Pharmaceutical Dissolution Testing

Pharmaceutical Dissolution Testing

Edited by

Jennifer Dressman
Johann Wolfang Goethe University
Frankfurt, Germany

Johannes Krämer
Phast GmbH
Homburg/Saar, Germany

Taylor & Francis
Taylor & Francis Group

Boca Raton London New York Singapore

Published in 2005 by
Taylor & Francis Group
6000 Broken Sound Parkway NW, Suite 300
Boca Raton, FL 33487-2742

Printed in the United States of America on acid-free paper
10 9 8 7 6 5 4 3 2 1

International Standard Book Number-10: 0-8247-5467-0 (Hardcover)
International Standard Book Number-13: 978-0-8247-5467-9 (Hardcover)

Library of Congress Cataloging-in-Publication Data

Catalog record is available from the Library of Congress

T&F informa

Taylor & Francis Group
is the Academic Division of T&F Informa plc.

Visit the Taylor & Francis Web site at
http://www.taylorandfrancis.com

This book is dedicated to dissolution scientists the world over, and to our spouses, Torsten and Heike, without whose support this work would not have been possible.

Preface

Over the last 20 years, the field of dissolution testing has expanded considerably to address not only questions of quality control of dosage forms but additionally to play an important role in screening formulations and in the evolving bioequivalence paradigm. Through our participation in various workshops held by the FIP, AAPS, and APV, it became clear to us that there is an international need for a book covering all aspects of dissolution testing, from the apparatus through development of methodology to the analysis and interpretation of results. *Pharmaceutical Dissolution Testing* is our response to this perceived need: a book dedicated to the equipment and methods used to test whether drugs are released adequately from dosage forms when administered orally. The focus on orally administered dosage forms results from the dominance of the oral route of administration on the one hand, and our desire to keep the book to a practicable length on the other hand.

Dissolution tests are used nowadays in the pharmaceutical industry in a wide variety of applications: to help identify which formulations will produce the best results in the clinic, to release products to the market, to verify batch-to-batch reproducibility, and to help identify whether changes made to formulations or their manufacturing procedure after marketing approval are likely to affect the performance in the clinic. Further, dissolution tests can sometimes be implemented to help determine whether a generic version of the medicine can be approved or not.

The book discusses the different types of equipment that can be used to perform the tests, as well as describing specific information for qualifying equipment and automating the procedures. Appropriate design of dissolution tests is put in the framework of the gastrointestinal physiology and the type of dosage form being developed. Although the discussion in this book is focused on oral dosage forms, the same principles can obviously be applied to other routes of administration. As important as the correct design of the test itself is the appropriate analysis and interpretation of the data obtained. These aspects are addressed in detail in several chapters, and suggestions are made about how to relate dissolution test results with performance in the patient (in vitro–in vivo correlation). To reflect the growing interest in dietary supplements and natural products, the last chapter is devoted to the special considerations for these products.

We would like to thank all of the authors for their valuable contributions to this work, which we trust will provide the dissolution scientist with a thorough reference guide that will be of use in all aspects of this exciting and ever-evolving field.

Jennifer Dressman
Johannes Krämer

Contents

Contributors

Cynthia K. Brown Eli Lilly and Company, Indianapolis, Indiana, U.S.A.

William E. Brown Department of Standards Development, United States Pharmacopeia, Rockville, Maryland, U.S.A.

Steffen M. Diebold Leitstelle Arzneimittelüberwachung Baden–Württemberg, Regierungspräsidium Tübingen, Tübingen, Germany

Stephen Dobro Product Testing and Validation, Zymark Corporation, Hopkinton, Massachusetts, U.S.A.

Jennifer Dressman Institute of Pharmaceutical Technology, Biocenter, Johann Wolfgang Goethe University, Frankfurt, Germany

Colm Farrell GloboMax, A Division of ICON plc, Marlow, Buckinghamshire, U.K.

Jayachandar Gajendran Phast GmbH, Biomedizinisches Zentrum, Homburg/Saar, Germany

Lee Timothy Grady Phast GmbH, Biomedizinisches Zentrum, Homburg/Saar, Germany

Vivian A. Gray V. A. Gray Consulting, Incorporated, Hockessin, Delaware, U.S.A.

Athanassios Iliadis Department of Pharmacokinetics, Mediterranean University of Marseille, Marseille, France

Kilian Kelly Department of Pharmaceutical Sciences, Strathclyde Institute for Biomedical Studies, University of Strathclyde, Glasgow, Scotland, U.K.

Sandra Klein Institute of Pharmaceutical Technology, Biocenter, Johann Wolfgang Goethe University, Frankfurt, Germany

Johannes Krämer Phast GmbH, Biomedizinisches Zentrum, Homburg/Saar, Germany

Frieder Langenbucher BioVista LLC, Riehen, Switzerland

Eleftheria Nicolaides Laboratory of Biopharmaceutics & Pharmacokinetics, National & Kapodistrian University of Athens, Athens, Greece

Christos Reppas Laboratory of Biopharmaceutics & Pharmacokinetics, National & Kapodistrian University of Athens, Athens, Greece

Myriam Rochdi GloboMax, A Division of ICON plc, Marlow, Buckinghamshire, U.K.

Vinod P. Shah Office of Pharmaceutical Science, Center for Drug Evaluation and Research, Food and Drug Administration, Rockville, Maryland, U.S.A.

Theresa Shepard GloboMax, A Division of ICON plc, Marlow, Buckinghamshire, U.K.

V. Srini Srinivasan Dietary Supplements Verification Program (DVSP), United States Pharmacopeia, Rockville, Maryland, U.S.A.

Ralf Steinmetz Phast GmbH, Biomedizinisches Zentrum, Homburg/Saar, Germany

Erika Stippler Phast GmbH, Biomedizinisches Zentrum, Homburg/Saar, Germany

Mira Symillides Laboratory of Biopharmaceutics & Pharmacokinetics, National & Kapodistrian University of Athens, Athens, Greece

Maria Vertzoni Laboratory of Biopharmaceutics & Pharmacokinetics, National & Kapodistrian University of Athens, Athens, Greece

Dale VonBehren Pharmaceutical Development and Quality Products, Zymark Corporation, Hopkinton, Massachusetts, U.S.A.

Clive G. Wilson Department of Pharmaceutical Sciences, Strathclyde Institute for Biomedical Studies, University of Strathclyde, Glasgow, Scotland, U.K.

Martin Wunderlich Institute of Pharmaceutical Technology, Biocenter, Johann Wolfgang Goethe University, Frankfurt, Germany

1

Historical Development of Dissolution Testing

JOHANNES KRÄMER, LEE TIMOTHY GRADY, and JAYACHANDAR GAJENDRAN

Phast GmbH, Biomedizinisches Zentrum, Homburg/Saar, Germany

INTRODUCTION

Adequate oral bioavailability is a key pre-requisite for any orally administered drug to be systemically effective. Dissolution (release of the drug from the dosage form) is of primary importance for all conventionally constructed, solid oral dosage forms in general, and for modified-release dosage forms in particular, and can be the rate limiting step for the absorption of drugs administered orally (1). Physicochemically, "Dissolution is the process by which a solid substance enters the solvent phase to yield a solution" (2). Dissolution of the drug substance is a multi-step process involving heterogeneous reactions/interactions between the phases of the solute–solute and solvent–solvent phases and at the solute–solvent interface (3). The heterogeneous reactions that constitute the overall mass transfer process may be categorized as (i) removal of the solute from the solid phase, (ii) accomodation of the solute in the liquid phase, and (iii) diffusive and/or convective transport of the solute away from the solid/liquid interface into the bulk phase. From the dosage form perspective, dissolution of the active pharmaceutical ingredient, rather than disintegration of the dosage form, is often the rate determining step in presenting the drug in solution to the absorbing membrane. Tests to characterize the dissolution behavior of the dosage form, which per se also take disintegration characteristics into consideration, are usually conducted using methods and apparatus that have been standardized virtually worldwide over the past decade or so, as part of the ongoing effort to harmonize pharmaceutical manufacturing and quality control on a global basis.

The history of dissolution testing in terms of the evolution of the apparatus used was reviewed thoroughly by Banakar in 1991 (2). This chapter focuses first on the pharmacopeial history of dissolution testing, which has led to mandatory dissolution testing of many types of dosage forms for quality control purposes, and then gives a detailed history of two newer compendial apparatus, the reciprocating cylinder and the flow-through cell apparatus. The last section of the chapter provides some historical information on the experimental approach of Herbert Strieker's group. His scientific work in combining permeation studies directly with a dissolution tester, is very much in line with the Biopharmaceutic Classification System (BCS), but was published more than two decades earlier than the BCS (4) and can therefore be viewed as the forerunner of the BCS approach.

FROM DISINTEGRATION TO DISSOLUTION

Compressed tablets continue to enjoy the status of being the most widely used oral dosage form. Tablets are solid oral dosage forms of medicinal substances, usually prepared with the aid of suitable pharmaceutical excipients. Despite the advantages offered by this dosage form, the problems associated with formulation factors remain to some extent enigmatic to the pharmaceutical scientist. In the case of conventional (immediate-release) solid oral drug products, the release properties are mainly influenced by disintegration of the solid dosage form and dissolution of drug from the disintegrated particles. In some cases, where disintegration is slow, the rate of dissolution can depend on the disintegration process, and in such cases disintegration can influence the systemic exposure, in turn affecting the outcome of both bioavailability and bioequivalence studies. The composition of all compressed conventional tablets should, in fact, be designed to guarantee that they will readily undergo both disintegration and dissolution in the upper gastrointestinal (GI) tract (1). All factors that can influence the physicochemical properties of the dosage form can influence the disintegration of the tablet and subsequently the dissolution of the drug. Since the 1960s, the so-called "new generation" of pharmaceutical scientists has been engaged in defining, with increasing chemical and mathematical precision, the individual variables in solid dosage form technology, their cumulative effects and the significance of these for in vitro and in vivo dosage form performance, a goal that had eluded the previous generation of pharmaceutical scientists and artisans.

As already mentioned, both dissolution and disintegration are parameters of prime importance in the product development strategy (5), with disintegration often being considered as a first order process and dissolution from drug particles as proportional to the concentration difference of the drug between the particle surface and the bulk solution. Disintegration usually reflects the effect of formulation and manufacturing process variables, whereas the dissolution from drug particles mainly reflects the effect of solubility and particle size, which are largely properties of the drug raw material, but can also be influenced significantly by processing and formulation. It is usually assumed that the dissolution of drug from the surface of the intact dosage form is negligible, so tablet disintegration is key to creating a larger surface area from which the drug can readily dissolve. However, tablet disintegration in and of itself may not be a reliable indicator of the subsequent dissolution process, so the tablet disintegration tests used as a quality assurance measure may or may not be a an adequate indicator of how well the dosage form will release its active ingredient in vivo. Only where a direct relationship between disintegration and dissolution has been established, can a waiver of dissolution testing requirements for the dosage form be considered (6).

Like disintegration testing, dissolution tests do not prove conclusively that the dosage form will release the drug in vivo in a specific manner, but dissolution does come one step closer, in that it helps establish whether the drug can become available for absorption in terms of being in solution at the sites of absorption. The period 1960–1970 saw a proliferation of designs for dissolution apparatus (7). This effort led to the adoption of an official dissolution testing apparatus in the United States Pharmacopeia (USP) and dissolution tests with specifications for 12 individual drug product monographs in the pharmacopeia. These tests set the stage for the evolution of dissolution testing into its current form.

DISSOLUTION METHODOLOGIES

The theories applied to dissolution have stood the test of time. Basic understanding of these theories and their application are essential for the design and development of sound dissolution methodologies as well as for deriving complementary statistical and mathematical techniques for unbiased dissolution profile comparison (3).

In the 1960s and 1970s, there was a proliferation of dissolution apparatus design. With their diverse design specifications and operating conditions, dissolution curves obtained with them were often not comparable and it was gradually realized that a standardization of methods was needed, which would enable correlation of data obtained with the various test apparatus. As a result, the National

Formulary (NF) XIV and USP XVIII and XIX (8) standardized both the apparatus design and the conditions of operation for given products. With these tests, comparable results could be obtained with the same apparatus design, even when the apparatus was produced by different equipment manufacturers.

PERSPECTIVE ON THE HISTORY OF COMPENDIAL DISSOLUTION TESTING

> ... it would seem that prompt action of certain remedies must be considerably impaired by firm compression. ... the composition of all compressed tablets should be such that they will readily undergo disintegration and solution in the stomach. [C. Caspari, "*A Treatise on Pharmacy*," 1895, Lea Bros., Philadelphia, 344.]

Tableting technology has had more than a century of development, yet the essential problems and advantages of tablets were perceived in broad brush strokes within the first years. Compression, powder flow, granulation, slugging, binders, lubrication, and disintegration were all appreciated early on, if not scientifically, at least as important considerations in the art of pharmacy. Industrial applications of tableting were not limited to drugs but found broad application in the confectionery and general chemical industry as well. Poor results were always evident and, already at the turn of the 20th century, some items were being referred to as "brickbats" in the trade.

With the modern era of medicine, best dated as starting in 1937, tablets took on new importance. Modern synthetic drugs, being more crystalline, were generally more amenable to formulation as solid dosage forms, and this led to greater emphasis on these dosage forms (9). Tableting technology was still largely empirical up to 1950, as is evidenced by the literature of the day. Only limited work was done before 1950, on drug release from dosage forms, as opposed to disintegration tests, partly because convenient and sensitive chemical analyses were not yet available. At that time, dissolution discussions mainly revolved around the question of whether the entire content could be dissolved and was mostly limited to tablets of simple, soluble chemicals or their salts.

The first official disintegration tests were adopted in 1945 by the British Pharmacopoeia and in 1950 by the USP. Even then, it was recognized that disintegration testing is an insufficient criterion for product performance, as evidenced by the USP-NF statement that "disintegration does not imply complete solution of the tablet or even of its active ingredient." Real appreciation of the significance of drug release from solid dosage forms with regard to clinical reliability did not develop until there were sporadic reports of product failures in the late 1950s, particularly vitamin products. Work in Canada by Chapman et al., for example, demonstrated that formulations with long disintegration times might not be physiologically available. In addition, the great pioneering pharmacokineticist John Wagner demonstrated in the 1950s that certain enteric-coated products did not release drug during Gl passage and that this could be related to poor performance in disintegration tests.

Two separate developments must be appreciated in discussing events from 1960 onward. These enabled the field to progress quickly once they were recognized. The first was the increasing availability of reliable and convenient instrumental methods of analysis, especially for drugs in biological fluids. The second, and equally important development, was the fact that a new generation of pharmaceutical scientists were being trained to apply physical chemistry to pharmacy, a development largely attributable, at least in the United States, to the legendary Takeru Higuchi and his students.

Further instances in which tablets disintegrated well (in vitro) but were nonetheless clinically inactive came to light. Work in the early 1960s by Campagna, Nelson, and Levy had considerable impact on this fast-dawning consciousness. By 1962, sufficient industrial concern had been raised to merit a survey of 76 products by the Phamaceutical Manufacturers of America (PMA) Quality Control Section's Tablet Committee. This survey set out to determine the extent of drug dissolved as a function of drug solubility and product disintegration time. They found significant problems, mostly occurring with drugs of less than 0.3% (30 ug/mL) solubility in water, and came within a hair of recommending that dissolution, rather than disintegration, standards be set on drugs of less than 1% solubility.

Another development that occurred between 1963 and 1968 that continues to confabulate scientific discussions of drug release and dissolution testing was the issue of generic drug approval. During this period, drug bioavailability became a marketing, political, and economic issue. At first, generic products were seen as falling short on performance. However later it turned out that the older formulations, that had been marketplace innovators, were often short on performance compared to the newly formulated generic products.

To better compare and characterize multi-source (generic) products, the USP-NF Joint Panel on Physiological Availability was set up in 1967 (Table 1) under Rudolph Blythe, who already had led industrial attempts at standardization of drug release tests. Discussions of the Joint Panel led to adoption, in 1970, of an official apparatus, the Rotating Basket, derived from the design of the late M. Pernarowski, long an active force in Canadian pharmaceutical sciences. A commercial reaction flask was used for cost and ruggedness. The monograph requirements were shepherded by William J. Mader, an industrial expert in analysis and control, who directed the American Pharmaceutical Association (APhA) Foundation's Drug Standards Laboratory. William A. Hanson prepared the first apparatus and later commercialized a series of models.

The Joint Panel proposed no in vivo requirements, but individual dissolution testing requirements were adopted in 12 compendial monographs. USP tests measured the time to attain a specified amount dissolved, whereas NF used the more workable test for the amount dissolved at a specified time. Controversy with respect to equipment selection and methodology raged at the time of the first official dissolution tests. As more laboratories entered the field, and experience (and mistakes!) accumulated, the period 1970–1980 was one of intensive refinement of official test methods and dissolution test equipment.

Table 1 USP Timeline from 1945–1999

1945–1950	Disintegration official in Brit Pharmacon and USP
1962	PMA Tablet Committee proposes 1% solubility threshold
1967	USP and NF Joint Panel on Physiological Availability chooses dissolution as a test chooses an apparatus
1970	Initial 12 monograph standards official
1971–1974	Variables assessment; more laboratories, three Collaborative Studies by PMA and Acad. Pharm. Sci
1975	First calibrator tablets pressed; First Case default proposed to USP
1976	USP Policy—comprehensive need; calibrators Collaborative Study
1977	USP Guidelines for setting Dissolution standards
1978	Apparatus 2—Paddle adopted; two Calibrator Tablets adopted
1979	New decision rule and acceptance criteria
1980	Three case Policy proposed; USP Guidelines revised; 70 monographs now have standards
1981	Policy adopted January, includes the default First Case, monograph proposals published in June
1982	Policy proposed for modified-release dosage forms
1984	Revised policy adopted for modified-release forms
1985	Standards now in nearly 400 monographs; field considered mature; Chapter <724> covers extended-release and enteric-coated
1990	Harmonization: apparatus 4—Flow- through adopted; Apparatus 3 Apparatus 5, 6, 7 fortransdermal drugs
1995	Third Generation testing proposed—batch phenomenon; propose reduction in calibration test number
1997	FIP Guidelines for Dissolution Testing of Solid Oral Products; pooled analytical samples allowed
1999	Enzymes allowed for gelatin capsules reduction from 0.1 N to 0.01 N Hcl

Later, a second apparatus was based on Poole's use of available organic synthesis round-bottom flasks as refined by the St. Louis laboratory. Neither choice of dissolution equipment proved to be optimal, indeed, it may have been better if the introduction of the two apparatus had occurred in the reverse order. With time, the USP would go on to offer a total of seven apparatuses, several of which were introduced primarily for products applied to the skin.

At the time, the biopharmaceutical problems, such as with low-solubility drugs, both in theoretical terms and in actual clinical failures were already well recognized. The objective of the Joint Panel was to design tests which could determine whether tablets dissolved within a reasonable volume, in a commercial flask. In those days, drugs were often prescribed in higher doses, so the volume of the dissolution vessels in terms of providing an adequate volume to enable complete dissolution of the dose had to be taken into design consideration. Over the last 35 years there has been a trend to develop more potent drugs, with attendant decrease in doses required (with notable exceptions, especially anti-infectives). For example, an antihypertensive may have been dosed at 250 mg, but newer drugs in the same category coming onto the market might be dosed as low as 5 mg. Subsequently, there has been a change in the amount of drug that needs to get dissolved for many categories of drugs. Nevertheless, a few monographs (e.g., digoxin tablets) have always presented a challenge to design of dissolution tests. The following factors exemplify typical problems associated with the development of dissolution tests for quality control purposes:

1. The need to have a manageable volume of dissolution medium.
2. The development of less-soluble compounds as drugs (resulting in problems in achieving complete dissolution in a manageable volume of medium).
3. Insufficient analytical sensitivity for low-dose drugs, especially at higher media volumes (as illustrated in the USP monograph on digoxin tablets).

It should be remembered that in 1970, when drug-release/dissolution tests first became official through the leadership of USP and NF, marketed tablets or capsules in general simply did not have a defined dissolution character. They were not formulated to achieve a particular dissolution performance, nor were they subjected to quality control by means of dissolution testing. Moreover, the U.S. Food and Drug Administration (FDA) was not prepared to enforce dissolution requirements or to even to judge their value.

The tremendous value of dissolution testing to quality control had not yet been established, and this potential role was perceived in 1970 only dimly even by the best placed observers. Until the early 1970s, discussions of dissolution were restricted to the context of in vivo–in vitro correlation (IVIVC) with some physiologic parameter. The missing link between the quality control and IVIVC aims of dissolution testing was that dissolution testing is sensitive to formulation variables that might be of biological significance because dissolution testing is sensitive in general to formulation variables.

Between 1970 and 1975, it became clear that dissolution testing could also play a role in formulation research and product quality control. Consistent with this new awareness of the value of dissolution testing in terms of quality control as well as bioavailability, USP adopted a new policy in 1976 that favored the inclusion of dissolution requirements in essentially all tablet and capsule monographs. Thomas Medwick chaired the Subcommittee that led to this policy. Due to lack of industrial cooperation, the policy did not achieve full realization. Nevertheless, by July 1980 the role of dissolution in quality control had grown to appeareance in 72 monographs, most supplied by USP's own laboratory under the direction of Lee Timothy Grady, and FDA's laboratory under the direction of Thomas P. Layloff. USP continued to adopt further dissolution apparatus designs (Fig. 1) and refine the methodology between 1975 and 1980, as shown in Table 1.

Over the years, dissolution testing has expanded beyond ordinary tablets and capsules—first to extended-release and delayed-release (enteric-coated) articles, then to transdermals, multivitamin and minerals products, and to Class Monographs for non-prescription drug combinations. (*Note*: at the time, "sustained-release" products were being tested, unofficially, in the NF Rotating Bottle apparatus).

Tablets and capsules that became available on the market in the above time frame often showed 10–20% relative standard deviation in amounts dissolved. The FDA's St. Louis Laboratories results on about 200 different batches of drugs

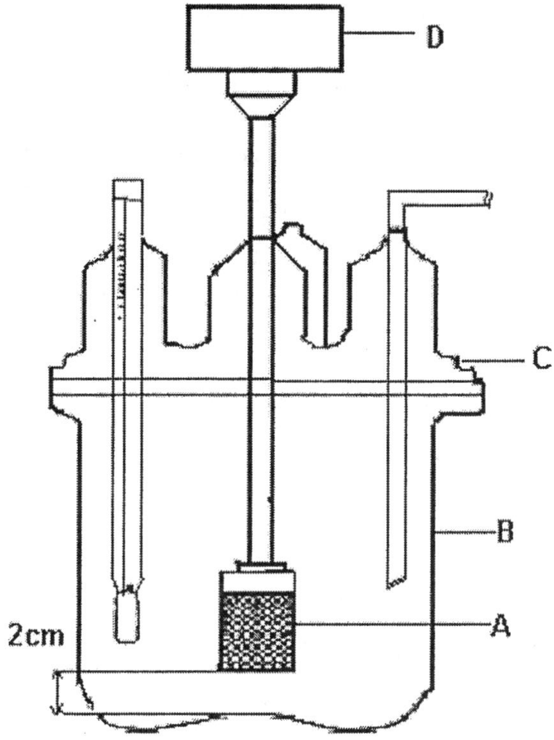

Figure 1 Rotating basket method. *Source*: From Ref. 10.

available showed that variation tend to be greatest for slowly dissolving drugs. Newer formulations, developed using dissolution testing as one of the aids to product design, are much more consistent. Another early problem in dissolution testing was lab-to-lab disagreement in results. This problem was essentially resolved when testing of standard "calibrator" tablets were added to the study design, for which average dissolution values had to comply with the USP specifications to qualify the equipment in terms of its operation. Every calibrator batch produced since the inaugauration of calibrators has been subjected to a Pharmaceutical Manufactorers of America (PMA)/Pharmaceutical Research and Manufacturers of America (PhRMA) collaborative study to determine acceptance statistics. Originally, calibrators were adopted to pick up the influence on dissolution results due to vibration in the equipment, failures in the drive chains and belts, and operator error. In fact, perturbations introduced in USP equipment are usually detected by at least one of the two types of calibrators (prednisone or salicylic acid tablets). Although the calibrators were not adopted primarily to test either deaeration or temperature control, they proved to be of value here, too. As a follow-up, the USP developed general guidelines on deaeration early in the 1990s, presently favoring a combination of heat and vacuum. In the late 1990s, the number of tests to qualify an apparatus was halved. Yet even today, an apparatus can fail the calibrator tablet tests, since small individual deviations in the mechanical calibration and operator error can combine to produce out of specification results for the calibrator. Thus, the calibrators are an important check on operating procedures, especially in terms of consistency between labs on an international basis.

In addition to the increasing interest in dissolution as a quality control procedure and aid to development of dosage forms, bioavailability issues continued to be raised throughout the 1970–1980 period, as clinical problems with various oral solid products dissolution and bioavailability continued to crop up. Much of the impetus behind the bioavailability discussions came from the issue of bioequivalence of drugs as this relates to generic substitution. In January 1973, FDA proposed the first bioavailability regulations. These were followed in January 1975 by more detailed bioequivalence and bioavailability regulations, which became final in February 1977. A controversial issue in these regulations proved to be the measurement of the rate of absorption. The 1975 revision proposal was the first to contain the concept of an in vitro bioequivalence requirement, which reflected the growing awareness of the general utility of dissolution testing at that time.

A major wave of generic equivalents were introduced to the U.S. market following the Hatch–Waxman legislation in the early 1970s and ANDA applications to the FDA provided the great majority of IVIVC available to USP for non-First Case standards setting during the following years.

From the USP perspective, digoxin tablets became and remained the benchmark for the impact of dissolution on bioavailability. It is a life-saving and maintaining drug, has a low therapeutic index, is poorly soluble, has a narrow absorption window (due to p-glycoprotein exotransport) and it is formulated using a low proportion of drug:excipients due to its high potency. Correlation between dissolution and absorption was first shown for digoxin in 1973. The official dissolution standard that followed was the watershed for the entire field. It is interesting to note that clinical observations for digoxin tablets were made in only few patients. Similarly, the original concerns of John Wagner over prednisone tablets were based on observations in just one patient. The message from these experiences is that decisive bioinequivalences can be picked up even in very small patient populations.

At the time the critical decisions were made, it seemed that diminished bioavailability could usually be linked to formulation problems. Scientists recognized early that when the rate of dissolution is less than the rate of absorption, the dissolution test results can be predictive of correlation with bioavailability or clinical outcome. At that time, there was little recognition that intestinal and/or hepatic metabolism mattered, an exception being the phenothiazines. So the primary focus was on particle size and solubility. Observations with prednisone, nitrofurantoin, digoxin and other low-solubility drugs were pivotal to decision making at the time, since the dissolution results could be directly linked to clinical data. Scientists recognized that it is not the solubility of the drug alone that is critical, but that the effective surface area from which the drug is dissolving also plays a major role, as described by the Noyes–Whitney equation, which describes the flux of drug into solution as a mathematical relationship between these factors.

In the mid-70s, it was a generally expressed opinion that there could be as many as 100 formulation factors that might affect bioavailability or bioequivalence. In fact, most of the documented problems centered around the use of the hydrophobic magnesium stearate as a lubricant or use of a hydrophobic shellac subcoat in the production of sugar-coated tablets. At that time, products were also often shellac-coated both for elegance and for longer shelf life. In addition, inadequate disintegration was still a problem, often related to disintegrant integrity and the force of compression in the tableting process. All four of these factors are sensitive to dissolution testing. Wherever there was a medically significant problem, a dissolution test was able to show the difference between the nonequivalent formulations and this is, in general, still true today.

In addition to the scientific aspects, much of the discussion around dissolution and bioequivalence really was and is a political, social, and economic argument. Because of reluctance on the part of the pharmaceutical industry to cooperate with USP, a default standard was proposed to the USP in 1975. This proposal called for 60% dissolved at 20 min in water, testing individual units in the official apparatus and was based on observations by Bill Mader and Rudy Blythe in 1968–1970, who had demonstrated that one could start getting meaningful data at 20 min, consistent with typical disintegration times in those days. In 1981, a USP Subcommittee pushed forward the default condition, resulting in an explosion in the number of dissolution tests from 70 to 400 in 1985, a five-fold increase in four years! Selection of a higher amount dissolved, 75%, made for tighter data, whilst the longer test time, 45 min, was chosen because it gave formulators some flexibility in product design to improve elegance, stability, and/or to reduce friability—in other words, a lot of considerations not directly linked to dissolution. Subsequently, industrial cooperation improved, and later the FDA Office of Generic Drugs and the USP established a cooperation, with the FDA supplying both dissolution and bioavailability data and information to USP.

Experience has demonstrated that where a medically significant difference in bioavailability has been found among supposedly identical products, a dissolution test has been efficacious in discriminating among them. A practical problem has been the converse, that is, dissolution tests are sometimes too discriminating, so that it is not uncommon for a clinically acceptable product to perform poorly in an official dissolution test. In such cases, the Committee of Revision has been mindful of striking the right balance: including as many acceptable products as possible, yet not setting forth dissolution specifications that would raise scientific concern about bioequivalence.

COMPENDIAL APPARATUS

The USP 27, NF22 (11) now recognizes seven dissolution apparatus specifically, and describes them and, in some cases allowable modifications, in detail. The choice of the dissolution apparatus should be considered during the development of the dissolution methods, since it can affect the results and the duration of the test. The type of dosage form under investigation is the primary consideration in apparatus selection.

Apparatus Classification in the USP

Apparatus 1 (rotating basket)
Apparatus 2 (paddle assembly)
Apparatus 3 (reciprocating cylinder)
Apparatus 4 (flow-through cell)
Apparatus 5 (paddle over disk)
Apparatus 6 (cylinder)
Apparatus 7 (reciprocating holder)

The European Pharmacopoeia (Ph. Eur.) has also adopted some of the apparatus designs (12) described in the USP, with some minor modifications in the specifications. Small but persistent differences between the two have their origin in the fact that the American metal processing industry, unlike the European, uses the imperial rather than the metric system. In the European Pharmacopeia, official dissolution testing apparatus for special dosage forms (medicated chewing gum, transdermal patches) have also been incorporated (Table 2 provides an overview of apparatus in Ph. Eur.).

Of all these types, Apparatus 1 and 2 are the most widely used around the world, mostly because they are simple, robust, and adequately standardized apparatus designs, and

Table 2 Apparatus Classification in the European Pharmacopoeia (2002) for Different Dosage Forms

For solid dosage forms	Paddle apparatus
	Basket apparatus
	Flow-through apparatus
For transdermal patches	Disk assembly method
	Cell method
	Rotating cylinder method
For special dosage forms	Chewing apparatus (medicated Chewing gums), Figure 2a
	Flow-through apparatus, Figure 2b

are supported by a wider experience of experimental use than the other types of apparatus. Because of these advantages, they are usually the first choice for in vitro dissolution testing of solid dosage forms (immediate as well as controlled/modified-release preparations). The number of monographs found in the USP for Apparatus 2 now exceeds that of apparatus 1. The description of these apparatus can be found in the USP dissolution testing, Chapter <711>(11) and Ph. Eur, Chapter <2.9> (12).

Generally speaking, it was intended that Apparatus 1, 2, 3, and 4 of the USP could all be used to evaluate all dosage forms, irrespective of the drug or the type of dosage form to be tested. Nowadays, with a wide variety of dosage forms being produced, most notable being the multiplicity of special dosage forms such as medicated chewing gums, transdermal patches, implants, etc. on the market, the USP dissolution Apparatuses 1 and 2 do not cover all desired dissolution studies. For these dosage forms, the term "drug release testing" is used instead of "dissolution." Figure 2a shows a special apparatus for the release of drug from medicated chewing gums.

Reciprocating Cylinder

The reciprocating cylinder was proposed by Beckett and coworkers (13) and its incorporation into the USP followed in 1991. The idea to generate a new test method came from a

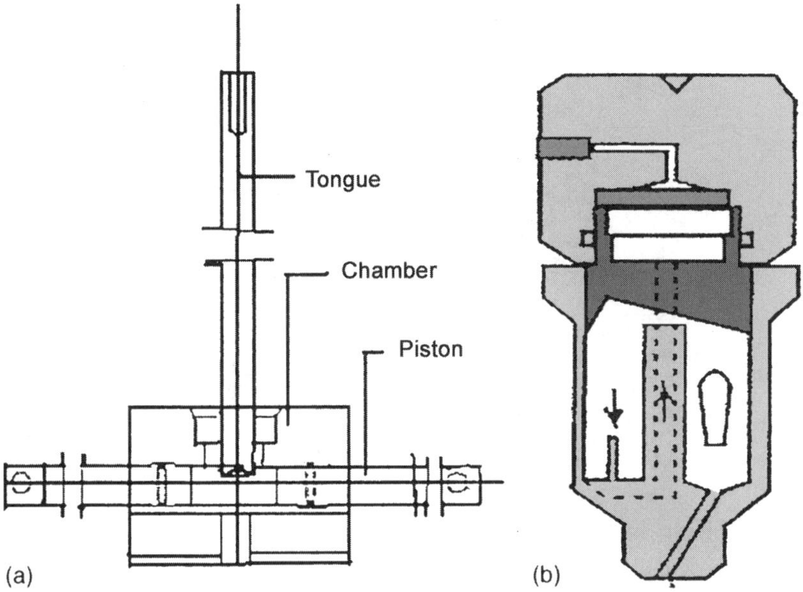

Figure 2 (a) Apparatus for the determination of drug release from medicated chewing gums and (b) flow-through cell for semi-solid products.

presentation at the International Pharmaceutical Federation (FIP) Conference in 1980 (U.S. Pharmcopeial Convention). In this presentation, problems with the dissolution results from USP Apparatuses 1 and 2, which may be affected physical factors like shaft wobble, location, centering, deformation of the baskets and paddles, presence of the bubbles in the dissolution medium, etc. were enumerated. It was agreed at the conference that major problems could arise in the acceptance of pharmaceutical products in international trade due to the resultant variations in the dissolution data (13). A team of scientists working under Beckett's direction in London, UK, subsequently developed the reciprocating cylinder, which is often referred to as the "Bio-Dis." Although primarily designed for the release testing of extended-release products, USP apparatus 3 may be additionally be used for the dissolution testing of IR products of poorly soluble drugs (14). In

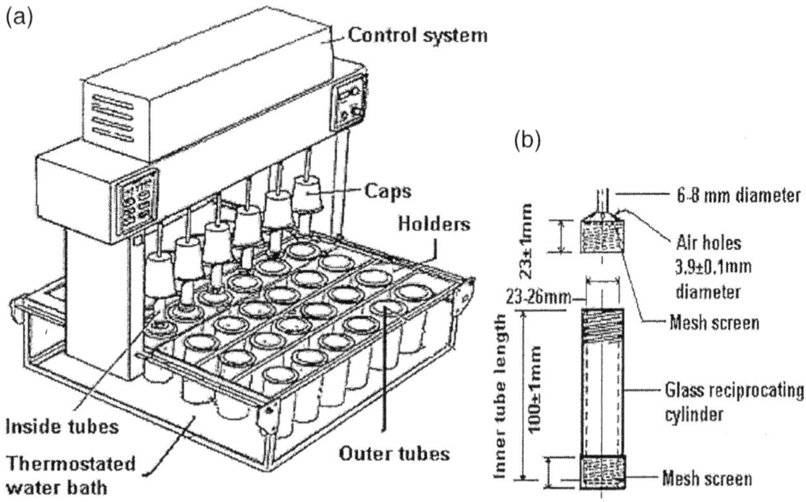

Figure 3 (a) The reciprocating cylinder apparatus (Bio-Dis) and (b) reciprocating cell.

terms of design, the apparatus is essentially a modification of the USP/NF disintegration tester (Fig. 3).

Principle and Design

The development of USP Apparatus 3 was based on the recognition of the need to establish IVIVC, since the dissolution results obtained with USP Apparatuses 1 and 2 may be significantly affected by the mechanical factors mentioned in the preceding section. The design of the USP Apparatus 3, based on the disintegration tester, additionally incorporates the hydrodynamic features from the rotating bottle method and provides capability agitation and media composition changes during a run as well as full automation of the procedure. Sanghvi et al. (15) have made efforts to compare the results obtained with USP Apparatus 3 and USP Apparatus 1 and 2. Apparatus 3 can be especially useful in cases where one or more pH/buffer changes are required in the dissolution testing procedure, for example, enteric-coated/sustained-release dosage forms, and also offers the advantages of mimicking the changes in physiochemical conditions and extraordinarily strong mechanical forces experienced by the drug products in the mouth or at certain locations in the GI tract, such as the pylorus and the ileocecal valve.

Apparatus 3 is currently commercially available with seven columns of six rows, each row consisting of a set of cylindrical, flat bottomed glass outer vessels, a set of reciprocating inner cylinders and stainless steel fittings (Fig. 3a and b). The screens are made of suitable materials designed to fit the top and bottom of the reciprocating cylinders. Operation involves the agitation, in dips per minute (dpm), of the inner tube within the outer tube. On the upstroke, the bottom tube in the inner tubes moves upward to contact the product and on the down stroke the product leaves the mesh and floats freely within the inner tube. Thus, the mechanics subject the product being tested to a moving medium.

The USP Apparatus 3 is considered as the first line apparatus in product development of controlled-release preparations, because of its usefulness and convenience in exposing products to mechanical as well as a variety of physicochemical conditions which may influence the release of products in the GI tract (13). The particular advantage of this apparatus is the technically easy and problem free use of test solutions with different pH values for each time interval. It also avoids cone formation for disintegrating (immediate release) products, which can be encountered with the USP apparatus 2. Ease of sampling, automation, and pH change during the test run, make it the method of choice in comparison to the rotating bottle apparatus, although both can lead to good correlations for extended-release formulations (16).

An additional advantage of apparatus 3 includes the feasibility of drug-release testing of chewable tablets. Chewable tablets for human use do not contain disintegrants, so they need to undergo physiological grinding (i.e., chewing) prior to dissolution. However, requirements concerning their biopharmaceutical quality are similar or identical to those for conventional immediate-release tablets. The use of compendial devices such as either stirred systems like the basket and the paddle apparatus or the flow-through cell apparatus were found not to provide suitable results for proper product characterization of chewable tablets. Pre-treatment by trituration to simulate mastication is not desirable because of the lack of standardization for this manual procedure. Furthermore, for safety reasons, it must be established that even when the unchewed tablets are swallowed, it would still release the active ingredient. The action produced by the reciprocating cylinder carries the chewable tablet being tested through a moving medium. The hydrodynamic forces in this apparatus were found to be stronger in comparison to Apparatus 1 and 2 (3). The results showed that 5 dpm (dips per min) in apparatus 3 is equivalent to 50 rpm in Apparatus 2. Hence, higher dip rates are creating forces that may not be achieved by the use of the paddle instrument but which are highly desired to mimic human masticatory forces.

Further experiments were performed to evaluate the suitability of the reciprocating cylinder apparatus to discriminate dissolution properties of different Pharmaceuticals including chewable tablets containing calcium carbonate (18). The oscillatory movement of USP Apparatus 3 operated at 20 dpm exhibited a high mechanical stress on the formulations. The results (19) were discussed at the Royal British Pharmaceutical Society (RBPS)/FIP Congress in September 1999 and later included as a recommendation in the FIP/AAPS guidelines (20). The use of USP Apparatus 3 to characterize the drug release behavior of chewable tablets represents the state of the art, but there are also some concerns about the carry over and the effect of surface tension retarding complete drainage of the test fluid during the "hold" period between rows (21).

Flow-Through Cell

The USP Apparatus 4, also known as the flow-through cell, was introduced and extensively studied by Langenbucher (22). In the open loop configuration, this system offers the advantage of unlimited medium supply, which is of particular interest for the dissolution of poorly soluble drugs. The idea to develop a flow-through cell method dates back more than 45 years. As early as 1957, a flow-through cell method with a

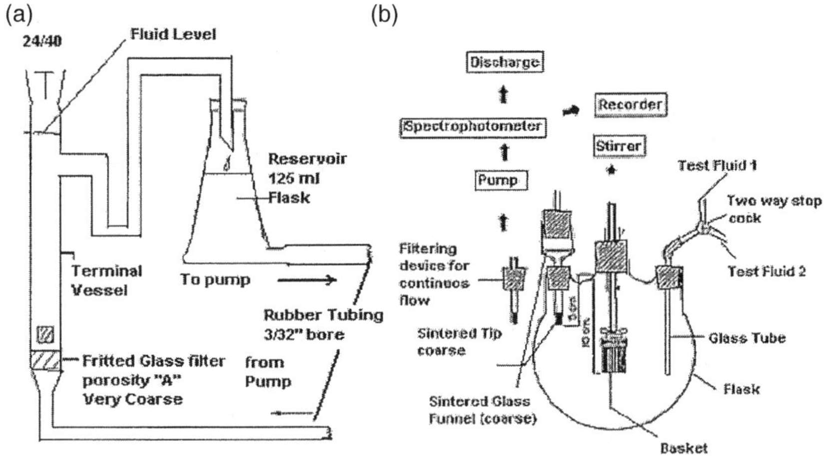

Figure 4 (a) Assembly for testing timed-release preparations. Redrawn from a letter typewritten on USP paper in 1957. *Source*: From Ref. 23. (b) Continuous flow dissolution apparatus. *Source*: From a 1968 publication by Pemarowski.

closed (limited) liquid volume was developed by the FDA (Fig. 4a) and discussed by both the PMA and the USP. In 1968, Pemarowski published a "continuous flow apparatus" which could supply an unlimited volume of liquid, as shown in Figure 4b. This design could have become an early version of the flow-through method, but instead became the forerunner of the basket method of USP. It had already been incorporated into the two semiofficial compendia, the German Arzneimittel Codex (1983) and the French "Pro Pharmacopoeia" (23). The flow-through cell was finally included officially in the USP as Apparatus 4, in a Supplement to USPXXII, in1990, even though little experience with the method had been accumulated at the time.

The flow-through cell is applicable not only for the determination of the dissolution rate of tablets and sugar-coated tablets, but has also been applied to suppositories, soft-gelatin capsules, semisolids, powders, granules, and implants. A small volume cell containing the sample solution is subjected to a continuous stream of dissolution media. The dissolution

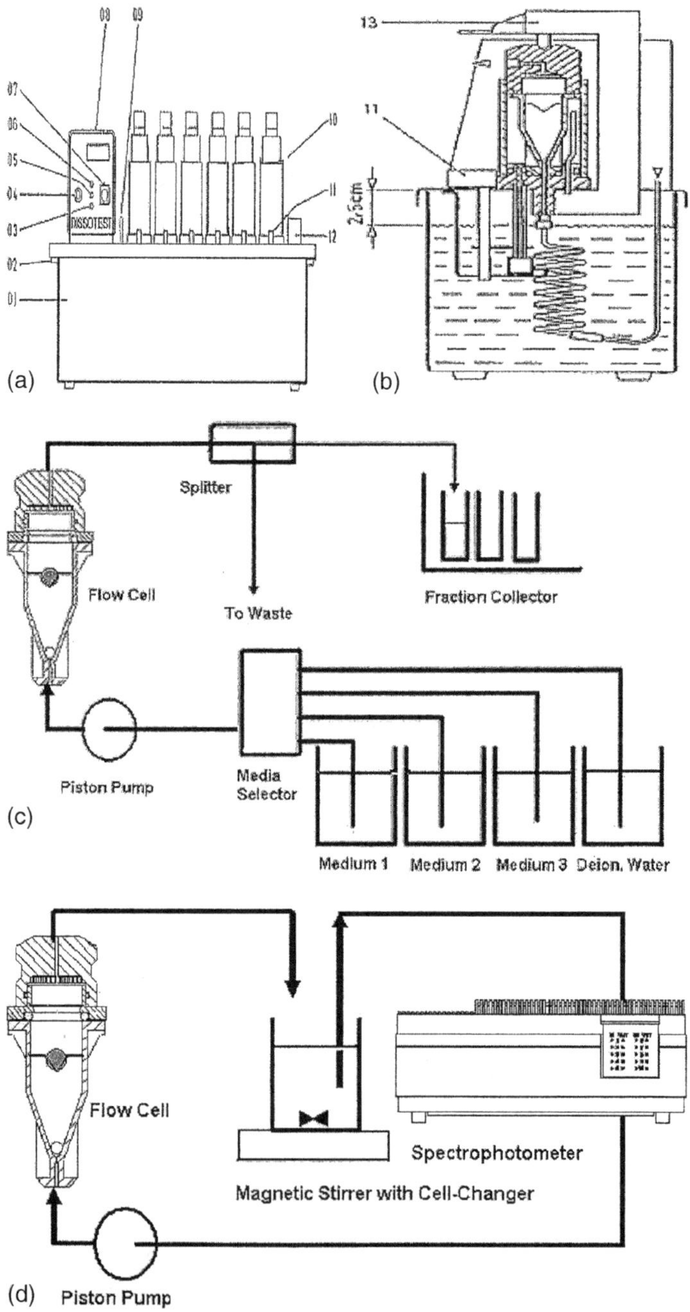

Figure 5 (*Caption on Facing Page*)

medium flows through the cell from bottom to top of the cell. The special pulsating movement of the piston pump obviates the need for further stirring and/or shaking elements. A filtration device at the top of the cell quantitatively retains all undissolved material and provides a clear solution for subsequent quantitative analysis of the compound dissolved. The set-up is illustrated in Figure 5. Unlike the closed systems, with their limited and constant volume of dissolution medium, the flow-through cell system is usually operated as an open loop, i.e., new dissolution medium is continuously introduced into the system. The experimental design of the closed systems results in cumulative dissolution profiles, as shown in Figure 5c. With the open systems, all drug dissolved is instantaneously removed along the flow of the dissolution medium, see Figure 5d. The results are therefore generated in the form of dissolution rates, i.e., fraction dissolved per time unit. The results obtained from tests in the flow-through system therefore need to be transformed in order to present the data in the usual form, i.e., dissolution profiles of cumulative amount dissolved vs. time. Use of devices to maintain temperature control, positioning of the specimen in the cell, and the possible need to adjust the flow rate are additional points which may need to be incorporated into the test design.

A common feature of widely used apparatus like the paddle or basket method is their limited volume. Typical volumes used in these systems range from about 500 to 4000 mL, limiting their use for very poorly soluble substances. Theoretically at least, open systems may be operated with infinite volumes to complete the dissolution of even very poorly soluble com-

Figure 5 (*Facing Page*) (a) and (b). General assemblage of a six-channel flow-through cell apparatus Dissotest. 01. Trough, 02. Bolt, 03. Alarm lamp, 04. Temperature control Knob, 05. Push Button for reference temperature value, 06. Signal Lamp, 07. Switcher, 08. Circulating thermostat, 09. Level Indicator, 10. Dissolution Unit, 11. Stopcocks, 12. Connecting bar, 13. Tensioning lever. *Source*: From Ref. 18. (c) Flow-through cell—open system. (d) Flow-through cell—closed system.

pounds. With these systems, the analytical limit of quantification and the preparation and cost of large volumes of dissolution medium represent practical limitations to attain 100% release. Some of the advantages of the flow-through cell apparatus include provision of sink conditions, the possibility of generating rapid pH changes during the test, continuous sampling, unlimited solvent volume, minimizing downtime between tests (since the cells can be prepared and loaded with samples independent of tests in progress), ability to adapt test parameters to physiological conditions, retention of undissolved particles within the cell, without the need for an additional step of filtration or centrifugation, and availability of specific sample cells depending on the type of dosage form, as illustrated in Figure 6. In summary, the flow-through cell is widely regarded as a promising instrument for formulations such as suppositories, implants and other sustained-release dosage forms as well as immediate-release dosage forms of poorly soluble compounds and continues to grow in terms of acceptance and application in the pharmaceutical industry.

QUALIFICATION OF THE APPARATUS

Due to the nature of the test method, "quality by design" is an important qualification aspect for in vitro disolution test equipment. The suitability of the apparatus for the dissolution/drug-release testing depends on both the physical and chemical calibrations which qualifies the equipment for further analysis. Besides the geometrical and dimensional accuracy and precision, as described in USP 27 and Ph.Eur., any irregularities such as vibration or undesired agitation by mechanical imperfection are to be avoided. Temperature of the test medium, rotation speed/flow rate, volume, sampling probes, and procedures need to be monitored periodically.

Apparatus Suitability Test

In addition to the mechanical calibration briefly described in the preceding section, another important aspect of qualification and validation is the "apparatus suitability test." The

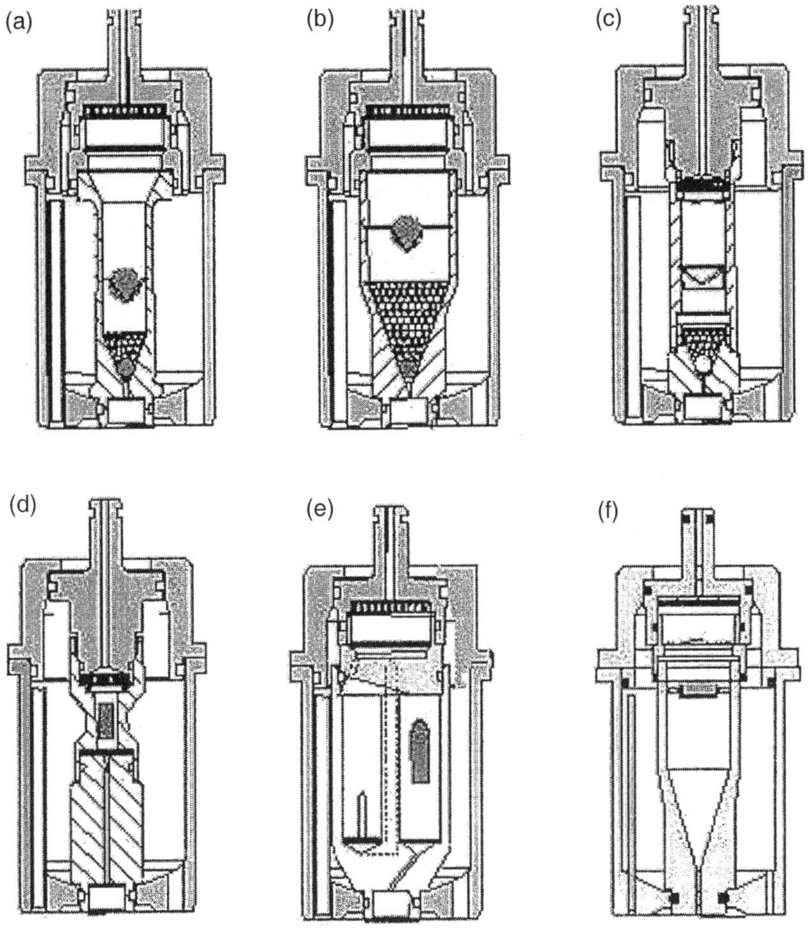

Figure 6 Different cell types for dissolution testing using the flow-through system. Type (a) tablet cell (12 mm), (b) tablet cell (22.6 mm), (c) cell for powders and granulates, (d) cell for implants, (e) cell for suppositories and soft gelatin capsules, (f) cell for ointments and creams.

use of USP calibrator tablets (for Apparatus 1 and 2 disintegrating as well as non-disintegrating calibrator tablets are used) is the only standardized approach to establishing apparatus suitability for conducting compendial dissolution tests and has been generally able to identify system or operator failures. Suitability tests have also been developed for Apparatus 3, using specific calibrators and the aim is to generate a set of calibrators for each and every compendial dissolution test apparatus.

Apparatus suitability tests are recommended to be performed not less than twice per year per equipment and after any equipment change, significant repair, or movement of the accessories. Thus, critical inspection and observation of test performance during the test procedure are required. Validation of the analytical procedure, including assessment of precision, accuracy, specificity, detection limit, quantification limit, linearity and range, applied in the dissolution testing, when using either automated or manual tesing, has to comply with "Validation of Analytical Procedures" (24) and "Validation of Compendial Methods" (25) (< 1225 > , USP27).

DESCRIPTION OF THE SARTORIUS ABSORPTION MODEL

The Sartorius Absorption Model (26), which served as the forerunner to the BCS, simulates concomitant release from the dosage form in the GI tract and absorption of the drug through the lipid barrier. The most important features of Sartorius Absorption Model are the two reservoirs for holding different media at 37°C, a diffusion cell with an artificial lipid barrier of known surface area, and a connecting peristaltic pump which aids the transport of the solution or the media from the reservoir to the compartment of the diffusion cell. The set-up is shown in Figures 7a and b.

The two media typically used include Simulated Gastric Fluid (pH 1–pH 3) and Simulated Intestinal Fluid (pH 6–pH 7). The drug substance under investigation is introduced, and its uptake in the diffusion cell ("absorption") is governed by its hydrophilic–lipophilic balance (HLB). The absorption model proposed by Stricker (26) in the early 1970s therefore effectively took into consideration (in an experimental sense) all aspects considered by the theory of the BCS, which was introduced more than 20 years later.

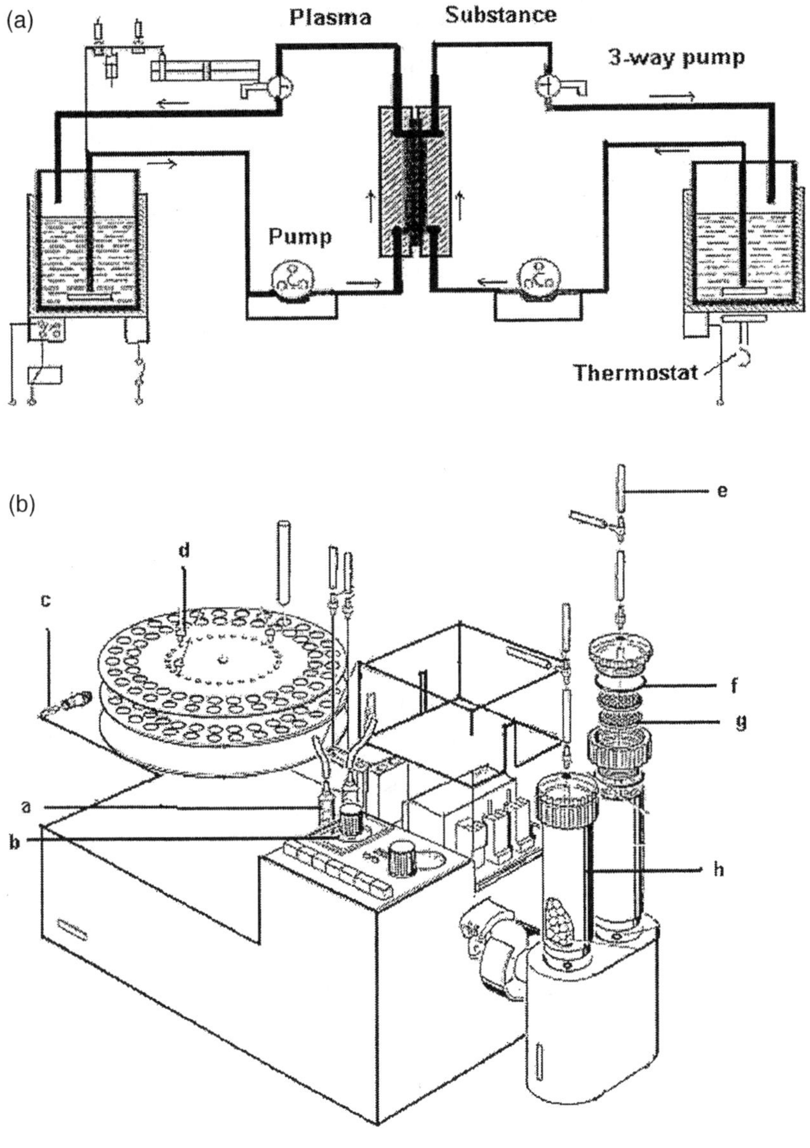

Figure 7 (a) Sartorius absorption model; (b) Sartorius dissolution model. a, Plastic syringe; b, timer; c, safety lock; d, cable connector; e, silicon tubes; f, silicon-O-rings; g, metal filter; h, polyacryl reaction vessel.

Biopharmaceutics Classification System

The introduction of the BCS in 1995 precipitated a tremendous surge of interest in dissolution and dissolution testing methodologies. Amidon et al. (4) devised the BCS to classify drugs based on their aqueous solubility and intestinal permeability. The BCS characteristics (solubility and permeability), together with the dissolution of the drug from the dosage form, takes the major factors that govern the rate and extent of drug absorption from dosage forms into account. According to current BCS criteria (2004), drugs are considered highly soluble when the highest dose strength of the drug substance is soluble in less than 250 mL water over a pH range of 1–6.8 and considered highly permeable when the extent of absorption in humans is determined to be greater than 90% of the administered dose.

According to the BCS, drug substances are classified as follows (20):

Class 1 Drugs: High solubility–High permeability;
Class 2 Drugs: Low solubility–High permeability;
Class 3 Drugs: High solubility–Low permeability;
Class 4 Drugs: Low solubility–Low permeability.

The FDA currently allows biowaivers (27) (drug product approval without having to show bioequivalence in vivo) for formulations that contain Class I drugs and can demonstrate appropriate in vitro dissolution (rapidly dissolving).

In Vitro Dissolution Testing Model

The principles of dissolution testing as an indication of in vivo performance had also been addressed in the experimental models proposed by Stricker (28). Figures 8 and 9 depict the processes occurring during the transformation of the drug in the solid dosage form to drug in solution in the gastrointestinal environment. The vessels containing the Simulated Gastric Fluid and Intestinal Fluid and maintained at 37°C, are rotated at 1.2 rotations per minute (rpm). The dissolution of the dosage form is controlled by the flow properties of the media, mechanical forces induced by the "GI tract," the pH,

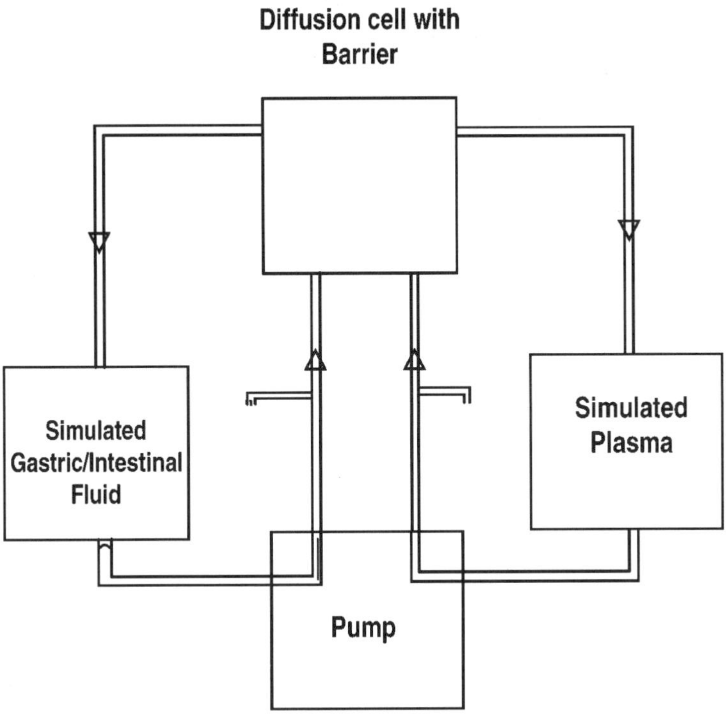

Figure 8 Scheme of in vitro absorption model according to Stricker. *Source*: From Ref. 28.

and the volume of the media. On the basis of absorption data, the operating parameters of Stricker's dissolution model were adjusted appropriately. Additional accessories like the dosing pump and the fraction sampler at various points in the model set-up were installed to facilitate a quantitative analysis. Using the Stricker model, it was possible to generate good IVIVC.

INTRODUCTION TO IVIVC

One challenge that remains in biopharmaceutics research is that of correlating in vitro drug-release profiles with the in vivo pharmacokinetic data. IVIVC has been defined by the

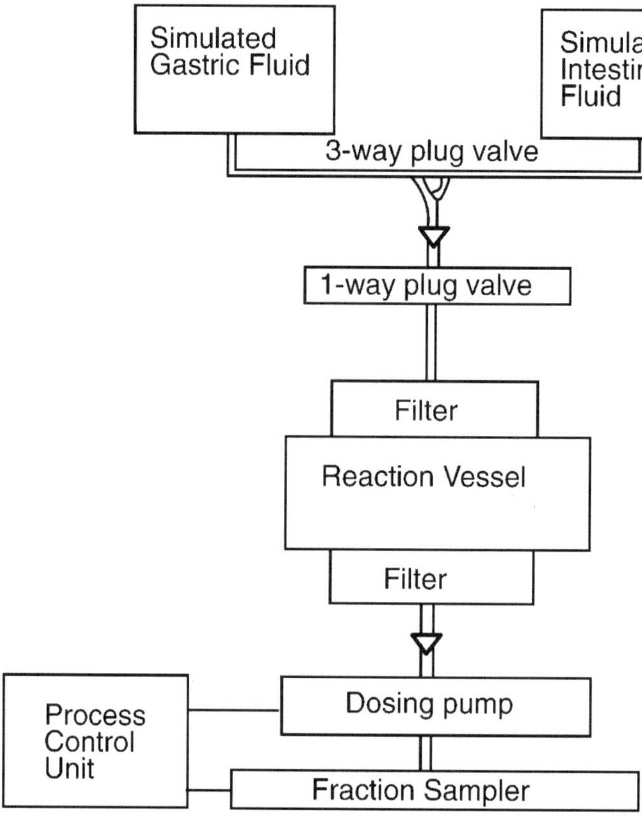

Figure 9 Scheme of in vitro dissolution model according to Stricker. *Source*: From Ref. 28.

FDA (29) as a "Predictive mathematical model describing the relationship between an in vitro property of the dosage form and an in vivo response." The concept behind establishing an IVIVC is that in vitro dissolution can serve as a surrogate for pharmacokinetic studies in humans, which may reduce the number of bioequivalence studies performed during the initial approval process as well as when certain scale-up and post-approval changes in the formulation need to be made. Obtaining a satisfactory correlation is, of course, highly dependent on the quality of the input variables. Though the dissolution testing gained official status in the USP in the early 1970s, it was questioned whether the dissolution data generated were sufficiently reliable to be used for IVIVC.

In case of pharmaceutical formulation development, the relation between the in vitro drug release from the dosage form and its in vivo biopharmaceutical performance needs to be within the acceptance criteria stated by the FDA guidance for industry. Lack of a relationship between the dissolution test results and in vivo behavior would lead to inappropriate control of the critical production parameters with the dissolution test methods and also confound biopharmaceutical interpretation of the dissolution test results. Therefore, in vitro specification limits should be set according to an established relationship between in vivo and in vitro results, best reached through a well-designed IVIVC. Relevant Guidances from the FDA reflect increasing consensus on in vitro–in vivo comparison techniques. Although some approaches deviate significantly from the standards, there is general agreement with the concept that in vitro systems should be developed which can distinguish between "good" and "bad" batches, ("good" in this context meaning "of acceptable and reproducible biopharmaceutical performance in vivo").

Two kinds of general relationships can be established between the in vitro dissolution and in vivo bioavailability: (1) IVIVC and (2) In vivo–in vitro associations. In the former, one or more in vivo parameters are correlated with one or more in vitro-release parameters of the product. In case of in vivo-in vitro associations, in vivo and in vitro performance of different formulations is in agreement, but a correlation does not exist per se. Situations can also exist where no correlation or association is possible between the in vitro and in vivo data (30). Regardless of which case applies, the extent of the relationships between the parameters must be clearly understood to arrive at a meaningful interpretation of the results (31). The procedures for comparing profiles and establishing an IVIVC are explained in detail in USP 27, Chapter <1088> and also addressed in Chapter 10 of this book. In the best case, IVIVC implies predictability of both similarity in and differences between in vitro and in vivo data in a symmetrical way, so that discrimination among formulations is even handed and the balance between patient and producer's risk is properly represented.

DISSOLUTION TESTING: WHERE ARE WE NOW?

The art and science of dissolution testing have come a long way since its inception more than 30 years ago. An appropriate dissolution procedure is a simple and economical method that can be utilized effectively to assure acceptable drug product quality and product performance (32). Dissolution testing finds application as a tool in drug development, in providing control of the manufacturing process, for batch release, as a means of identifying potential bioavailability problems and to assess the need for further bioequivalence studies relative to scale-up and post-approval changes (SUPAC) and to signal possible bioinequivalence of formulations (33). In the case of drug development, it is used to guide formulation development and to select an appropriate formulation for in vivo testing. With respect to quality assurance and control, almost all solid oral dosage forms require dissolution testing as a quality control measure before a drug product is introduced and/or released into the market. The product must meet all specifications (test, methodology, acceptance criteria) to allow batch release. Dissolution profile comparison has additionally been used extensively in assessing product sameness, especially when post-approval changes are made. Decades of extensive study and collaborative testing have increased the precision of test methodology greatly, leading to increasingly stringent protocols being used to optimize the repeatability of experimental results. It has also been recognized that the value of the test is significantly enhanced when the product performance is evaluated as a function of time. With the evolution and advances in the dissolution testing technology, the understanding of scientific principles and the mechanism of test results, a clear trend has emerged, wherein dissolution testing has moved from a traditional quality control test to a surrogate of in vitro bioequivalence test (34), which is generally referred as a biowaiver. This represents a shift in the dissolution thought process and a new regulatory perspective on dissolution.

A recent and important further development has been initiated by the research group of Dressman and Reppas (1) who introduced the concept of using more biorelevant dissolution media, FaSSIF and FeSSIF media. FaSSIF stands for Fasted State Simulated Intestinal Fluid and FeSSIF for Fed State Simulated Intestinal Fluid. These fluids consist of ingredients that provide physicochemical properties similar to the content of the human GIT. Their composition is given in Table 3 (see also Chapter 5). A practical feature of these physiologically based dissolution testing procedures is that they use compendial devices in combination with the biorelevant dissolution media. The procedures thus provide a link between research-oriented dissolution testing, mainly for development purposes, with a strong capability for predicting in vivo performance of the drug and/or drug product and routine quality control dissolution testing of batches in the industry, which is performed with the primary goal of detecting non-bioequivalent batches. More than a mere academic project this technology was proven to be useful as a surrogate for bioavailability (BA)/bioequivalence (BE) studies. Most recently, the collaborative work of Stippler (35) and Dressman together with the WHO has resulted in the development of dissolution methods and specifications that permit not only

Table 3 Composition of FeSSIF and FaSSIF Media

Composition	Quantity required for 1 L basis	
	FaSSIF	FeSSIF
NaH_2PO_4	3.9 g	—
NaoH	pH 6.5 (qs)	pH 5 (qs)
Na taurocholate	3 mM	15 mM
Lecithin	0.75 mM	3.75 mM
NaCl	7.7 g	11.874 g
Acetic Acid	—	8.65 g

quality control but also biopharmaceutical assessment of a group of drugs on the WHO'S List of Essential Medicines.

REFERENCES

1. Dressman JB, Reppas C. In vitro–in vivo correlations for lipophilic, poorly water soluble drugs. Eur J Pharm Sci 2000; 11:73–80.
2. Banakar UV. Introduction, Historical Highlights, and the Need for Dissolution Testing. Pharmaceutical Dissolution Testing. 49. New York: Marcel Dekker, 1991:1–18.
3. Pillai V, Fassihi R. Unconventional dissolution methodologies. J Pharm Sci 1999; 88(9):843–851.
4. Amidon GL, Lennernas H, Shah VP, Crison JR. A theoretical basis for a biopharmaceutic drug classification: the correlation of in vitro drug product dissolution and in vivo bioavailability. Pharm Res 1995; 12(3):413–420.
5. Shah VP. Dissolution: a quality control test vs. a bioequivalence test. Dissol Technol 2001; 11(4):1–2.
6. ICH Topic Q6A. Note on Guidance Specifications: test procedures and acceptance criteria for new drug substances and new drug products: chemical substances. Oct 6, 1999.
7. Crist B. The History of Dissolution Testing: Dissolution Discussion Group (DDG); North Carolina 1999.
8. Carstensen JT, Fun lai TY, Prasad VK. DSP Dissolution IV: comparison of methods. J Pharm Sci 1978; 67(9):1303–1307.
9. Grady TL. Perspective on the History of Dissolution Testing. Vice President and Director Emeritus, United States Pharmacopeia. Rockville, MD.
10. The National Formulary XIV (NF XIV). American Pharmaceutical Association, Washington, DC, General Tests, 1975; 892–894.
11. United States Pharmacopoeia 27 (USP 27); National Formulary 22 (NF 22). United States Pharmacopeial Convention, Rockville. MD 2003. <724> Drug Release:2157–2165.

12. European Pharmacopoeia 4th ed; European directorate for the quality of medicines, Council of Europe, France, 2002.
13. Borst I, Ugwu S, Beckett AH. New and extended applications for USP drug release apparatus 3. Dissol Technol 1997; 4(1):1–6.
14. Lawrence X, Jin T, Wang, Ajaz S, Hussain. Evaluation of USP Apparatus 3 for dissolution testing of immediate release products. AAPS Pharm Sci 2002; 4(1):1.
15. Sanghvi PP, Nambiar JS, Shukla AJ, Collins CC. Comparison of three dissolution devices for evaluating drug release. Drug Dev Ind Pharm 1994; 20(6):961–980.
16. Esbelin B, Beyssac E, Aiache JM, Shiu GK, Skelly JP. A new method of dissolution in vitro, the "Bio-Dis" apparatus: comparison with the rotating bottle method and in vitro: in vivo correlations. J Pharm Sci 1991; 80(10):991.
17. Kraemer J. Chewable Tablets and Chewing Gums. Workshop on Dissolution Testing of Special Dosage Forms, Frankfurt, March 05, 2001 (oral presentation).
18. Kraemer J. Untersuchungen zur In vitro Freisetzung und ihre Praediktiven Eigenschaften, Proc. 11. ZL-Experttreffen: Bioverfuegbarkeitsstudien zu mineralstoffen, Eschborn, Oct. 07, 1994.
19. Kraemer J, Stippler E. Chewable Tablets and Chewing Gums. Proceedings of the Royal British Pharmaceutical Society/FIP: Dissolution Testing of Special Dosage Forms, London, Sep. 02–03, 1999.
20. Siewert M, Dressman JB, Cynthia KB, Shah VP. FIP/AAPS guidelines for dissolution/in vitro release testing of novel/special dosage forms. Pharm Ind 2003; 65(2):129–134.
21. Hanson WA. Handbook of Dissolution Testing. Alternative Methods—reciprocating cylinder. Vol.2. Eugene, OR: Aster Publishing Corporation, 1991:42–45.
22. Langenbucher F. In vitro assessment of dissolution kinetics: description and evaluation of a column-type method. J Pharm Sci 1969; 58(10):1265–1272.

23. Langenbucher F, Benz D, Kuerth W, Moeller H, Otz M. Standardized flow-cell method as an alternative to existing pharmacoepoeial dissolution testing. Pharm Ind 1989; 51(11): 1276–1281.

24. FIP: Guidelines for dissolution testing of solid oral products. Joint report of the section for official laboratories and medicines control services and the section of Industrial pharmacists of the FIP. Dec: 1996.

25. United States Pharmacopoeia 27 (USP 27): National Formulary 22 (NF 22). United States Pharmacopeial Convention, Rockville. MD 2003; < 1225 > Validation of Compendial Methods: 2662–2625.

26. Stricker H. Die Arzneistoffresorption im Gastrointestinaltrakt-ln vitro-Untersuchung Lipophiler Substanzen. Pharm Ind: 1973; 35(1):13–17.

27. U.S. Department of Health and Human Services Food and Drug Administration Center for Drug Evaluation and Research (CDER). Guidance for Industry: Waiver of In Vivo Bioavailability and Bioequivalence Studies for Immediate Release Solid Oral Dosage Forms Based on a Biopharmaceutics Classification System. 2000.

28. Stricker H. Die In-vitro-Untersuchung der "Verfugbarkeit von Arzneistoffen" im Gastrointestinaltrakt. Pharm Tech 1969; 11:794–799.

29. Shah VP, Williams RL. In vivo and in vitro correlations: scientific and regulatory perspectives. Generics Bioequivalence 2000; 6:101–110.

30. Extended Release Solid Oral Dosage Forms: Development, Evaluation and Application of In vitro/In vivo Correlations. Center for Drug Evaluation and Research (CDER) FDA 1997.

31. United States Pharmacopoeia 27 (USP 27): National Formulary 22 (NF 22). United States Pharmacopeial Convention, Rockville, MD 2003; < 1088 > In vitro and In vivo Evaluation of Dosage forms: 2334–2339.

32. Shah VP, Williams RL. Roles of dissolution testing: regulatory, industry and academic perspectives: role of dissolution testing in regulating pharmaceuticals. Dissol Technol 1999; 8(3):7–10.

33. Gohel MC, Panchal MK. Refinement of lower acceptance value of the similarity Factor F2 in comparison of dissolution profiles. Dissol Technol 2002; 9(1).

34. Shah VP. Dissolution: a quality control test vs. a Bioequivalence test. Dissol Technol 2001; 11(4).

35. Stippler E. Bioequivalent dissolution test methods to assess bioequivalence of drug products. Ph.D. dissertation, Johann Wolfgang Goethe University, Frankfurt am Main, 2004.

2

Compendial Testing Equipment: Calibration, Qualification, and Sources of Error

VIVIAN A. GRAY
V. A. Gray Consulting, Incorporated,
Hockessin, Delaware, U.S.A.

INTRODUCTION

During the dissolution test, the hydrodynamic aspects of the fluid flow in the vessel have a major influence on the dissolution rate (1). Therefore, the working condition of the equipment is of critical importance. In this chapter, the qualification and calibration of the equipment referred to in the two USP General Chapters related to dissolution, < 711 > Dissolution and < 724 > Drug Release (2), will be discussed. Sources of error when performing dissolution tests and using dissolution equipment will be examined in detail later in the chapter.

QUALIFICATION

To ensure that equipment is fit for its intended purpose, there is a series of qualifying steps that the analyst or vendor should apply to analytical instrumentation (3,4). Equipment can be evaluated through a series of tests or procedures designed to determine if the system meets an established set of specifications governing the accepted operating parameters. The successful completion of such tests justifies that the system operates and performs as expected. There are four components of instrument qualification: design, installation, operational, and performance.

A. When developing a dissolution method, the design qualification is built into the apparatus selection process. The dosage form and delivery system process will dictate at least initially the equipment of choice. For example, the first choice for a beaded product may be United States Pharmacopeia (USP) Apparatus 3, which is designed to confine the beads in a screened-in cylinder.
B. The installation qualification consists of the procedures used to verify that an instrument has been assembled in the appropriate environment and is functioning according to pre-defined set of limits and tolerances. The data should be documented throughout the procedure, especially the hardware installation. Safety issues should be addressed. For example, setting up the fully automated dissolution equipment requires the proper plumbing, hot water source and pressure, electrical wiring and voltage, and drainage capability. Dissolution equipment should be installed on a stable bench top, free of environmental sources of vibration.

C. During operational qualification the analyst or vendor would assess if the equipment works as specified, generating appropriately documented data. The procedures will verify that the instrument's individual operational units are functioning within a given range or tolerance, reproducibly. For the dissolution apparatus, the water bath temperature and spindle assembly and shaft rpm speed would be obvious operational parameters.

D. Performance qualificationis conducted to ensure that the system is in a normal operating environment producing or performing designated set of tasks within the established specifications. In dissolution testing, the physical parameters such as centering, wobble, height of paddle or basket attached to shaft, speed, and temperature are performance qualifications. However, most important is the equipment performance with a known product, in many cases this is the calibration procedure using the calibrator tablets supplied by USP.

QUALIFICATION OF NON-COMPENDIAL EQUIPMENT

In dissolution testing of novel dosage forms, non-compendial equipment may be used. Some examples of non-compendial equipment are the rotating bottle, mini paddle, mega paddle (5), peak vessel, diffusion cells, chewing gum apparatus, and unique cell designs for USP Apparatus 4. In all cases, compendial equipment should be the first choice and there should always be justification, including data, showing why official equipment is not suitable.

Methods

If the equipment is a commercial product, the installation and operational qualifications can be obtained from the equipment vendor. This would include the vendor specifications and tolerances for the equipment. If it is an in-house design, then the process becomes more difficult. The first objective would be to look for adjustments and moving parts. Obtain a baseline of operational parameters, such as agitation rate (rpm), dip speed, flow rate, temperature, alignment, and/or volume control. After enough historical data have been obtained, examine the data for reproducibility, assessing the variability of the various components. If the analyst is satisfied that the equipment performs consistently, then choose ranges or limits based on this data. Then develop a per-run performance checklist based on these parameters.

Calibration

Non-compendial equipment, and in some cases compendial apparatus (Apparatus 4, for example), do not have calibrator tablets. In this case, an in-house calibrator tablet can be designated. This should be a product that is readily available with a large amount of reproducible historical data generated on the equipment. Evaluation of mechanical parameters such as agitation rate, volume control, alignment, etc. may be sufficient in some cases, circumventing the need to develop a calibrator tablet. However, it should be determined if there is some unique aspect of the equipment that can only be detected using a calibrator tablet. Currently, with Apparatus 1 and 2, vibration and vessel irregularities must be detected with the USP calibrator tablets, as there are no other practical measuring tools available to the analyst.

Hydrodynamics

The dissolution fluid flow characteristics should consist of a predictable pattern that is free of irregularities or variable turbulence. Observations of the product dissolution behavior are critical when choosing a dissolution apparatus. If there are aberrant or highly variable data that can be attributed to the apparatus, then it may be unsuitable for that product.

Other Considerations

When using non-compendial equipment, the transferability to another site or laboratory should be considered. Non-compendial equipment for quality control testing or at a contract laboratory could present problems of ruggedness. Therefore, ruggedness should be thoroughly evaluated before considering transferring product testing to another site, which uses a similar piece of equipment. For non-compendial as with compendial equipment, it is necessary to have adequate documentation, often with a log book, to keep track of maintenance, problems, repairs and product performance. Regular calibration, mechanical and/or chemical, should be documented and an appropriate time interval between calibrations determined. A standard operating procedure on operation, maintenance and calibration should be included. In addition, training and training documentation is critical. Further, the cleaning of all equipment parts is important, with special attention paid to parts that may be hard to clean and lead to contamination or residue build up.

COMPENDIAL APPARATUS

Apparatus 1 and 2

The USP Dissolution General Chapter <711> describes the basket (Apparatus 1) and paddle (Apparatus 2) in detail. There are certain variations in usage of the apparatus that occur in the industry and are allowed with proper validation. The literature contains a recommendation for a new USP general chapter for dissolution testing (6). In this article, guidance for method validation and selection of equipment is described. It may be a useful guide when showing equipment equivalence to compendial equipment.

Calibration or Apparatus Suitability Test

In <711>, there is a paragraph titled the Apparatus Suitability test. In this paragraph, the use of the USP calibrator tablets (Fig. 1) is required. There is some debate as whether

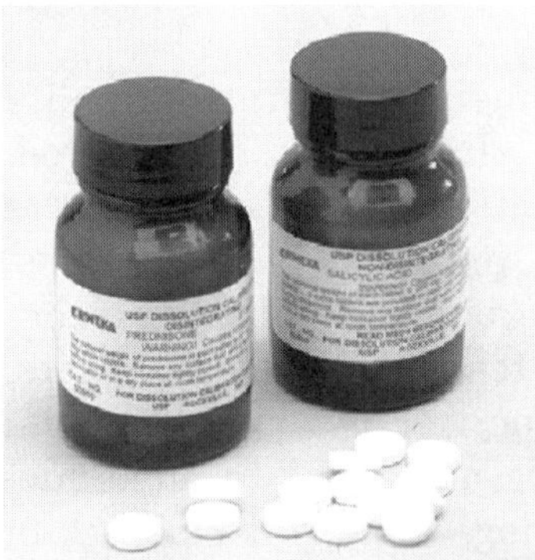

Figure 1 USP calibrator tablets, prednisone and salicylic acid. (Courtesy of Erweka, GmbH, Heusenstamm, Germany.)

the calibrator tablets are misnamed, since the tablets do not correct or adjust any parameter. During calibration, the analyst is given a set of ranges that need to be met by each calibrator tablet. The results of the calibration tell the analyst whether the apparatus is suitable. The calibrator tablets have a long history (7). The major reason for the calibrator tablets, and this remains a major reason for them today, is the ability of the tablets to pick up vibration effects. The Dissolution Committee within Pharmaceutical Research Manufacturers of America (PhRMA) formerly known as Pharmaceutical Manufacturers of America (PMA) conducted the collaborative studies that determined the aforementioned ranges for the initial USP calibrator tablets. These collaborative studies included 20–30 laboratories that performed dissolution tests on the calibrator tablets using both the basket and paddle dissolution apparatus at different speeds. This procedure is still followed today for new batches of calibrator tablets and the results of the studies are published in the Pharmacopeial Forum (PF) of the USP to inform the scientific community how the range specifications are obtained and show the detailed statistical analysis (8). Within the PhRMA Dissolution Committee, there was a Dissolution Calibration Subcommittee. This subcommittee's purpose was to examine the dissolution bath calibration and look for ways to reduce testing without relaxing the standards for operating the equipment. For example, mechanical calibration was studied thoroughly as an alternative to using the calibrator tablet testing (9,10).

Heating Jacket

A water-less bath method is stated in <711> as an alternative way to heat the vessels other than a conventional water bath (11). As shown in Figure 2, the vessels are heated with a water jacket and are not submerged into a water bath. With this bath, as with all testers that use the basket apparatus, when the basket shaft with the basket is introduced into the vessel medium, the temperature will drop slightly. There-

Figure 2 Water bath-less dissolution testing equipment. (Courtesy of Distek, Inc., North Brunswick, New Jersey, U.S.A.)

Figure 3 Peak vessel. (Courtesy of VanKel, a member of the Varian, Inc. Life Science Business, Cary, North Carolina, U.S.A.)

fore, equilibration or stabilization of the vessel medium temperature is necessary before beginning the run.

Peak Vessel

This vessel is designed to eliminate "mounding or coning" by having a cone molded into the bottom of the glass vessel, see Figure 3. The peak vessel is non-compendial, but may have utility with products that contain dense excipients that can have a tendency to cone rather than disperse freely inside the vessel (12).

Clip and Clipless Baskets

Two types of basket shafts are commercially available to the analyst. One type has an O-ring inset in the disk at the end of the shaft with the basket fitting snuggly around the O-ring. The other has three clips attached to the disk at the end of the shaft. The basket is attached by fitting between the clips and the disk. The latter design is described in <711>. The two designs are shown in Figure 4. A recent study (13) compared

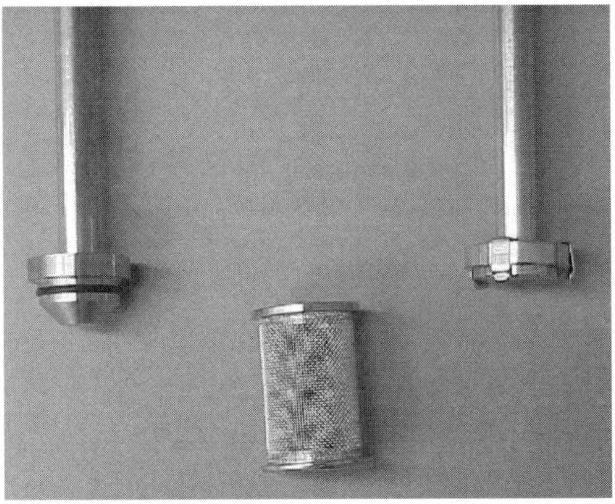

Figure 4 Two basket attachment designs: On the left is the O-ring design and on the right is the three-pronged USP Apparatus 1 design.

these two types of basket shafts using the two USP calibrator tablets, prednisone and salicylic acid, and three development products. The study concluded that there was no difference between the two basket shaft types for the three development products and USP salicylic acid tablets. However, the USP prednisone calibrator tablets did show a significantly different dissolution rate, with a higher dissolution rate using the clipped basket shaft design. The clipped basket shaft is the official USP design; however, there are some drawbacks to this design. The clips protrude and disturb the fluid flow in the vessel. In addition, the clips can weaken over time and cause the basket to be attached too loosely to the shaft—increasing the chance for wobble. Further, when using robotic dissolution testers, a robotic arm can remove the O-ring-type basket more efficiently.

Since the O-ring style is not an official design, the analyst should show that it does not give results different from the clipped shafts when testing the product. As part of validation, the two basket shaft types should be compared and equivalence shown. If the types do not give comparable results, there could be problems with technology transfer. In addition, if a

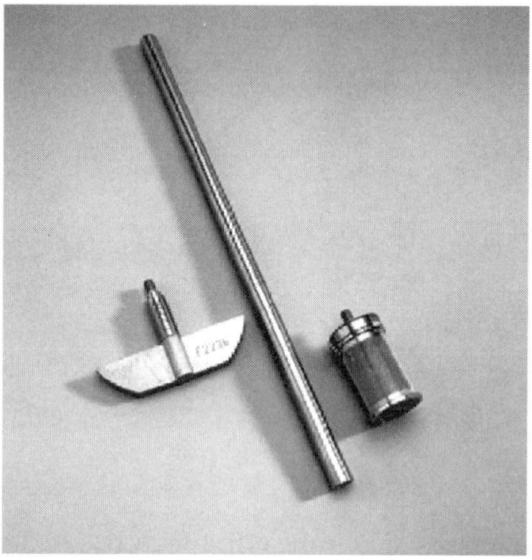

Figure 5 Detachable basket and paddle apparatus device. (Courtesy of Erweka, GmbH, Heusenstamm, Germany.)

regulatory agency performs the dissolution test on a product using the USP procedure, the results obtained could be different.

Single Entity, Including Two-Part Detachable Shaft Design

In Figure 5, an example of the two-part detachable design is shown. As <711> states, the assembly must be firmly

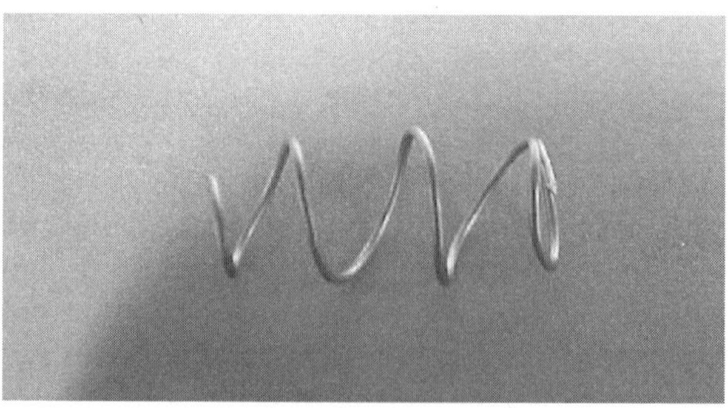

Figure 6 A hand-made sinker.

engaged during the test. If this aspect is satisfied then no particular equivalence validation needs to occur. During calibration this apparatus using this two-part design would be assessed for significant wobble.

Sinkers

Sinkers are used for floating or sticking of dosage forms. The description of sinkers in <711> is brief and not detailed. An example of a hand-made USP sinker is shown in Figure 6. In Figure 7, the sinker described in the Japanese Pharmacopoeia (JP) is pictured, but several other sinkers are available commercially. Since <711> contains the statement that other validated sinkers may be used, any of these designs could be considered.

Deaeration

The compendium contains a note in <711> that requires that air bubbles be removed if they change the results of the test. The suggested method found as a footnote in <711> uses heat followed by filtration under vacuum. There is a plethora of methods for deaeration (14), an earlier method was to boil and cool the medium. There are also several varieties of automated deaeration equipment. The mechan-

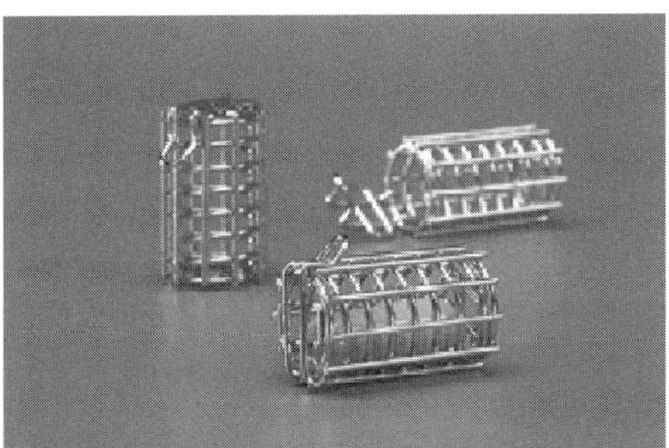

Figure 7 The sinker required in the Japanese Pharmacopeia. (Courtesy of VanKel, a member of the Varian, Inc. Life Science Business, Cary, North Carolina, U.S.A.)

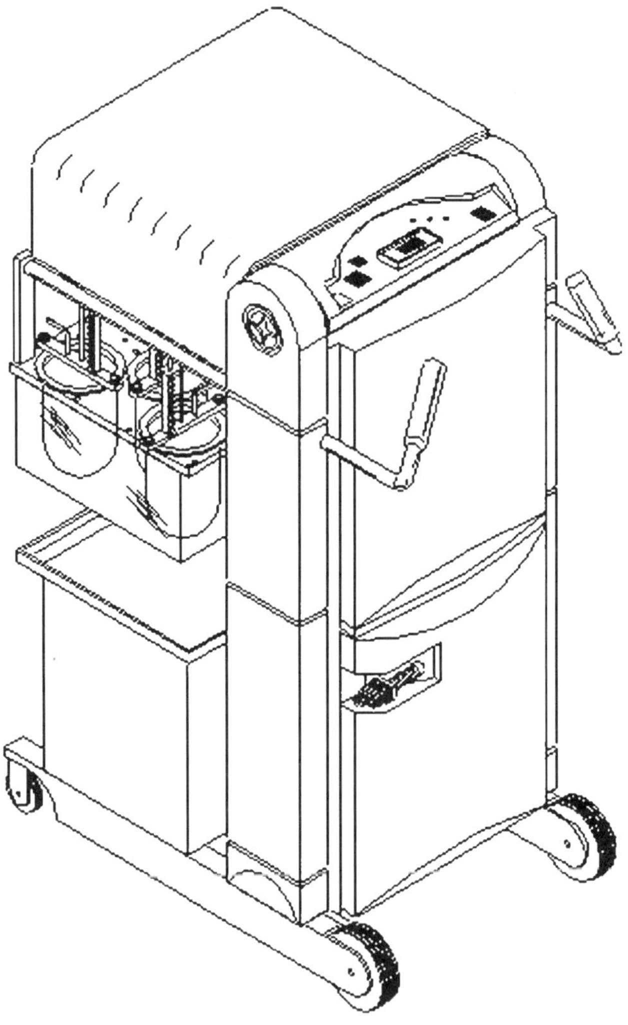

Figure 8 Deaeration equipment. (Courtesy of Hanson Research Corporation, Chatsworth, California, U.S.A.)

ism for the equipment shown in Figure 8 uses a thin film vacuum; that is, pre-heated dissolution media is slowly injected through a spray-disbursing nozzle into a closed vessel. As the media is sprayed, vacuum is applied to remove gasses. The closed chamber will fill to a pre-adjusted volume

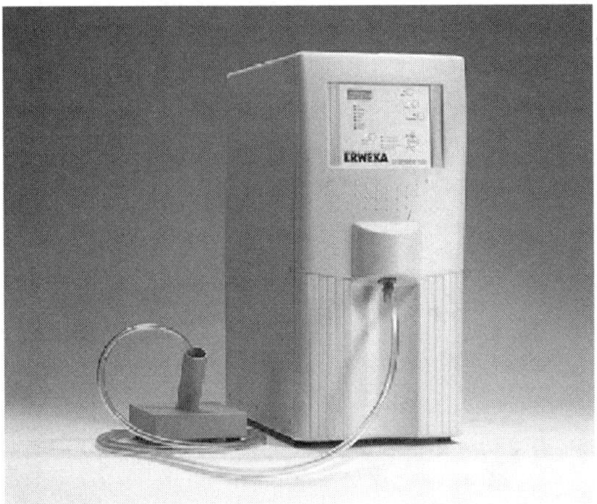

Figure 9 Deaeration equipment. (Courtesy of Distek, Inc., North Brunswick, New Jersey, U.S.A.)

level (typically 900 mL) and then, media is subsequently dispensed into the dissolution flasks. With the equipment shown in Figure 9, the media is filtered, heated and degassed under vacuum, and precisely dispensed in individual volumes into each vessel.

Automated Sampling

Modification of the apparatus to accomplish automation is allowed by <711>. One example is hollow shaft sampling as illustrated in Figure 10 (15). This method is theoretically within the stated sampling location of the text of <711>, although there may be question about the concentration of sample surrounding the shaft. This and other sampling techniques, for example in-residence probes, are convenient sampling tools but should be properly validated.

Apparatus 3

We have now started to discuss the equipment in the USP Drug Release General Chapter (<724>). The reciprocating cylinder, as shown in Figure 11, has special utility for beaded products along with the capability of changing medium by

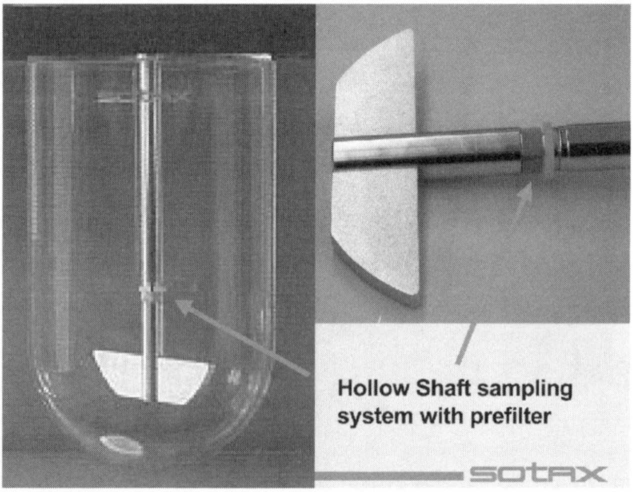

Figure 10 Hollow shaft autosampler. (Courtesy of Sotax Corporation, Horsham, Pennsylvania, U.S.A.)

removing the dosage unit and placing it in another pH medium. This apparatus has been found to be useful for both immediate and controlled-release products (16).

Calibration

This equipment has one calibrator tablet: a single tablet product, chlorpheniramine extended-release tablets (drug-release calibrator, single unit). It has been found that this equipment is not particularly sensitive to vibration and has reliable and consistent operation (17).

Apparatus 4

The flow-through cell is especially useful for dissolution rate-limited products, where sink conditions may be hard to obtain (18,19). The operation of the flow-through cell is illustrated in Figure 12. A closer look at the tablet holders is shown in Figure 13. This particular apparatus can be utilized as either a closed or open system. In Figure 14, the closed system mode,

USP 3 (Bio-Dis)

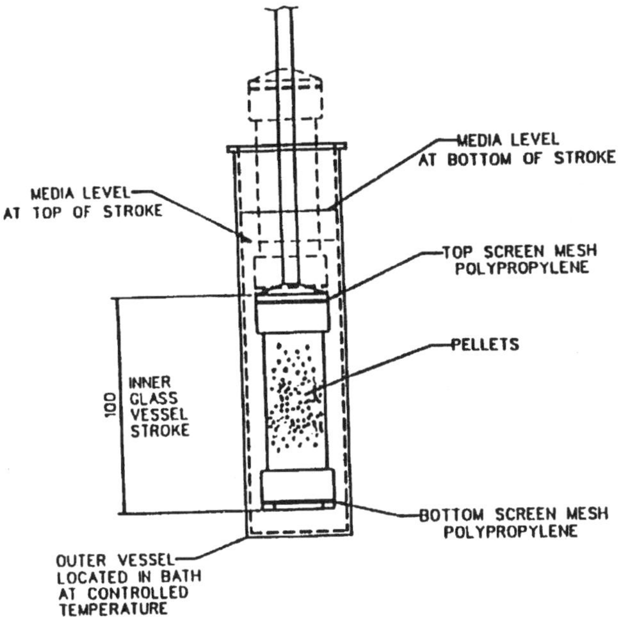

Figure 11 Apparatus 3. (Courtesy of VanKel, a member of the Varian, Inc. Life Science Business, Cary, North Carolina, U.S.A.)

including on-line ultraviolet sampling using flowcells, is illustrated. Notice that there is no part of the equipment design that allows for waste lines or sampling ports. The system would conserve medium, continuing to recycle the testing liquid. The open system mode, which is typical in dissolution testing, is shown in Figure 15. With the flow-through cell design, this system uses a copious amount of medium for the test, especially if the test is continued for many hours.

Calibration

The performance of the apparatus has been studied using the USP prednisone and salicylic acid tablets (20), but to date there are no official calibrator tablets for Apparatus 4. As

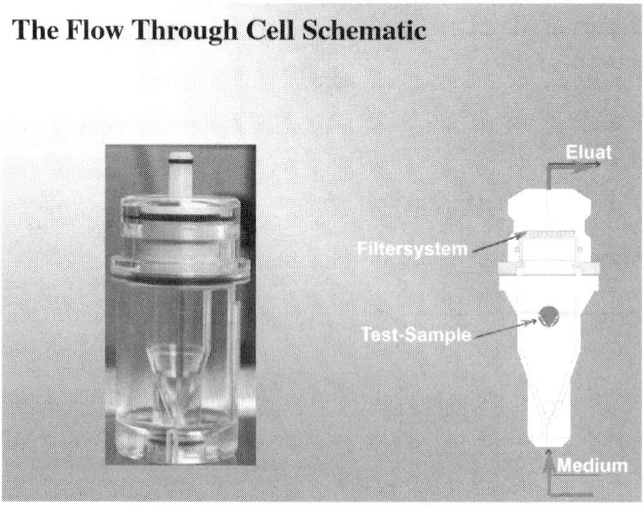

Figure 12 Schematic of Apparatus 4. (Courtesy of Sotax Corporation, Horsham, Pennsylvania, U.S.A.)

mentioned previously, the critical instrument parameters should be measured and limits or ranges set. For this equipment, flow rate is the most critical factor. The medium must also deaerated.

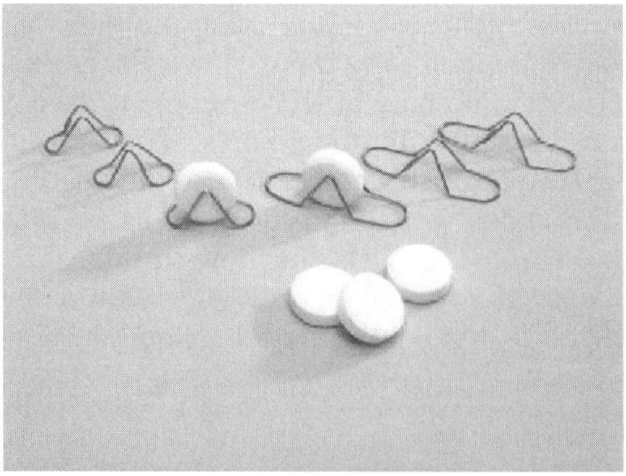

Figure 13 Apparatus 4 tablet holders. (Courtesy of Erweka, GmbH, Heusenstamm, Germany.)

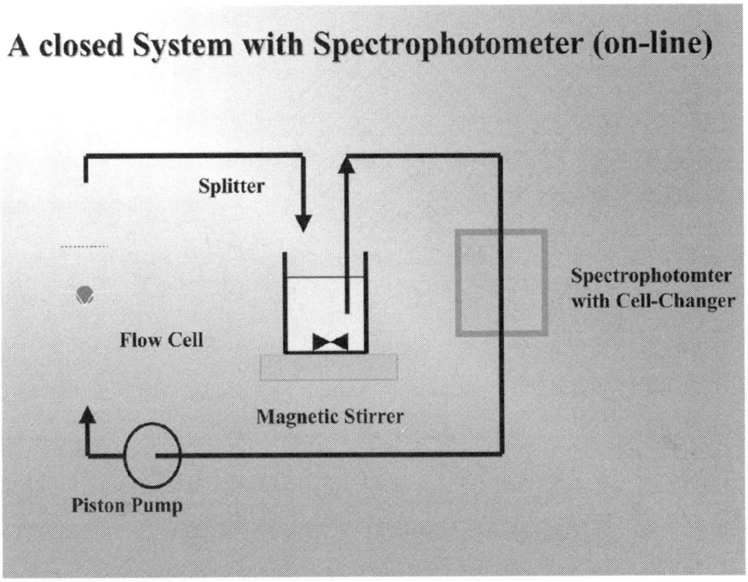

Figure 14 Schematic of the Apparatus 4 as a closed system. (Courtesy of Sotax Corporation, Horsham, Pennsylvania, U.S.A.)

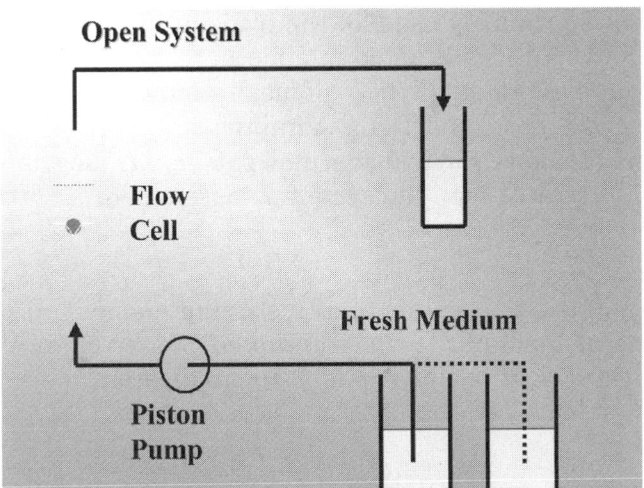

Figure 15 Schematic of the Apparatus 4 as an open system. (Courtesy of Sotax Corporation, Horsham, Pennsylvania, U.S.A.)

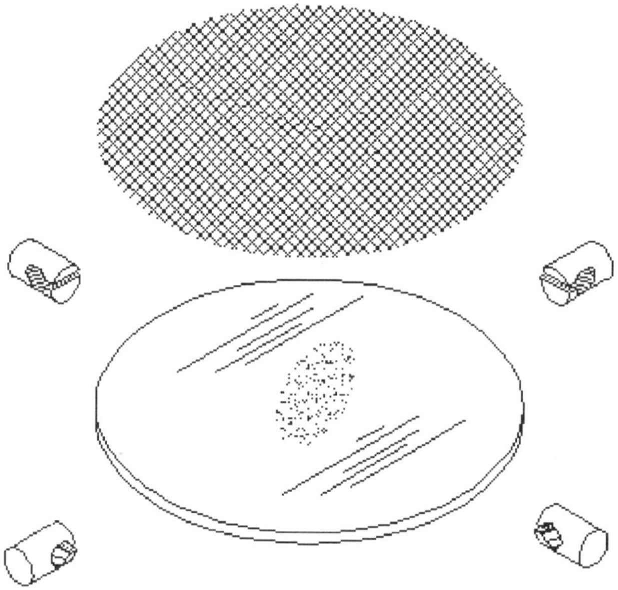

Figure 16 The watchglass–patch–polytef mesh sandwich. (Courtesy of Hanson Research Corporation, Chatsworth, California, U.S.A.)

Apparatus 5

This apparatus is primarily used for the transdermal patch. A variation of the apparatus is noted in a footnote in <724> . It is called the watchglass–patch–polytef mesh sandwich , and is favored by the US Food and Drug Administration (FDA) as the equipment of choice for transdermal patches. A diagram in Figure 16 illustrates how the system is assembled.

Calibration

This apparatus uses the paddle as the stirring element in a typical volume of medium. If the equipment passes calibration for Apparatus 2, it is suitable for this application.

Apparatus 6

The rotating cylinder is shown in Figure 17. It also in used for transdermal patches and can be lengthened for larger patches using an adapter.

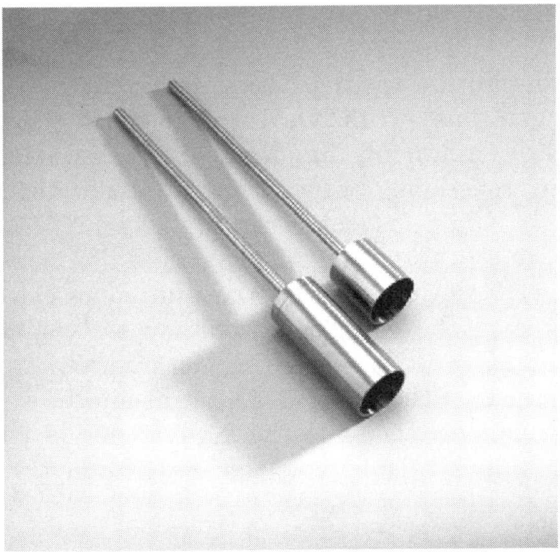

Figure 17 Apparatus 6. (Courtesy of Erweka, GmbH, Heusenstamm, Germany.)

Calibration

If the equipment passes the calibration for basket and paddles, then it can be assumed that the spindle assemblies, motor, and drive belt are functioning properly. The analyst may be able to test the wobble using equipment that assesses the run out measurement for the basket.

Apparatus 7

This apparatus has many design configurations, some applying to transdermal patches and others to oral dosage forms, in particular the osmotic pump extended-release tablet.

Calibration

There are no calibrator tablets available for this apparatus. The approach to performance qualification would be as outlined previously, that is, to determine the critical parameters, which in this case will include dip rate and volume control.

SOURCES OF ERROR

When performing dissolution testing, there are many ways that the test may generate erroneous results. The testing equipment and its environment, handling of the sample, formulation, in situ reactions, automation and analytical techniques can all be the cause of errors and variability. The physical dissolution of the dosage form should be unencumbered at all times. Certain aspects of the equipment calibration process may show these errors as well as close visual observation of the test. The essentials of the test are accuracy of results and robustness of the method. Aberrant and unexpected results do occur, however, and the analyst should be well trained to examine all aspects of the dissolution test and observe the equipment in operation.

Drug Substance Properties

Knowledge of drug properties, especially solubility in surfactants or as a function of pH, is essential. One could anticipate precipitation of the drug as the pH changes in solution, or if release from the dosage form leads to supersaturation of the test media. Be aware that preparation of a standard solution may be more difficult than expected. It is customary to use a small amount of alcohol to dissolve the standard completely. A history of the typical absorptivity range of the standard can be very useful to determine if the standard has been prepared properly.

Drug Product Properties

Highly variable results indicate that the method is not robust, and this can cause difficulty in identifying trends and effects of formulation changes. Two major causal factors influence variability: mechanical and formulation. Mechanical causes can arise from the dissolution conditions chosen. Carefully observe the product as it dissolves. An apparatus or speed change may be necessary. The formulation can have poor content uniformity, additionally, reactions and/or degradation may be occurring in situ. The film coating may cause sticking to the vessel walls. Upon aging, capsule shells are known for pellicle formation and tablets may become harder or softer, depending upon the excipients and drug interaction with moisture, which in turn may affect the dissolution and disintegration rate.

Equipment

Major components of dissolution equipment are the tester (including typically, but not limited to, spindle assemblies, belt, motor, tension adjuster, and circulator pump and hoses), water bath, paddles, baskets and shafts, vessels, samplers, and analyzers. Mechanical aspects, such as media temperature, paddle or basket speed, shaft centering and wobble, and vibration can all have a significant impact on the dissolution of the product. Mechanical and chemical calibration should therefore be conducted periodically, usually every 6 months, to ensure that the equipment is working properly.

In <711>, there is a requirement for the analyst to perform the apparatus suitability test using USP calibrator tablets. USP calibrator tablets come with certificates identifying appropriate ranges. The apparatus suitability test is designed to detect sources of error associated with improper operation and inadequate condition of the equipment (9,10,21). Two calibrators are used; USP prednisone tablets, 10 mg, and USP salicylic acid tablets, 300 mg. Use of each of these types of calibrator tablets involves calibrator-specific considerations. The salicylic acid tablets should be brushed before using to remove fine particles. This task should be performed in a hood to avoid breathing the irritating dust. Use whole tablets, and check whether the tablets are chipped or nicked. Since this tablet dissolves through erosion and is pure compressed salicylic acid, minor chips or nicks have no significant effect on the dissolution rate, if large chunks are missing results may be affected. The buffer should be prepared according to USP Reagent (Buffers) section.

The prednisone tablets use deaerated water as the medium. There are numerous methods for deaeration of medium (14,22). As mentioned above, there are also automated methods available. The method described in <711> uses heat, filtration, and vacuum. Helium sparging is also a typical method for deaeration. The level of dissolved oxygen and other gases is related to the presence of bubbles. Bubbles are common and will cause problems in non-deaerated medium. In <711>, it is stated that bubbles can interfere with dissolution test results and should be avoided. Dissolved air can slow down dissolution by creating a barrier; either adhering to the tablet surface or to basket screens, or particles can cling to bubbles on the glass surface of the vessel or shafts. Dissolution tests should always be performed immediately after deaeration. It is best not to have the paddle rotating before adding the tablet, as paddle movement will reaerate the medium.

When preparing standard solutions, be sure to dry the reference standard properly, preferably on the day of use. Care should be taken to ensure that the drug powder is completely dissolved. In the case of prednisone reference standard, the powder becomes very hard upon drying, making it slower to dissolve. Dissolving the powder first in a small amount of alcohol helps to overcome this problem.

Vibration interference is a common problem with dissolution equipment (23). Careful leveling of the top plate and lids is critical. Within the spindle assembly, the bearings can become worn and cause vibration and wobble of the shaft. The drive belts should be checked for wear and dirt. The tension adjustments for the belt should be optimized for smooth operation. Surging of spindles, though difficult to detect without closely scrutinizing the tester operation, can cause spurious results. Vessels need to be locked in place so that they are not moving with the flow of water in the bath.

External vibration sources might include other equipment on bench tops, such as shakers, centrifuges, or sonicators. Local construction in the area or within the building is a common, though often overlooked, source of vibration. The testers should not be near hoods or significant airflow sources. Additionally, heavy foot traffic and door slamming should be avoided.

These days, the water bath itself is rarely a source of vibration because the design has been changed to eliminate noisy circulators near the bath. Measuring the temperature of the medium in all the vessels, rather than just one, can assure the temperature uniformity. The bath water level should always be maintained at the top of the vessels to ensure uniform heating of the medium. Lastly, the water bath should contain clean water so that observations of the dissolution test can be performed clearly and easily.

Close inspection of USP Apparatus 1 and 2 before use can help identify sources of error. Obviously, dimensions should be as specified. In cases of both baskets and paddles, shafts must be straight and true. The paddles are sometimes partially coated with Teflon. This coating can peel and partially shed from the paddle, causing flow disturbance of hydrodynamics within the vessel. Paddles can rust and become nicked or dented; this can adversely affect dissolution hydrodynamics and be a source of contamination. Thorough cleaning of the paddles is also important, to preclude carry over of drug or medium.

The baskets need special care and examination. They can become frayed, misshapen, or warped with use. Screen mesh size may change over time, especially when used with acidic medium. Baskets are especially prone to gelatin or excipient build up if not cleaned immediately after use.

Vessels have their own set of often-overlooked problems. Vessels are manufactured from large glass tubing. Then the vessel bottom is individually rounded. Depending upon techniques of the heating/shaping process, irregular surfaces can occur and the uniformity of vessel bottom roundness can vary. Cheaply made vessels are notorious for this problem. Close examination of vessels when newly purchased is very important, as surface irregularity can cause dissolution results to differ significantly. Another common problem with vessels is residue build up either from oily products or sticky excipients. Insoluble product, not rinsed well from previous testing, can also cause contamination. Vessels become scratched and etched after repeated washing with wire brushes and should be discarded. Lids need to be in place to prevent evaporation. As mentioned before, vessels should be locked down to avoid vibration.

Off center shafts are often critical factors in failed calibration, especially with the USP prednisone calibrator tablets.

In assessing calibration failure, one should examine the system, changing one parameter at a time. Repeated testing until passing results are obtained is strongly discouraged, as it does not address the underlying problem. If aberrant results are obtained with just one vessel, only this position needs to be retested. But if adjustments are made to the tester, the entire calibration procedure must be conducted for all positions. Good manufacturing practices dictate that all adjustments should be documented and that all maintenance recorded.

Method Considerations

The best way to avoid errors and data "surprises" is to put a great deal of effort into selecting and validating methods. There are many good references on method selection and validation (6,24,25). Some areas of testing are especially troublesome. Sample introduction can be tricky and, unfortunately at times, not easy to perform reproducibly. Products can have a dissolution rate that is "position dependent." For example, if the tablet is off-center, the dissolution rate may be higher due to shear forces. Or if it is in the center, coning may occur and the dissolution rate will go down. Film-coated tablets can be sticky and pose problems related to tablet position in the vessel. Little can be done except to use a basket (provided there is no gelatinous or excipient build up) or a sinker.

Suspensions can be introduced in a variety of ways. Some examples are to manually use syringes or pipettes, pour from a tared beaker, or automate delivery using calibrated pipettes. Each method has its own set of limitations, although automated methods may show less variability. Mixing of the suspension sample will generate air bubbles; therefore, the mixing time of suspension samples must be strictly uniform to reduce erroneous or biased results.

The medium is a critical component of the test that can cause problems. One cause of inaccurate results may be that too great a volume of medium has been removed, through multiple sampling without replacement, in which case sink conditions may no longer prevail.

Surfactants can present quite a cleaning problem, especially if the concentration is high (over 0.5%). In the sampling lines, surfactants such as sodium lauryl sulfate may require many rinsing to assure total elimination. The same is true with carboys and other large containers. This particular surfactant has other limitations, as quality can vary depending upon grade and age and the dissolving effect can consequently change, depending upon the surface-active impurities and electrolytes (26). The foaming nature of surfactants can make it very difficult to effectively deaerate. Some pumps used in automated equipment simply are not adapted to successful use with surfactants. One caution when lowering a basket into surfactant medium is that surface bubbles can adhere to the bottom of the basket and decrease the dissolution rate substantially. When performing HPLC analysis using surfactants as the medium, several sources of error may be encountered. The autoinjectors may need repeated needle washing to be adequately cleansed. Surfactants, especially cetrimide, may be too viscous for accurate delivery. Surfactants can affect column packing to a great degree, giving extraneous peaks or poor chromatography. Basic media, especially above 8 pH, may cause column degradation.

Observations

One of the most useful tools for identifying sources of error is close observation of the test. A well-trained analyst can pinpoint many problems because he or she understands the cause and effect of certain observations. Accurate, meaningful dissolution occurs when the product dissolves without disturbance from barriers to dissolution, or disturbance of vessel hydrodynamics from any source. The particle disintegration pattern must show freely dispersed particles. Anomalous dissolution usually involves some of the following observations: floating chunks of tablet, spinning, coning, mounding, gumming, swelling, capping, "clam shell" erosion, off-center position, sticking, particles adhering to apparatus or vessel walls, sacs, swollen/rubbery mass, or clear pellicles. Along with good documentation, familiarity with the dissolution behavior of a product is essential in quickly identifying changes in stability or changes associated with a modification of the formulation. One may notice a change in the size of the dissolving particles, excipients floating upward, or a slower erosion pattern. Changes in the formulation or an increase in strength may produce previously unobserved basket screen clogging. If contents of the basket immediately fall out and settle to the bottom of the vessel, a spindle assembly surge might be the cause. If the medium has not been properly deaerated, the analyst may see particles clinging to vessel walls. The presence of bubbles always indicates that deaeration is necessary.

Sinkers are defined in USP as "not more than a few turns of a wire helix. . . ." Other sinkers may be used, but the analyst should be aware of the effect different types of sinkers may have on mixing (27). Sinkers can be barriers to dissolution when the wire is wound too tightly around the dosage unit.

Filters are used on almost all analyses; many types or different materials are used in automated and manual sampling. Validation of the pre-wetting or discard volume is critical for both the sample and standard solutions. Plugging of filters is a common problem, especially with automated devices and with Apparatus 4.

Manual sampling techniques can introduce error by virtue of variations in strength and size of the human hand, from analyst to analyst. As a result, the pulling velocity through the filter may vary considerably. Too rapid a movement of liquid through the filter can compromise the filtration process itself.

Automation

While automation of dissolution sampling is very convenient and laborsaving, errors often occur with these devices because the analysts tend to overlook problem areas. Sample lines are often a source of error for a variety of reasons: unequal lengths, crimping, wear beyond limits, disconnection, carryover, mix-ups or crossing, and inadequate cleaning.

The volume dispensed, purged, recycled, or discarded should be routinely checked. Pumping tubes can wear out through normal use or repeated organic solvent rinsing and may necessitate replacement.

The use of flow cells may generate variability in absorbance readings. Air bubbles can become caught in the cell, either introduced via a water source containing bubbles or by air entering inadvertently into poorly secured sample lines. Flow rate and dwell time should be evaluated so that the absorbance reading can be determined to have reached a steady plateau. Cells need to be cleaned frequently to avoid build up of drug, excipient, surfactant, or buffer salts from the dissolution medium.

Cleaning

The analyst should take special care to examine this aspect when validating the method. In many laboratories, where different products are tested on the same equipment, this is a critical issue that, if inadequately monitored, may be a cause of inspection failures.

Method Transfer

Problems occurring during transfer of methods can often be traced to not having used exactly the same type of equipment, such as baskets/shafts, sinkers, dispensing apparatus, or sampling method. A precise description of medium and standard preparation, including grade of reagents, may be useful. The sampling technique (manual vs. automated), and sample introduction, should be uniform.

REFERENCES

1. Mauger JW. Physicochemical and fluid mechanical principles applied to dissolution testing. Dissolution Technol 1996; 3(1):7–11.
2. USP 25/NF 20. Maryland: United States Pharmacopeial Convention, Inc., 2002.
3. Sigvardson KW, Manalo JA, Roller RW, Saless F, Wasserman D. Laboratory equipment qualification. Pharm Technol 2001; October:102–108.

4. Burgess C, Jones DG, McDowall RD. Equipment qualification for demonstrating the fitness for purpose of analytical instrumentation. Analyst 1998; 123:1879–1886.

5. Ross MS, Rasis M. Mega paddle—a recommendation to modify Apparatus 2 used in the USP general test for dissolution <711>. Pharm Forum 1998; 24(3):6351–6359.

6. Gray VA, Brown CK, Dressman JB, Leeson LJ. A new general information chapter on dissolution. Pharm Forum 2001; 27(6):3432–3439.

7. Morgan TA. History of dissolution calibration. Dissolution Technol 1995; 2(4):3–9.

8. PhRMA Dissolution Committee. The USP dissolution calibrator tablet collaborative study—an overview of the 1996 process. Pharm Forum 1997; 23(3):4198–4242.

9. PhRMA Subcommittee on Dissolution Calibration, Brune S, Bucko J, Emr S, Gray V, Hippeli K, Kentrup A, Whiteman D, Loranger M, Oates M. Dissolution calibrator: recommendations for reduced chemical testing and enhanced mechanical calibration. Pharm Forum 2000; 26(4):1149–1166.

10. Mirza T, Grady LT, Foster TS. Merits of dissolution system suitability testing: response to PhRMA's proposal on mechanical calibration. Pharm Forum 2000; 26(4):1167–1169.

11. Brinker G, Goldstein B. Bathless dissolution: validation of system performance. Dissolution Technol 1998; 5(2):7–14, 22.

12. Beckett AH, Quach TT, Kurs GS. Improved hydrodynamics for USP apparatus 2. Dissolution Technol 1996; 3(2):1–4.

13. Gray VA, Beggy M, Brockson R, Corrigan N, Mullen JA. A comparison of dissolution results using O-ring versus clipped basket shafts. Dissolution Technol 2001; 8(4):8–11.

14. Queshi SA, McGilveray IJ. Impact of different deaeration methods on the USP dissolution apparatus suitability test criteria. Pharm Forum 1994; 20(6):8565–8566.

15. Schauble T. A comparison of various sampling methods for tablet release tests using the stirrer method [USP apparatus 1 & 2]. Dissolution Technol 1996; 3(2):11–15.

16. Borst I, Ugwu S, Beckett AN. New and extended applications for USP drug release apparatus 3. Dissolution Technol 1997; 4(1):1–6.
17. Rohrs BR. Calibration of the USP 3 [reciprocating cylinder] dissolution apparatus. Dissolution Technol 1997; 4(2):11–18.
18. Nicolaides E, Hempenstall JM, Reppas C. Biorelevant dissolution tests with the flow-through apparatus. Dissolution Technol 2000; 7(1):8–11.
19. Looney TJ. USP apparatus 4 [flow through method] primer. Dissolution Technol 1996; 3(4):10–12.
20. Nicklasson M, Langenbucher F. Description and evaluation of the flow cell dissolution apparatus as an alternative test method for drug release. Pharm Forum 1990; 16(3):532–537.
21. Thakker KD, Naik NC, Gray VA, Sun S. Fine-tuning of dissolution apparatus for the apparatus suitability test using the USP dissolution calibrators. Pharm Forum 1980; 6(4):177–185.
22. Moore TW. Dissolution testing: a fast, efficient procedure for degassing dissolution medium. Dissolution Technol 1998; 3(2): 3–5.
23. Collins CC. Vibration—what is it and how does it affect dissolution testing. Dissolution Technol 1998; 5(4):16–18.
24. Rohrs BR. Dissolution method development for poorly soluble compounds. Dissolution Technol 2001; 8(3):6–12.
25. Leeson LJ. ANDA dissolution method development and validation. Dissolution Technol 1997; 4(1):5–9, 18.
26. Crison JR, Weiner ND, Amidon GL. Dissolution media for in vitro testing of water-insoluble drugs, effect of surfactant purity and electrolyte on in vitro dissolution of carbamazepine in aqueous solutions of sodium lauryl sulfate. J Pharm Sci 1997; 86(3):384–388.
27. Soltero RA, Hoover JM, Jones T, Standish M. Effects of sinker shapes on dissolution profiles. J Pharm Sci 1989; 78(1):35–39.

3

Compendial Requirements of Dissolution Testing—European Pharmacopoeia, Japanese Pharmacopoeia, United States Pharmacopeia

WILLIAM E. BROWN
Department of Standards Development,
United States Pharmacopeia,
Rockville, Maryland, U.S.A.

PHARMACOPEIAL SPECIFICATIONS

A pharmacopeia is a collection of recommended specifications and other information for therapeutic products, including drug substances (active ingredients), excipients, dosage forms (also called preparations), and other articles. One function of a pharmacopeia is to provide a uniform and public basis on which to evaluate these therapeutic products, which are used in the practice of medicine and pharmacy. Ingredients and products that fall short of these specifications can be judged unsuitable for commerce. The authority of such a collection is given through the particular regulatory mechanism of the country, as in the United States or Japan, or in a multinational region, as for Europe. The existence of such a body of information allows its citation outside of its originating environment. Thus, reference may be found in the regulations of countries thousands of miles from the primary national or regional audience. Note that for the purposes of this chapter, the International Conference on Harmonization of Technical Requirements for the Registration of Pharmaceuticals (ICH) definition of a specification as "a list of tests, associated analytical procedures, and acceptance criteria" will be used (1).

HISTORICAL BACKGROUND AND LEGAL RECOGNITION

European Pharmacopoeia

The states of Europe have a deep history of pharmacopeial activity that even now is evidenced in publications by the United Kingdom, Denmark, Sweden, Spain, and Russia that date from the late 18th century. European unification as a modern process saw the creation of a common drug standard in 1964. The European Pharmacopoeia (EP) grew out of subsequent discussions within the European Economic Committee to establish a common set of rules and guidelines for the quality of drugs among the member states.

The Convention Number 50 of the European Treaty Series of the Council of Europe gives the European Pharmacopoeia legal recognition to provide harmonized specifications for medicinal substances or pharmaceutical preparations within the member states. Within the signatory countries, existing national requirements may be superceded as the EP standards are implemented (2).

Alterations to the content of the EP are first presented for public review in the quarterly, PharmEuropa, which was first published in 1988 and the EP is updated accordingly via quarterly supplements. The fourth edition of EP appeared in 2003.

Japanese Pharmacopoeia

Established in 1886, the Pharmacopoeia of Japan (JP) is published by the Ministry of Health and Welfare. It received legal recognition in 1960 through Article 41 of the Pharmaceutical Affairs Law and is administered by the Committee on the Japanese Pharmacopoeia of the Central Pharmaceutical Affairs Council. The experts serving on scientific panels represent Japanese Trade Organization members. As with other pharmacopeias, it presents official standards that form the basis for regulating the qualities and attributes of drugs. The inclusion of materials in this book is based on their importance to medical practice as evidenced by the frequency of prescription or particular clinical importance. Any interested individual or organization may submit materials in support of the inclusion of additional information or revision of JP monographs (3). The specifications given in JP monographs are mandatory for the particular drug. Furthermore, all drugs and drug products involved in licensing in Japan are subject to the general test methods, such as dissolution, given in the JP.

Revision of the JP is preceded by an announcement in the Japanese Pharmacopoeial Forum (JPF). Public comment is reviewed and if appropriate, accommodated, before the change is made official via the JP or its supplement. The JPF was established in 1992 and is published quarterly in January, April, July, and October. Currently, JPXIII (1996) is official and is updated via supplements approximately every 2 years.

United States Pharmacopeia

The U.S. Pharmacopeial Convention currently meets every 5 years. The first meeting of the Convention was in 1820 and was attended by a group of 11 physicians interested in providing unified information on therapeutic products available at the time (4). Although recognized within national law, it represents the only non-governmental national pharmacopeia. The content of the USP is the responsibility of the Council of Experts, a volunteer body elected for a 5-year term by the USP Convention. The USP Convention represents state associations and schools of medicine and pharmacy, national and international associations and governmental agencies (5).

The USP was combined as a compendium with the National Formulary (NF) in 1975. Currently, the USP gives information regarding substances considered as having active medicinal properties while pharmaceutically inactive necessities are described in NF. The combined *United States Pharmacopeia and National Formulary* (*USP–NF*) is legally recognized under the U.S. Federal Food, Drug and Cosmetic Act.

The USP–NF is revised annually with two intervening supplements. As of the writing of this chapter, USP27–NF22 (2004) was official. Revision proposals are presented under authority of the Council of Experts in Pharmacopeial Forum, published bimonthly.

NECESSITY FOR COMPENDIAL DISSOLUTION TESTING REQUIREMENTS

Dissolution testing has become an important component of the assessment of the quality of solid oral dosage forms and oral suspensions. The basic procedures for these oral dosage forms have been extended to transdermal delivery systems as well. The release rate for modified-release oral dosage forms adds a level of sophistication to the concept of dissolution testing, setting acceptance criteria at multiple time points.

The relationship between manufacturing variables and therapeutic action of compressed oral dosage forms was noted early in the history of mass-produced medicines. Caspari (6), in the late 19th century, recommended that a tablet have a composition that promotes disintegration and subsequent solution in the stomach to avoid impairment of its therapeutic value. The implementation of a disintegration procedure to verify this important quality attribute can be found in the major pharmacopeias of the mid-20th century. The British Pharmacopoeia included a general disintegration standard in 1945. USP incorporated disintegration as a general test procedure in 1950 using the Stoll–Gershberg apparatus that had previously been employed in the evaluation of quality of drug products by the U.S. Army–Navy Procurement Agency. Yet problems in therapeutic action with products meeting the disintegration standard were reported in the literature. Campagna et al. reported problems with prednisone tablets meeting the *USP XVI* standards for assay (strength) and disintegration. Comparison of the dissolution rate between tablets that were known to be clinically active and the problem product indicated that the dissolution rates in vitro exhibited a rank–order correlation. With this observation, Campagna et al. (7) suggested that the dissolution performance in vitro of an oral dosage form might be used as an estimate of the efficacy of a product. Early studies of aspirin tablets demonstrated that ready disintegration did not necessarily correlate with prompt dissolution (8). Noticeable increase in the exposed surface area is therefore not an irrefutable metric for acceptable performance. Performance is better measured by the solution formed by the active contents in a physiologically relevant solvent. Clearly, a dissolution test could provide greater prediction of the ability of a dosage form to deliver its active contents than a disintegration test and could thus form the basis for the control of this important manufacturing quality attribute.

INTRODUCTION AND IMPLEMENTATION OF COMPENDIAL DISSOLUTION TEST REQUIREMENTS

USP

USP recognition of the need to control the in vitro dissolution performance of oral products by some level of compendial requirement was evidenced by the formation of a joint USP–NF panel on physiological availability in 1967. The USP and NF separately introduced dissolution procedures to drug products in 1970. Each compendium originally included dissolution tests in six monographs. As indicated above, at that time USP and NF were individual publications but would be combined in 1975. By 1980, the number of monographs with a dissolution test had grown to 72. This followed a 1976 policy statement that dissolution tests would be adopted for all tablets and capsules with a few exceptions. Emphasis was to be placed on products containing low-solubility drug substances, while it was thought unnecessary to implement dissolution standards for products such as antacids and stool softeners whose action did not require systemic absorption. Since the USP or any other facility would necessarily lack the resources to determine dissolution test conditions and criteria for each official product, the Executive Committee of Revision determined to use whatever resources could be made available in the effort (9).

Early optimism about the possibility of in vitro–in vivo correlation was tempered by the need for a performance test that would yield reproducible results (10). Even though not necessarily correlated to bioavailability, dissolution requirements were seen as useful in controlling variables in formulation or processing. Thus, from the start, sources of variability in the results were seen as factors to be minimized in any proposed compendial method.

A proposal to merely publish the official standards, allowing any apparatus to be used in regulatory filing to meet the standard, met with opposition by the USP (11). Clearly, the compendial standard required a specific procedure to allow the demonstration of compliance.

The desire of USP experts for contributed dissolution procedures for most official immediate-release solid oral dosage forms was not fulfilled. In 1980, a policy giving a framework for the comprehensive application of a dissolution test procedure was formulated. The policy recognized three classes of products for which the dissolution test could be applied with increasing brand-linked specificity. First Case conditions were intended for the most general class where either the basket at 100 rpm or the paddle at 50 rpm was used with from 500 to 1000 mL of water and a value of 75% of label claim for the active ingredient (strength) was specified to be released in 45 min of testing. Testing by First Case conditions was to be applied to all official USP solid oral dosage forms. In the case of affected products where application of First Case conditions was not appropriate and where no evidence of bioavailability problems existed, deviation from strict adherence to the medium, apparatus/speed, test time, or acceptance criterion would be considered. Such a departure was termed Second Case and would apply to all preparations conforming to the monograph. Where data indicated that bioavailability was a concern for articles not already conforming to First Case conditions, a separate test could be applied that considered available clinical information. In such a Third Case, in vivo data were viewed as paramount (12).

Initially, USP did not extend a dissolution requirement to non-immediate-release products. The USP recognized two categories of modified-release dosage forms, where intentional alteration of the formulation or process contributed to a dissolution profile for which the First Case dissolution would not be appropriately applied. The first category included extended-release dosage forms that allowed a two-fold reduction in dosing frequency. The second category, termed delayed-release, was associated with release at a time other than promptly upon administration. Delayed-release products are typified by enteric-coated products, where release is inhibited in the gastric environment but can be prompt once the product is exposed to the higher pH of the small intestine.

The application of dissolution or drug-release testing to extended-release dosage forms followed the approach given for immediate-release forms. For Case One, the test procedure for First Case was applied with times adapted to the fractions of the dosing interval. At 25% of the dosing interval, a range of 20–50% of the labeled content was to be released, 50% of the interval would find 45–75% of the labeled content released and not less than 75% of the label was to be in solution a the full dosing interval. Where either the properties of the active or of the product did not permit the application of First Case test conditions or the in vitro release occurred in a time period that was less than the dosing interval, Case Two would apply and with appropriate justification, alternative procedures or criteria could be considered. For those products, the particular procedure and acceptance criteria would be given in the individual monograph. Case Three was applied where differences among the products available from several manufacturers prevented the application of a single procedure with acceptance criteria. Monographs where Case Three is applied will have multiple drug-release tests numbered in order of USP Committee approval. Affected products are required to state the number of the test on the label to allow confirmation of compliance to the appropriate test (13).

USP 27 (2004) contains 185 capsule monographs representing 121 monographs with dissolution test and 15 other monographs with a drug-release test. Out of 527 tablet monographs, 346 contain a dissolution test while 21 cited a drug-release test (14).

British Pharmacopoeia

As an example of a national standard that has played a notable role in the evolution of dissolution testing, the process by which the British Pharmacopoeia (BP) adopted dissolution testing is given here. It should be noted that while much of the contents of the BP are identical with the EP in agreement with the ongoing process to harmonize drug regulations in the European community, the EP itself does not provide any specific methods for dissolution testing in individual drug monographs. Consequently, the dissolution tests in the BP are often applied throughout Europe (and, for that matter, the whole world) for product quality control.

The need to develop compendial standards for dissolution for capsules and tablets containing poorly soluble drug products was noted by the BP in 1973. By 1980, the British Pharmacopoeia Commission had identified a list of drug products included in the 1973 BP for which the development of a dissolution standard was necessary. The list included products for which clinical problems were associated with bioavailability, bioequivalence had been questioned, or where the properties of the drug substance indicated that dissolution might be a concern.

Implementation of dissolution testing by BP was in a tiered program similar to that employed at the time by USP. For the first category, products would conform to 75% release in 45 min. Where the drug had a narrow therapeutic index and should not release too rapidly, was known to exhibit a brief plasma half-life, or have site-specific absorption, additional testing to satisfy the need for greater control would be considered. Dissolution tests were included in 1980 for 14 tablet and four capsule monographs (15,16).

The 2002 BP has 73 capsule monographs with dissolution applied for 29. In the same edition, 351 tablet monographs can be found with 103 of them giving a dissolution method (17).

Japanese Pharmacopoeia

A dissolution test was first described in the JP in 1981 (18). General rules for capsules and tablets stated that the requirements of the disintegration test must be met unless otherwise specified. Several specific capsule and tablet monographs included new dissolution tests.

In the intervening years, the increase in specifications for oral dosage forms dissolution has been less dramatic. The 14th edition of the Japanese Pharmacopoeia (2002) has included additional dissolution tests for tablets and capsules. Out of a total of 61 tablet monographs, dissolution tests are included in 32. From four capsule monographs, one dissolution test is given (19).

European Pharmacopoeia

A general chapter giving the dissolution test for solid oral dosage forms was first described in the EP in 1991 (20). As mentioned above, the EP has no product monographs in which to elaborate specific dissolution procedures.

HARMONIZATION

With the USP as the pioneer, much of the overall approach to dissolution has been by the application of similar test procedures to locally available products. Regional differences in the specifications for otherwise similar oral dosage forms were inevitable. While regional differences among specification for the hundreds of individual oral dosage forms will likely continue into the future, the harmonization of the general dissolution test has developed to a fairly high degree. The areas of harmonization for the general dissolution test are: apparatus, procedure, and acceptance criteria.

Periodic discussions among the EP, JP, and USP, with the World Health Organization as observer, facilitate compendial harmonization. This association is known as the Pharmacopeial Discussion Group (PDG). The PDG has prioritized the harmonization effort for individual general test chapters based originally on those identified within ICH Q6A (1). Dissolution is prominent on the PDG work agenda.

Any proposal for harmonization must be presented for public comment in each of the pharmacopeial journals, Pharmeuropa (EP), Japanese Pharmacopoeial Forum (JP), and Pharmacopeial Forum (USP). This was accomplished early in 2003 (21–23). Comments were collated and further PDG discussions conducted. Any agreement will be presented again, prior to implementation. The PDG harmonization process can be found as General Information Chapter <1196> in USP 27 (24).

REFERENCES

1. International Conference on Harmonization. Guidance on Q6A specifications: test procedures and acceptance criteria for new drug substances and new drug products: chemical substances. Fed Reg 2000; 65(251):83,041–83,063.

2. Artiges AF. Pharmacopeial standards: European Pharmacopoeia. In: Swarbrick J, Boylan JC, eds. Encyclopedia of Pharmaceutical Technology. Vol. 12. New York: Marcel Dekker, 1995:53–72.

3. Uchiyama M. Pharmacopeial standards: Japanese Pharmacopoeia. In: Swarbrick J, Boylan JC, eds. Encyclopedia of Pharmaceutical Technology. Vol. 12. New York: Marcel Dekker, 1995:73–79.
4. Anderson L, Higby GJ. The Spirit of Volunteerism—A Legacy of Commitment and Contribution. The United States Pharmacopeial Convention, Inc., Rockville, Maryland, USA, 1995.
5. USP. USP26–NF21. The United States Pharmacopeial Convention, Inc., Rockville, Maryland, USA, 2003:2871–2872.
6. Caspari CA. Treatise on Pharmacy. Lea Bros, Philadelphia, 1895.
7. Campagna FA, Cureton G, Mirigian RA, Nelson E. Inactive prednisone tablets USP XVI. J Pharm Sci 1963; 52:605–606.
8. Levy G, Hayes BA. Physicochemical basis of the buffered acetylsalicylic acid controversy. N Engl J Med 1960; 262(21): 1053–1058.
9. USP. USP policy statement on dissolution requirements. Pharm Forum 1976; 2(1):85–86.
10. Tingstad JE. J Pharm Sci 1973; 62(7):VI.
11. Banes D. J Pharm Sci. 1973; 62(7):VI.
12. USP. USP policy on dissolution standards. Pharm Forum 1981; 7(4):1225.
13. USP. USP policy on modified-release dosage forms. Pharm Forum 1983; 9(3):2999–3001.
14. USP. USP27–NF22. The United States Pharmacopeial Convention, Inc., Rockville, Maryland, USA, 2003.
15. British Pharmacopoeial Commission. Solution rate. In: British Pharmacopoeia 1973 Addendum 1975. London: Her Majesty's Stationary Office, 1975:xii, xix.
16. British Pharmacopoeial Commission. Dissolution test for tablets and capsules. In: British Pharmacopoeia 1980. London: Her Majesty's Stationary Office, 1980:A114.
17. British Pharmacopoeial Commission. London: The Stationary Office, 2002.

18. Committee on JP. Dissolution test. In: The Pharmacopoeia of Japan. 10th ed 1981. English Version. Tokyo: Society of Japanese Pharmacopoeia, 1982:729–733.

19. JP. The Japanese Pharmacopoeia. 14th ed. English Version. Tokyo: Society of Japanese Pharmacopoeia, 2001.

20. EP. Dissolution test for solid oral dosage forms. In: European Pharmacopoeia. 2nd ed. Fifteenth Fascicule. Sainte-Ruffine: Maisonneuve S. A., 1991:v.5.4–1– v.5.4–8.

21. EP. Dissolution. Pharm Eur 2003; 15(1):191–198.

22. Secretariat of Japanese Pharmacopoeial Forum. Dissolution. J Pharm Forum 2002; 11(4):623–641.

23. USP. <711> Dissolution. Pharm Forum 2002; 28(6):1972–1987.

24. USP. <1196> Pharmacopeial harmonization. In: USP27-NF22. The United States Pharmacopeial Convention, Inc., Rockville, Maryland, USA, 2003:2608–2612.

4

The Role of Dissolution Testing in the Regulation of Pharmaceuticals: The FDA Perspective

VINOD P. SHAH
Office of Pharmaceutical Science, Center for Drug Evaluation and Research, Food and Drug Administration, Rockville, Maryland, U.S.A.

INTRODUCTION

Over the last quarter century the dissolution test has emerged as a most powerful and valuable tool to guide formulation development, monitor the manufacturing process, assess product quality, and in some cases to predict in vivo performance of solid oral dosage forms. Under certain conditions, the dissolution test can be used as a surrogate measure for bioequivalence (BE) and to provide biowaivers, assuring BE of the product. Dissolution test has turned out to be a critical test for measuring product performance. Generally, dissolution testing of solid oral dosage form is carried out by the basket (USP Apparatus 1) or paddle (USP Apparatus 2) method under mild agitation (100 rpm with the basket or 50–75 rpm with the paddle), in an aqueous buffer in the pH range 1.2–6.8. Dissolution samples are analyzed at 15 min intervals for immediate-release (IR) products or at hourly intervals for extended-release products until at least 85% dissolution is achieved. For water-insoluble drug products, small amounts of surfactants are often employed to achieve sink conditions.

Dissolution is also used to identify bioavailability (BA) problems and to assess the need for further BE studies relative to scale-up and post-approval Changes (SUPAC), where it functions as a signal of bioinequivalence. In vitro dissolution studies for all product formulations investigated (including prototype formulations) are encouraged, particularly if in vivo absorption characteristics can be defined for the different product formulations. With such efforts, it may be possible to achieve an in vitro/in vivo correlation. When an in vitro correlation or association is available, the in vitro test can serve not only as a quality control (QC) specification for the manufacturing process, but also as an indicator of in vivo product performance.

Several in vitro tests are currently employed to assure drug product quality. These include purity, potency, assay, content uniformity, and dissolution specifications. For a pharmaceutical product to be consistently effective, it must meet all of its quality test criteria. When used as a QC test, the in vitro dissolution test provides information for marketing authorization. The dissolution test forms the basis for setting specifications (test, methodology, acceptance criteria) to allow batch release into the market place. Dissolution tests also provides a useful check on a number of physical characteristics, including particle size distribution, crystal form, etc., which may be influenced by the manufacturing procedure. In vitro dissolution tests and QC specifications should be based on the in vitro performance of the test batches used in in vivo studies or on suitable compendial specifications. For conventional-release products, a single-point dissolution test is commonly used as a compendial specification. However, a two-point test or a profile is suggested for characterizing the dosage form. For extended-release products, a three to four-point dissolution test is recommended as a routine QC test. The dissolution test or the drug-release test is also employed for evaluating other non-oral (special) dosage forms such as topicals and transdermals, suppositories, implants, etc. It is anticipated that the drug-release test for these products will also be of value in assuring drug product quality.

For the test to be useful, the dissolution test should be simple, reliable and reproducible, and should be able to discriminate between different degrees of in vivo product performance. The value of the test is significantly enhanced when product performance is evaluated as a function of time, i.e., when the dissolution profile is determined rather than a single-point determination. Increasingly, dissolution profile comparison is used for assuring product sameness under SUPAC-related changes and for granting biowaivers. Thus, an increasing role of dissolution is seen in regulating the quality of pharmaceutical drug products.

DISSOLUTION-RELATED FDA GUIDANCES

Because of the importance of dissolution, FDA has developed dissolution-related guidances that provide information and recommendations on the development of dissolution test methodology, setting dissolution specifications, and the regulatory applications of dissolution testing (1,2). In addition, it provides information with respect to when a single-point dissolution test is adequate as a QC test and when two points or a dissolution profile is needed to characterize the drug product. A procedure for establishing a predictive relationship between dissolution and in vivo performance and setting specifications for extended-release drug products is also discussed (2). A recent FDA guidance on biowaiver based on Biopharmaceutics Classification System (BCS) suggests that documentation of BE via dissolution studies is appropriate for orally administered IR drug products which are highly soluble, highly permeable, and rapidly dissolving (3). The FDA dissolution-related guidances are:

- Guidance for Industry: Dissolution Testing of Immediate Release Solid Oral Dosage Form, August 1997.
- Guidance for Industry: Extended Release Solid Oral Dosage Forms: Development, Evaluation and Application of In Vitro/In Vivo Correlations, September 1997.
- Guidance for Industry: Waiver of In Vivo Bioavailability and Bioequivalence Studies for Immediate-Release Solid Oral Dosage Forms Based on a Biopharmaceutics Classification System, August 2000. (BCS Guidance).

A recent FDA guidance on Bioavailability and Bioequivalence Studies for Orally Administered Drug Products—General Considerations (4) provides "how to" information for conducting BA and BE studies, defines proportionally similar formulations, and provides provision for biowaivers for lower strength(s) of IR as well as modified-release (MR) drug products. The guidance lowers regulatory burden without sacrificing product quality. The general BA and BE guidance and BCS guidance clearly establish a trend whereby the dissolution test has moved from traditional QC test to a surrogate in vitro BE test. Figure 1 for IR and Figure 2 for MR dosage forms summarize the BE and dissolution requirements as discussed in this guidance.

A dissolution profile or at least a two-point determination should be used to characterize the in vitro performance of an IR drug product. Because a MR dosage form is a more complex formulation, three to four dissolution time points are needed to characterize the product. In addition, SUPAC guidances also rely on dissolution testing and profile comparison to assure product sameness between pre- and post-approval change for drug products. In order to avoid subjective evaluation of dissolution profile comparison, FDA has adopted a simple method to compare dissolution profiles, termed the similarity factor, f_2. The pharmaceutical industry has used this approach extensively to assure product sameness for changes in manufacturing site (SUPAC-related changes).

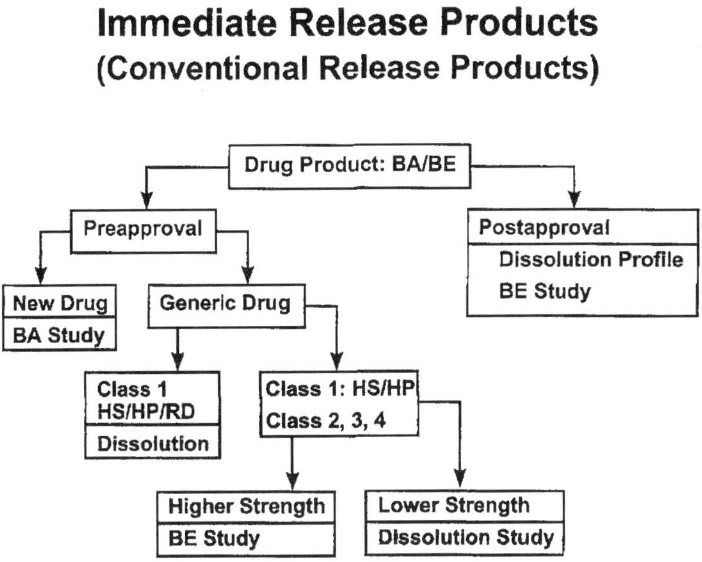

Figure 1 The IR dosage forms.

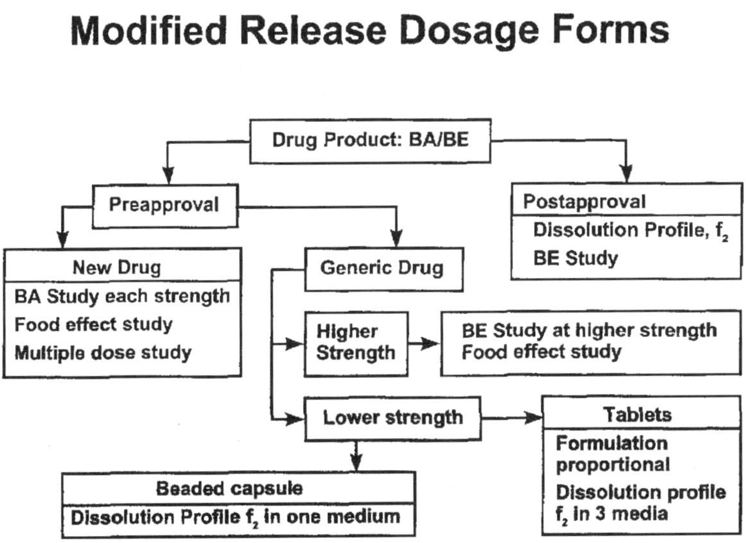

Figure 2 The MR dosage forms.

CHANGES IN DISSOLUTION SCIENCE PERSPECTIVES

As more experience and knowledge is gained in understanding of the dissolution science and mechanism, the dissolution test has undergone a shift in its application and value. The current regulatory perspective on dissolution is depicted in Figure 3. In this new era of dissolution, dissolution tests can be used not only for QC but also as a surrogate marker for BE test, as outlined in a recent BCS guidance (3). The possibility of using dissolution testing as a tool for providing biowaivers has considerably enhanced the value of the test. The BCS guidance takes into account three major factors, dissolution, solubility, and intestinal permeability, which govern the rate and extent of drug absorption from IR solid dosage forms. The BCS provides a scientific framework for classifying drug substances based on aqueous solubility and intestinal permeability, and in combination with dissolution data, provides a rationale for biowaiver of IR drug products. In addition, the General Bioavailability and Bioequivalence Guidance (4) allows biowaivers for lower strength(s) of IR as

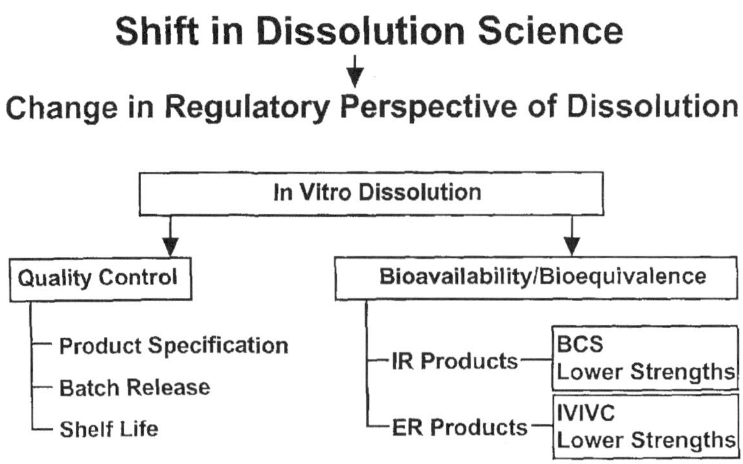

Figure 3 Current regulatory perspective on dissolution.

well as MR drug products based on formulation proportionality and dissolution profile comparison. These changes in BE requirements, moving away from in vivo study requirement in certain cases and relying more on dissolution test, clearly establish a change in dissolution testing applications. In all cases where the dissolution test is used as a BE test, an anchor with a bioavailable product is established or a rational for waiving in vivo studies is provided. Further, the reliance on dissolution testing can be extended to improve drug product quality in developing countries. In several instances, biowaivers can be justified on the basis of a dissolution profile comparison with a reference product.

DISSOLUTION-BASED BIOWAIVERS—DISSOLUTION AS A SURROGATE MARKER OF BE

The BCS provides a new perspective to the dissolution testing (3,5). It provides scientific rationale to lower regulatory burden and justifies a biowaiver under certain circumstances. It is based on aqueous solubility and intestinal permeability of the drug substance and dissolution of the drug product. When combined with the dissolution of the drug product, the BCS takes into account three major factors that govern the rate and extent of drug absorption from IR solid dosage forms namely dissolution, solubility, and intestinal permeability. It classifies the drug substance (and therefore the drug product) into four classes, class 1: high solubility/high permeability (HS/HP), class 2: low solubility/high permeability (LS/HP), class 3: high solubility/low permeability (HS/LP) and class 4: low solubility/low permeability (LS/LP). BCS takes into consideration GI physiological factors such as pH, gastric fluid volume, gastric emptying, intestinal transit time, etc and permeability factors (5). According to the BCS guidance:

- the drug substance is considered highly soluble when the highest dose strength is soluble in 250 mL or less of aqueous media over the pH range of 1–7.5;

- the drug substance is considered highly permeable when the extent of drug absorption in humans is determined to be 90% or more of an administered dose based on a mass balance determination or in comparison to an intravenous reference dose; and
- an IR drug product is considered rapidly dissolving when 85% or greater of the labeled amount of the drug substance dissolves within 30 min, using basket method (Apparatus I) at 100 rpm or paddle method (Apparatus II) at 50 rpm in a volume of 900 mL or less in each of the following media: (i) 0.1 N HCl or simulated gastric fluid USP without enzymes (ii) a pH 4.5 buffer and (iii) a pH 6.8 buffer or simulated Intestinal Fluid USP without enzymes.

The BCS also predicts the possibility of obtaining an in vitro/in vivo correlation. Justification of a biowaiver is based on a combination of the BCS classification of the drug substance and a drug product dissolution profile comparison. In all these instances, an anchor with a bioavailable product is established. Specifically, to obtain a biowaiver for an IR generic product:

- the reference product should belong to Class 1, HS/HP;
- the test and reference drug products should dissolve rapidly (85% or greater in 30 min or less) under mild test conditions in pH 1.2, 4.5, and 6.8 and
- the test product and the reference product should meet the profile comparison criteria under all test conditions.

Dissolution-based biowaivers for generic IR and MR drug products are discussed in the General BA and BE Guidance (4).

For IR Products,

1. A biowaiver is applicable for drug products meeting the BCS Class 1 criteria, HS/HP/RD (Rapid Dissolution).
2. A biowaiver is applicable for lower strength(s) when the highest strength is shown to be BE to the innova-

tor product and the formulation(s) of the generic product is (are) proportional to the highest strength and meets dissolution profile comparison criteria.

For MR products,

1. A biowaiver is applicable for beaded capsules when the lower strength differs only in number of beads of active drug and the dissolution profile is similar in the recommended dissolution test media and conditions.
2. A biowaiver is applicable for extended-release tablet formulations, where the lower strength(s) are compositionally similar to the highest strength and uses the same release mechanism and the dissolution profile is similar in pH 1.2, 4.5, and 6.8.

The biowaiver criteria described in BCS guidance (3) are regarded as very conservative. Discussions are underway to consider relaxing some of the requirements for biowaiver of the drug product. These dissolution-based biowaivers exemplify the role of dissolution in regulating pharmaceutical drug products.

DISSOLUTION/IN VITRO RELEASE OF SPECIAL DOSAGE FORMS

In the last decade, the application of dissolution testing has been extended to oral and non-oral "special" dosage forms, such as transdermal patches, semisolid preparations such as creams, ointments and gels, orally disintegrating dosage forms, suppositories, implants, microparticles, liposomes, etc. Can the principles and applications of dissolution/in vitro drug release be extended to these "special" dosage forms? Current scientific knowledge suggests that the drug release from the formulation is the crucial first step for the therapeutic activity of the drug product. Thus, the principles of dissolution, i.e., in vitro drug release from the special dosage forms can at least be used as a QC tool to assure batch-to-batch reproducibility. The goal of these in vitro release tests is analogous to that for solid oral dosage forms, i.e., to use the in vitro-release test as a regulatory tool to assure consistent product quality in the market place. A final report is out and would prefer to give the final reference report published by *FIP Dissolution Working Group* summarizes the current status of test procedures and developments in this area (6).

The in vitro drug release from semisolid preparations, creams, ointments, and gels can be determined using vertical diffusion cell system and synthetic membrane. The method is simple, rugged, and easily reproducible. The method is applicable to all creams, ointments, and gels (7). In vitro drug release from transdermal patches can be easily determined using simple modification of paddle method, paddle over disk method (8). This is also simple, rugged, reproducible, and applicable to all marketed transdermal patches. In several cases, modification of the paddle method is used for drug release of suppositories (6,9).

Going beyond the application of the in vitro-release test as a QC tool for special dosage forms to biowaivers and in vitro–in vivo correlations will require more research.

DISSOLUTION PROFILE COMPARISON

In recent years, FDA has placed more emphasis on dissolution profile comparison in the area of post-approval changes and biowaivers. Under appropriate test conditions, a dissolution profile can characterize the product more precisely than a single-point dissolution test. A dissolution profile comparison between (i) pre-change (reference) and post-change (test) products for SUPAC-related changes, or (ii) with different strengths of a given manufacturer, or (iii) comparison between manufacturers for BCS class 1 (HS/HP/RD) drug products, evaluates similarity in product performance, with poor results signaling bioinequivalence.

Among several methods investigated for dissolution profile comparison, the f_2 factor is the simplest and widely applicable (1). Moore and Flanner (10) proposed a model independent mathematical approach to compare the dissolution profile using two factors, f_1 and f_2.

$$f_1 = \{[_{t=1^n}|R_t - T_t|]/[_{t=1^n}R_t]\}\ 100$$

$$f_2 = 50\log\{[1 + (1/n)_{t=1}^{n}(R_t - T_t)^2]^{-0.5}\ 100\}$$

where R_t and T_t are the cumulative percentage dissolved at each of the selected n time points of the reference and test product, respectively. The factor f_1 is proportional to the average difference between the two profiles, where as factor f_2 is inversely proportional to the average squared difference between the two profiles, with emphasis on the larger difference among all the time points. The factor f_2 measures the closeness between the two profiles. Because of the nature of measurement, f_1 was described as a difference factor, and f_2 as a similarity factor (11). The similarity factor, f_2 (10–12), has been adopted by the FDA in its Guidances, since the regulatory interest is to know whether the dissolution profiles of the test and reference products are similar. When the two profiles are identical, $f_2 = 100$. A plot of f_2 values determined using computer-simulated average differences between the reference and test dissolution profiles indicated that an average difference of 10% at all measured time points between the two profiles results in a f_2 value of 50 (Fig. 4). FDA has set a public standard of f_2 value between 50 and 100 to indicate similarity between two dissolution profiles. (Further discussion of the advantages and limitations of the f_2 factor and other measures of profile similarity can be found in Chapter 13.)

For a dissolution profile comparison:

- At least 12 units should be used for each profile determination. Mean dissolution values can be used to estimate the similarity factor, f_2. To use mean data, the percentage coefficient of variation at the earlier point should not be more than 20% and at other time points should not be more than 10%.
- For circumstances where wide variability is observed, or a statistical evaluation of f_2 metric is desired, a bootstrap approach to calculate a confidence interval can be performed (8).

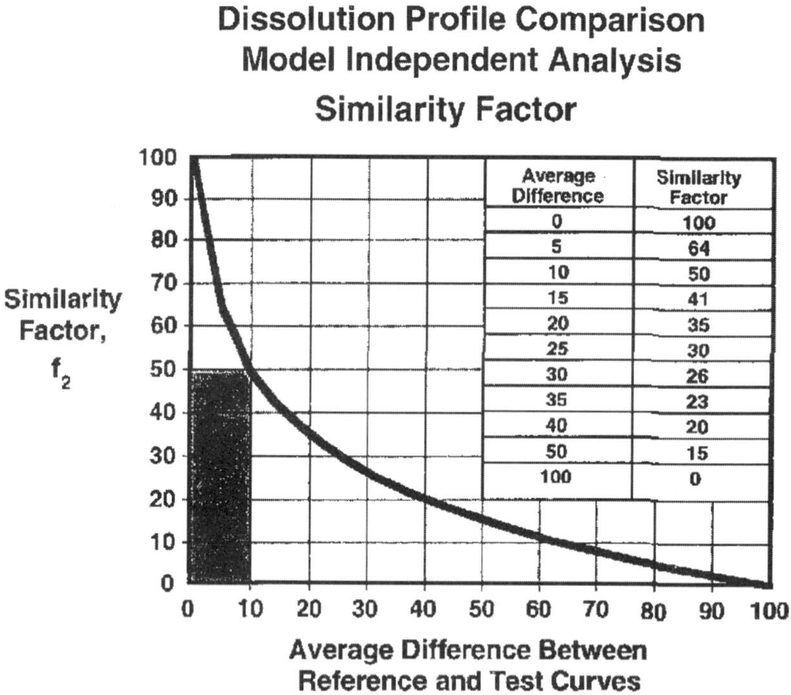

Figure 4 Dissolution profile comparison model independent analysis.

- The dissolution measurements of the two products (test and reference, pre- and post-change, two strengths) should be made under the same test conditions. The dissolution time points for both the profiles should be the same, e.g., for IR products 15, 30, 45, and 60 min, for extended-release products 1, 2, 3, 5, and 8 hr.
- Because f_2 values are sensitive to the number of dissolution time points, only one measurement should be considered after 85% dissolution of the product.
- For drug products dissolving 85% or greater in 15 min or less, a profile comparison is not necessary.

A f_2 value of 50 or greater (50–100) ensures sameness or equivalence of the two curves and, thus, the performance of the two products. From a public health point of view, and as a regulatory consideration, a conservative approach of $f_2 \geq 50$ is appropriate. The f_2 comparison metric with a value of 50 or greater is a conservative, but reliable basis for granting a biowaiver, and for assuring product and product performance sameness. A value below 50 may be acceptable based on additional information available about the drug substance and drug product. Additional research and data mining are needed to address the general question of what can be done if the f_2 value is <50.

FUTURE DIRECTIONS

One of the major efforts of the FDA is to reduce regulatory requirements and unnecessary in vivo testing, without sacrificing the quality of the product. The BCS guidance is a step in the right direction, but future extensions of the BCS remain a major challenge. Appropriate data need to be collected and evaluated before biowaiver extensions in other classes can be considered. Principles of BCS, especially solubility information, can be utilized in the selection of an appropriate dissolution medium. In addition, based on the BCS, the dissolution specification for class 1 drug products (HS/HP) can be set at 85% dissolution in 30 min to improve the quality of pharmaceutical products in the market place. A good knowledge and understanding of GI physiology, excipient effects on drug absorption and GI motility, and the use of biorelevant dissolution media may be useful in this evaluation. The dissolution test using a biorelevant dissolution medium may be especially helpful in product development, establishing in vitro–in vivo correlation, determining appropriate dissolution test media (particularly for drugs belonging to BCS class 2 and 4), and also in predicting food effects (13–15). The use of biorelevant dissolution media can serve as an excellent prognostic tool in these areas.

Further, there is an increased reliance on use of in vitro dissolution as a surrogate marker for in vivo blood level data. When dissolution is used as a QC test for IR products, it is generally a single-point dissolution test and is represented as $X\%$ dissolved in Y minutes. But when the dissolution test is used as a BE test, it is different: comparison of the dissolution profile with a bioavailable product is crucial.

The value of dissolution test can be further appropriately utilized in developing countries where it can be used as a "BE test." The question is raised: "Can dissolution test alone be used as a BE test for approval of IR products in developing countries"? Generally in developing countries, the technology and other resources are very limited to conduct an appropriate in vivo BE studies. Under these circumstances, appropriate dissolution studies, for e.g., profile comparison between the local generic product and the reference product in pH 1.2, 4.5, and 6.8 media under mild test conditions, e.g., basket method at 100 rpm or paddle method at 50 rpm, may be used to assure product quality. This appears to be a practical approach that can be easily considered and adopted for BE test in developing countries (16). The research in the area of dissolution/in vitro release test for non-oral (special) dosage forms will lead to its application as a QC test for batch-to-batch uniformity as well as other regulatory applications.

IMPACT OF DISSOLUTION TESTING

The art and science of dissolution testing have come a long way since its inception about 30 years ago. The procedure is well established, reliable, and reproducible. Application of dissolution testing as a QC test, to guide formulation development, to use as a manufacturing/process control tool and as a test for product sameness under SUPAC-related changes is well established. Increasingly, in vitro dissolution testing and profile comparison are relied on to assure product quality and performance and to provide a biowaiver. An appropriate dissolution test procedure is identified as a simple and economical method that can be utilized effectively in developing countries to assure acceptable drug product quality. An increasing role of dissolution in regulating pharmaceutical drug product quality is becoming clearly evident. The dissolution test is currently being used as a both QC test (generally single point for IR products and 3-to-4 points for extended-release products), as well as an in vitro BE test (generally dissolution profile and profile comparison). Since dissolution testing plays a different role when it is used as a QC test than when it is used as a surrogate for BE, the discussion and assessment of dissolution in these roles should be carefully separated.

REFERENCES

1. Guidance for Industry: Dissolution Testing of Immediate Release Solid Oral Dosage Form. Aug. 1997.
2. Guidance for Industry: Extended Release Solid Oral Dosage Forms: Development, Evaluation and Application of In Vitro/In Vivo Correlations. Sep. 1997.
3. Guidance for Industry: Waiver of In Vivo Bioavailability and Bioequivalence Studies for Immediate-Release Solid Oral Dosage Forms Based on a Biopharmaceutics Classification System. Aug. 2000.
4. Guidance for Industry: Bioavailability and Bioequivalence Studies for Orally Administered Drug Products—General Considerations. Oct. 2000.
5. Amidon GL, Lennernas H, Shah VP, Crison JR. A theoretical basis for a biopharmaceutics drug classification: the correlation of in vitro drug product dissolution and in vivo bioavailability. Pharm Res 1995; 12:413–420.
6. Siewert M, Dressman J, Brown CK, Shah VP. FIP/AAPS Guidelines for dissolution in vitro release testing of novel/special dosage forms. AAPS Pharm Sci Tech 2003; 4(1):43–52; Pharm Ind 2003; 65:129–134; Dissolut Technol 2003; 10(1):6–15.
7. Shah VP, Elkins JS, Williams RL. Evaluation of the test system used for in vitro release of drugs from topical dermatological drug products. Pharm Develop Technol 1999; 4: 377–385.
8. Shah VP, Tymes NW, Skelly JP. In vitro release profile of clonidinetransdermal therapeutic systems scopolamine patches. Pharm Res 1989; 6:346–351.

9. Gjellan K— Suppositories.
10. Moore JW, Flanner HH. Mathematical comparison of curves with an emphasis on in vitro dissolution profiles. Pharm Tech 1996; 206:64–74.
11. Shah VP, Tsong Y, Sathe P, Liu JP. In vitro dissolution profile comparison—statistics and analysis of the similarity factor, *f*2. Pharm Res 1998; 15:889–896.
12. Shah VP, Tsong Y, Sathe P, Williams RL. Dissolution profile comparison using similarity factor, *f*2. Dissolut Technol 1999; 6(3):15.
13. Dressman JB, Amidon GL, Reppas C, Shah VP. Dissolution testing as a prognostic tool for oral drug absorption: immediate release drug dosage forms. Pharm. Res 1998; 15:11–22.
14. Galia E, Nicolaides E, Horter D, Lobenberg R, Reppas C, Dressman JB. Evaluation of various dissolution media for predicting in vivo performance of class I and II drugs. Pharm. Res 1998; 15:698–705.
15. Lobenberg R, Kraemer J, Shah VP, Amidon GL, Dressman JB. Dissolution testing as a prognostic tool for oral drug absorption: dissolution behavior of glibenclamides. Pharm Res 2000; 17:439–444.
16. Shah VP. Dissolution: quality control test vs. bioequivalence test. Dissolut Technol 2001; 8(4):6–7.

5

Gastrointestinal Transit and Drug Absorption

CLIVE G. WILSON and KILIAN KELLY
Department of Pharmaceutical Sciences,
Strathclyde Institute for Biomedical Studies,
University of Strathclyde, Glasgow,
Scotland, U.K.

INTRODUCTION

The human gut has evolved over many thousands of years to provide an efficient system for the extraction of nutrients from a varied diet. Functionally, the gut is divided into a preparative and primary storage region (mouth and stomach), a secretory and absorptive region (the midgut), a water reclamation system (ascending colon), and finally a waste-product storage system (the descending and sigmoid colon regions and the rectum). The organization of the upper gut facilitates the controlled presentation of calories to the systemic circulation allowing the replete person to perform physical work, undergo social activities, and to go to sleep.

The physiology of the digestive process is less than convenient for the efficient absorption of many of the modern therapeutic entities that we wish to administer. For example, drug absorption can be highly dependent on gastrointestinal (GI) transit, with absorption kinetics in some cases varying hugely in different parts of the GI tract. This is due to factors such as the mechanical forces applied to the formulation as well as the nature of the mucosa, the available surface area, pH, and the presence of enzymes and bacteria. The influence of feeding and temporal patterns on GI transit is therefore of great relevance in attempting to optimize drug absorption.

Most of the work on GI transit published to date has utilized gamma scintigraphy studies. The use of gamma-emitting radionuclides for diagnostic imaging in nuclear medicine has been established for over three decades. Sophisticated gamma-ray detecting camera systems and high-speed computer links enable the clinical investigator to image different regions of the body and to quantify organ function. Parallel developments have occurred in the field of radiopharmaceuticals, and a wide range of products are available that will exhibit uptake within specific tissues following parenteral administration. The situation with regard to investigations of GI transit is much simpler: the chief requirement is to be able to label different components within the formulation or food and for the label to remain associated with the component in both strongly acidic and neutral conditions. From the pharmaceutical perspective, the most important recent advances have come in the applications of other imaging modalities such as magnetic resonance imaging (MRI) and magnetic moment imaging which, when combined with an appreciation of scintigraphic data and its interpretation, can help the pharmaceutical scientist to understand formulation behavior.

A review of GI transit and oral drug absorption can be organized in many ways, but a logical sequence is to start at the top and work down. In this review, techniques to study buccal and rectal delivery will not be covered, but a detailed description of these is available in a recent book (1).

ESOPHAGEAL TRANSIT

After the dosage form leaves the buccal cavity, which is a relatively benign environment, transit through the esophagus is normally complete *within five seconds*. However, this may be influenced by several factors, including the dosage form, exact mode of administration, posture, age, and certain pathologies (2). It has been known for many years that disorders of normal motility (dysphagia), left-sided heart enlargement or stricture of the esophagus can result in impaired clearance of formulations, which in turn could result in damage to the esophageal tissues. Radiological studies of an asymptomatic group of 56 patients, mean age 83 years, showed that a normal pattern of deglutition was present in only 16% of individuals (3). Oral abnormalities, which included difficulty in controlling and delivering a bolus to the esophagus following ingestion, were noted in 63% of cases. Structural abnormalities capable of causing esophageal dysphagia include neoplasms, strictures, and diverticula, although several workers have commented that only minor changes of structure and function are associated specifically with aging. The difficulty for elderly patients, therefore, appears to relate to neuromuscular mechanisms associated with the coordination of tongue, oropharynx, and upper esophagus during a swallow. In the past, researchers have suspected that reflux of gastric acid might contribute to esophageal damage; however, a recent study conducted by our group suggests that persistent gastroesophageal reflux does not predispose towards problems in the clearance of film-coated oval tablets (4).

In scintigraphic studies of transit rates of hard gelatin capsules and tablets, elderly subjects were frequently unable to clear the capsules (5,6). This appears to be due to the separation of the bolus of water and capsule in the oropharynx, resulting in a "dry" swallow. Capsule adherence occurred in the lower third of the esophagus, although subjects were unaware of sticking. The importance of buoyancy in capsule formulation has hitherto been ignored and may be an additional risk factor in dosing the elderly.

The issue of surface properties in tablets is also important and, surprisingly, small flat tablets can cause problems. In the development of a risedronate product, we needed to develop a procedure that was able to discriminate between alternative formulations. The key conditions necessary to differentiate among products with respect to the ease of swallowing was to dose the unit with one mouthful of water—30 mL. Using this procedure we demonstrated that small, uncoated, shallow convex-shaped tablets (9.5 mm diameter) were arrested in the esophagus more often than the final design of the formulation—an oval of 5.7 $\times$ 11.5 mm (2). In 5 out of 30 cases, esophageal transit of the smaller tablet was slower (6).

GASTRIC RETENTION

Our understanding of the behavior of dosage forms in the stomach has been gained largely from scintigraphic studies in which solid and liquid phases of a meal and formulations are labeled with different radionuclides, most often Tc-99m and In-111 (7,8). These two radionuclides can be distinguished according to the energy of their emissions and thus can be separately detected, even when both are present in the field of view. Such studies have demonstrated that retention times of formulations in the stomach are dependent on the size of the formulation (9) and whether or not the formulation is taken with a meal (10). Enteric-coated tablets dosed on an empty stomach are generally emptied from the stomach quite rapidly (<2 hr after ingestion), while after a heavy meal they may be retained for a considerable period of time (over 15 hr in some cases) (11). It is well established that, after eating a meal, the shape of the stomach changes and the upper part (the fundus) relaxes to accommodate the extra volume. There is a short lag phase before the mixing movements in the lower part of the stomach (the pyloric antrum) increase. There is, therefore, a sharp contrast between the activity in the top and bottom halves of the stomach.

Multi-particulate dosage forms will empty more slowly in the presence of food than in the fasted state. Since the dosage

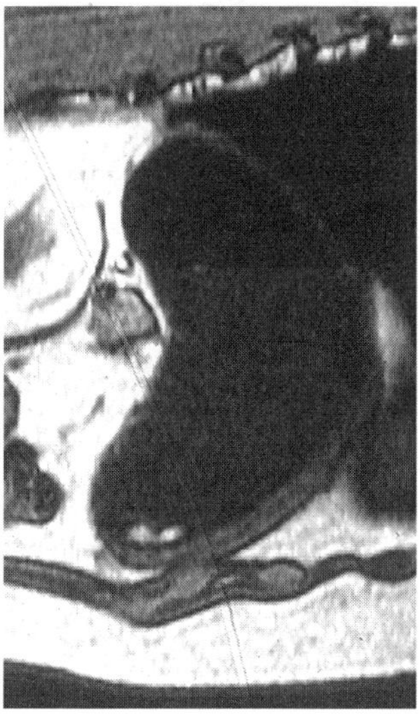

Figure 1 Oil-filled gelatin capsules dissolving on the floor of the stomach.

forms mix evenly with the food, their entry into the small intestine will be strongly influenced by the calorific density and bulk of the ingested meal (9). The rate of gastric emptying, therefore, determines the absorption behavior and is reasonably reproducible. In contrast, the absorption of drugs from larger, non-disintegrating solids and even small soft gelatin capsules is sometimes less predictable, and in these cases other, non-radionuclide measurements may aid in the understanding of the dosage form behavior.

As an example, we observed erratic performance of a soft gel formulation containing a poorly soluble drug when given with a high carbohydrate meal (a baguette). Reduction of dose size increased the variability and we had some difficulty in explaining these results. We, therefore, had to look for other imaging possibilities, including MRI. Using this technique, the differences in proton shift of gut contents and tissues can be used to explore the behavior of formulations in the GI tract, provided that movement artifacts can be minimized. At first, there were difficulties in obtaining good definition

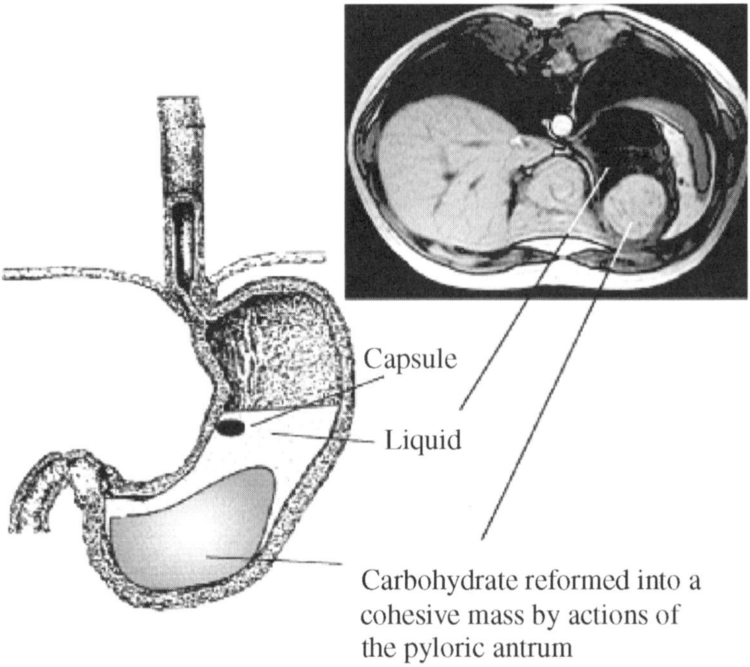

Figure 2 Magnetic resonance image showing the semisolid fraction of a sandwich-based meal lying in the stomach. A small capsule given soon after the meal floats on the liquid above the solid mass, becoming stuck in the gastric rugae in the body of the stomach or floats off ahead of the bulk of the gastric contents.

until it was found that rolling the subject into a prone position immobilized the stomach contents: in this position the pressure of the viscera causes mixing to abruptly cease and the liquid and solid phases separate in the stomach. The stasis produced by the maneuver allows the behavior of small objects to be clearly discriminated in the stomach as illustrated in Figure 1 in which two filled gelatin capsules can be seen in the greater curvature.

Using this same maneuver, the MRI clearly revealed the heterogeneity in the stomach associated with the baguette-based meal and helped to explain the variability associated with the formulation. Figure 2 shows the semisolid fraction of a sandwich-based meal lying in the stomach. Because the solid phase is not fully hydrated, it shows up as a bright doughnut-shaped solid against the liquid phase above it. Over a period of about 30 min to an hour, the solids gradually hydrate and the two phases are no longer distinct. During the early phase of digestion, the center of the lumen is relatively immobile and the secreted gastric juice flows around the food mass. This lack of homogeneity in the lumenal contents prevents efficient mixing and can have therapeutic consequences. For example, a small capsule given soon after the meal could either float on the liquid above the solid mass or float off ahead of the bulk of the gastric contents, resulting in quite different delivery patterns to the absorptive sites in the small intestine.

It is reasonable to expect that altering the balance between solids and liquids will affect emptying of both phases. The interaction is quite complex: Collins et al. (12) tried increasing the volume of the solid phase relative to the liquid, in meals containing either 100 or 400 g minced beef and a fixed amount of water. They showed that, with the larger meal, the lag phase increased from 31 to 56 min but that after this lag time the emptying of solid was accelerated. Furthermore, the larger meal retarded intragastric distribution and gastric emptying of the liquid (12). On the basis of this observation, it would be expected that an oral formulation given after a large meal would show a decreased rate of emptying. Scintigraphic studies show that the tablet is generally held in the fundus and may remain static as in the upper stomach stirring movements are sluggish or even absent.

Faas et al. (13) in Zurich were able to elucidate the cause of the observations made by Meyer and Lake (14), who showed a mismatch in delivery between the digestible fat fraction and the delivery of pancreatin from an enteric-coated pellet formulation. The study conducted by the Zurich group extended MRI observations on meal effects and homogeneity by studying meals which were homogenous, contained particulates or were highly heterogeneous (a hamburger-based meal with different amounts of water). They showed that the intragastric distribution of the marker was highly affected by the consistency of the meal, whereas the amount of co-ingested liquid had a small effect. A large fraction of the contents of the fundus did not come in contact with the marker, and in agreement with our earlier studies (15), it appears that the liquid phase moved around the consolidated solid phase.

For certain drugs, it is desirable to increase the rate of gastric emptying in order to speed up absorption and achieve a faster onset of action. Grattan et al. (16), and Rostami-Hodjegan and coworkers (17) reported that a novel acetaminophen (paracetamol) formulation containing sodium bicarbonate showed a shorter time to maximum serum concentration (t_{max}), in both the fed and fasted states, compared to conventional paracetamol tablets. These results can be explained on the basis of an old observation of Hunt and Pathak (18), who described a prokinetic effect of sodium bicarbonate, which was maximal with an isotonic solution. Given that the recommended dose of the new formulation, two tablets taken with 100 mL water, would produce an approximately isotonic solution of sodium bicarbonate, faster gastric emptying seemed a likely explanation for the faster absorption—at least in the fasted state. The new formulation was also shown to display faster in vitro dissolution compared to conventional tablets in 0.05 M HCl, using the USP II paddle apparatus at low stirrer speeds (10–40 rpm). Although the reason for this faster in vitro dissolution remained to be established, it was proposed that there may be a corresponding increase in the in vivo dissolution rate.

We suspected that the increased dissolution rate could be due to the altered hydrodynamic environment resulting from the release of gaseous carbon dioxide by the reaction of sodium bicarbonate with hydrochloric acid. According to the Noyes–Whitney equation, drug dissolution rate is inversely proportional to the thickness of the boundary diffusion layer at the surface of the tablet. Therefore, turbulence caused by gaseous carbon dioxide could effectively reduce the thickness of the diffusion layer and thus increase dissolution rate (see Chapter 6 for a discussion of the effects of turbulence on drug dissolution). In order to further investigate the influence of gaseous carbon dioxide on dissolution rate, our group carried out in vitro dissolution studies using carbonated and

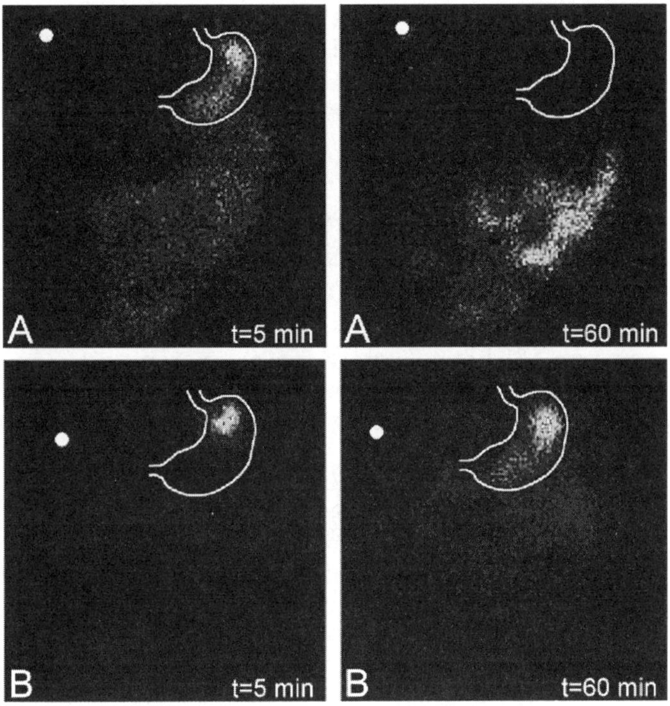

Figure 3 Representative scintigraphic images taken from a single volunteer following dosing with new paracetamol tablets containing sodium bicarbonate (A) and conventional tablets (B) in the fasted state.

de-gassed soda water as dissolution media with a stirrer speed of 30 rpm. There was no significant difference between the dissolution profiles of the conventional formulation in the de-gassed medium and in 0.05 M HCl. However, the carbonated medium increased the dissolution rate of the conventional formulation to such an extent that the dissolution profile was similar to that for the new formulation in 0.05 M HCl. This is consistent with the hypothesis that the increased dissolution rate of the new formulation in HCl is due to turbulence caused by the generation of gaseous carbon dioxide.

We also conducted a combined scintigraphy and pharmacokinetic study in healthy volunteers, which allowed comparison of the in vivo rates of disintegration and gastric emptying with the serum concentration vs. time profiles of the two formulations. We confirmed both faster disintegration and gastric emptying of the new formulation in both fed and fasted states, with the differences in gastric emptying being more pronounced in the fasted state and the differences in disintegration more pronounced in the fed state (19). As one might expect, the effect of food already present in the stomach appeared to impair the prokinetic effect of the sodium bicarbonate. Figure 3 shows representative scintigraphic images from an individual volunteer in the fasted state. After 5 min, the new tablets have largely disintegrated and some gastric emptying has already occurred, whereas the conventional tablets remain almost intact. After 60 min, gastric emptying of the new tablets is complete, while little emptying of the conventional tablets has occurred.

It has been established in many experiments that fat retards gastric emptying, although the presence of fat in the stomach is not the key issue. Much work has been done to establish the exact nature of this mechanism, and it has been known for many years that this effect is mediated through receptors in the small intestine (20). Studies in dogs using manometry and three-dimensional x-ray techniques established that the presence of fat in the upper intestine delays emptying by increasing resistance to flow through the pylorus (21). It has also been established that the hormone cholecystokinin (CCK) is at least partly responsible for this effect in humans (22). This leads to the possibility that fats could be used to retard the gastric emptying of drug formulations. Gröning and Heun (23,24) incorporated fatty acid salts in formulations of riboflavin and nitrofurantoin and showed an increase in both gastric residence time and drug absorption.

SMALL INTESTINE

In the small intestine, contact time with the absorptive epithelium is limited, and a small intestinal transit time (SITT) of 3.5–4.5 hr is typical in healthy volunteers. The Holy Grail of drug delivery would be to discover a mechanism that extended the period of contact with this area of the GI tract. Various approaches have been suggested, but a universal solution is not evident and data demonstrating phenomena that extend GI residence are often subject to controversy. Attempting to examine the effects of altering the contact time of a drug with the small intestine by treatment with metoclopramide or propantheline bromide has been a classical stratagem since the first observations on the effects of these compounds on the absorption of griseofulvin (25). More recently, Marathe et al. (26) examined the effects on metformin solutions labeled by addition of 99mTc-DTPA. Metformin absorption began when the solutions entered the small intestine and started to decline when the material reached the colon. In those cases where propantheline was used to greatly increase the residence time in the small intestine, absorption appeared to be complete prior to arrival at the colon.

Infusion of fat into the ileum has been shown to cause a lengthening of the SITT—a phenomenon known as the *ileal brake* (27,28). However, the effect is generally modest (causing a delay of 30–60 min) and attempts to exploit this mechanism in drug delivery have had limited success. Dobson et al. (29,30) studied the effect of co-administered oleic acid on the small intestinal transit of non-disintegrating tablets. They showed a delay in SITT in over half of all cases, and a doubling of SITT in some instances, but in the other cases SITT was either unaffected or even reduced. Lin et al. (31) have also showed slowed GI transit in patients with chronic diarrhea by administration of emulsions containing 0, 1.6, and 3.2 g of oleic acid. Small intestinal transit in *normal subjects* was measured at 102 ± 11 min, while the transit times in the *patients* treated with the three emulsions were, respectively, 29 ± 3, 57 ± 5 and 83 ± 5 min.

MOTILITY AND STIRRING IN THE SMALL INTESTINE

Muscular contractions in the wall of the small intestine have to achieve two objectives: first, stirring of the contents to

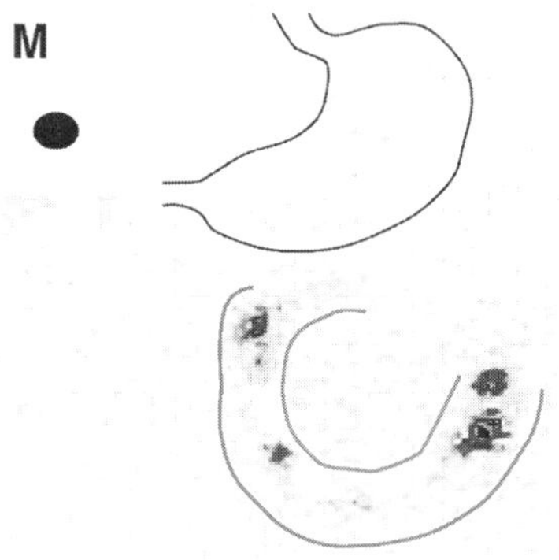

Figure 4 Gamma scintigraphic images of small intestinal transit of capsules showing periods of stasis during a 30 sec acquisition. M = exterior marker.

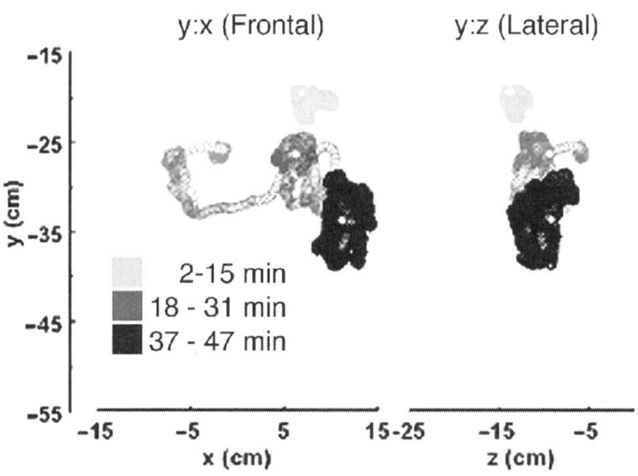

Figure 5 Magnetic moment images of an enteric-coated tablet containing a small amount of magnetized ferric oxide. Left-hand panel shows three sequences in a single volunteer viewed from the front. The right-hand panel shows the same sequences viewed from the top. (Courtesy of Prof. Dr. W. Weitschies.)

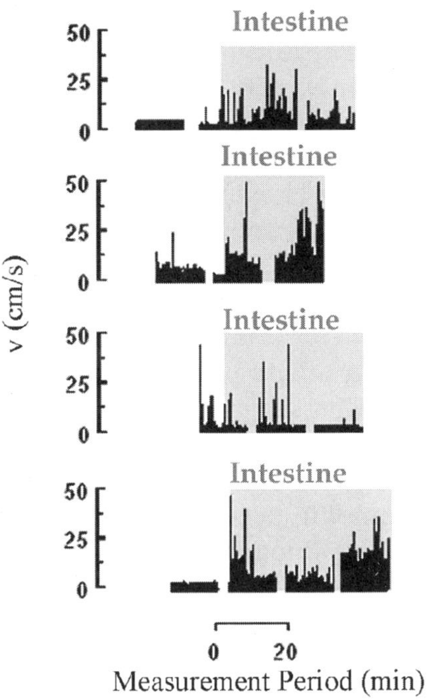

Figure 6 Differences in transit velocities in four subjects, before and after leaving the stomach. (Courtesy of Prof. Dr. W. Weitschies.)

increase exposure to enzymes and to bring the lumenally digested products close to the wall and second, propulsion of indigestible material towards the distal gut. To accomplish this, movements of the gut consist of a mixture of annular constricting activity together with peristaltic movements, which are of both long and short propagation types.

Gamma scintigraphy is not well suited to the study of real time movement, although Kaus et al. (32) applied the technique to measure the average transit rate through the jejunum and ileum of a Perspex capsule labeled with technetium-99 m. More recently, magnetic moment imaging has been used by several workers, in particular Professor Weitschies' group in Greifswald, to examine the pattern of movement of capsules through the GI tract. The technique involves the incorporation of a small amount of iron oxide into the formulation and detecting the tiny induced magnetic field against the Earth's magnetic field. The authors have used the technique to examine the manner in which the formulations move along the small intestine. This is typified by a series of hops and short periods of stasis as the periodic contractions push the capsule down the intestine. These movements gradually become weaker and weaker. In a gamma camera image periods of stasis can also be observed, as illustrated in Figure 4. Magnetic moment imaging provides much more detail, in part because imaging is carried out continuously or as a contiguous recording for short measurement periods. Figure 5 shows the passage of an enteric-coated tablet moving through the gut of a volunteer over three periods of time up to 47 min post-administration. The greater rate transit through the upper gut is clearly seen in the middle period—18–31 min—when the unit travels through the duodenum. Differences in applied agitation forces on the formulation in four volunteers are evident in Figure 6. Comparing formulations movements during the time the unit is in the stomach and in the upper intestine, as shown in Figures 4 and 5, suggests that the period of contact with the mucosa is low in these regions compared to further along the gut.

As might be expected, the presence of nutrients in the gut alters motility — drinking glucose solutions or Intralipid® increases contraction of the gut significantly. Both increase contractions to the same extent, with the duration of the increase dependent on caloric activity (33). The same group previously showed that increasing the viscosity of the gastric contents by administration of guar (5 g) delayed gastric emptying of the glucose load (300 kcal in 300 mL water) and produced a prolongation of the post-prandial contractile activity (34). The effect was seen when the guar was given with a meal, but not with water, suggesting that the guar effect is due to a slowed delivery of calories from the stomach and perhaps from the intestinal lumen.

Exposure of the intestinal cells to high concentrations of polyethylene glycol 2000 causes villus shortening, goblet cell capping, and destruction of the villus tip (35). The effects of smaller molecular weight materials were more extreme and

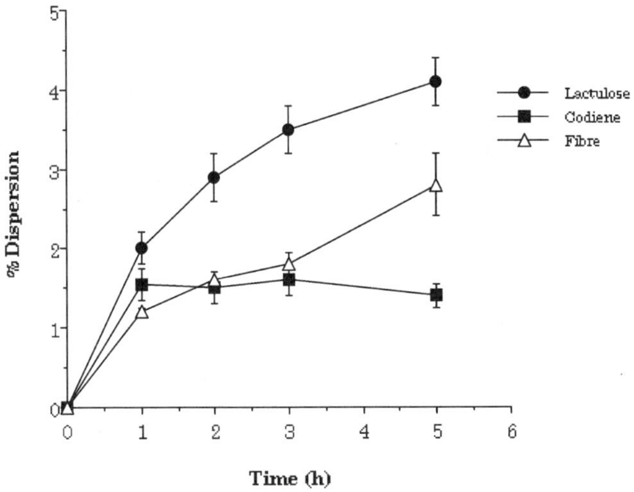

Dispersion of material following timed release in the colon using Pulsincap devices. Mean ± s.d.

Figure 7 Graph illustrating the dispersion of colonic contents of a Pulsincap released in the ascending bowel.

were not tolerated by the intestinal tissue. Contact with strong osmotically active agents would be expected to reverse water flux from the tissues and cause contractions. Basit et al. (36) recently reported a study in which a 150 mL orange juice drink containing 10 g PEG 400 was given with an immediate-release pellet formulation of ranitidine (150 mg). The control was the juice without PEG400 and the liquids were tagged with In-111 to allow measurement of transit. Mean small intestinal transit was decreased from 226 to 143 min and the absolute bioavailability of ranitidine decreased by a third.

COLONIC WATER

For most formulations, colonic absorption represents the only real opportunity to increase the interval between dosing. Transit through the lower part of the gut is quoted at around 24 hr, but in reality only the ascending colonic environment has sufficient fluid to facilitate dissolution. The supplementation of diet by fiber increases the water content of the colon—undigested insoluble fiber carries about 2 mL water per gram of dry weight (37) but effectiveness of fiber in easing functional constipation appears to require an additional intake of 1.5–2 L of extra fluid a day (38). Soluble fibers have a higher capacity for retaining water, at least in vitro, swelling more than 20 times their dry weight (39). The impact of this large amount of hydrogel on drug dispersion in the colon has not been investigated but remains a subject of considerable interest.

In the colon, water availability is low past the hepatic flexure, as the ascending colon is extremely efficient at water and electrolyte absorption. Release at the ileocaecal junction, before significant absorption of lumenal water has occurred, appears to provide satisfactory dispersion in the right colon. Recent evidence suggests that net absorptive water flux in the colon, in both the basal and postprandial state, appears to be augmented by intraluminal glucose (40). Further, changing the water content of the human colon by co-administering 20 g lactulose for three days markedly increases dispersion and dissolution in the transverse colon, as shown for subjects dosed with quinine sulfate in a colon-targeted device in Figure 7.

Motility changes in the colon can also be brought about by bacterial overgrowth and there is a school of thought which believes that patients with irritable bowel syndrome show symptoms which are similar to those of small intestinal bacterial colonization. It would be expected that the overgrowth would produce contraction and segmentation leading to stasis and pockets of gas in the bowel. Indeed, eradication of overgrowth with antibiotics appears to be associated with relief of symptoms in irritable bowel syndrome as judged by standard assessment criteria (41).

COLONIC GAS

In the cecum, the fermentation of any soluble fiber present produces short chain fatty acids (SCFA) and gas (largely carbon dioxide, but with small amounts of hydrogen and methane if the redox conditions are appropriate). In vitro fermentation studies of fiber with a human fecal innoculate show the amount of gas produced correlates approximately with SCFA production and varies with the fiber type. In the studies described by Campbell and Fahey (42), pectin produced the most gas during extended fermentation (108 mL/g^{-1}) whereas methylcellulose produced only 0.57 mL/g^{-1}. The same group has found considerable inter- and intra-subject variability in potential in vivo fermentation of pectin-containing vegetables (37), which may be due to the presence of other bacterial commensals. In fecal incubations from pigs fed probiotic bacteria (live lactobacilli), carbon dioxide production was reduced although hydrogen sulfide production was increased (43). When *Lactobacillus plantarum* was dosed to patients with irritable bowel syndrome, flatulence decreased and less pain was reported in the test vs. the placebo group (44).

The gas rises into the transverse colon and can form temporary pockets, which can restrict access of water to the formulation, particularly if the design does not permit uptake of water through the surface. For this reason, distal release of drug can be hampered by poor wetting/spreading and the reduced surface area, leading to restricted absorption.

Drugs that affect transit time would be expected to alter the normal flora and metabolic activity of the colonic lumen. Oufir et al. (45) investigated the effects of treatment with cisapride and loperamide on fecal flora and SCFA production. By doubling the transit time with loperamide, the concentration of SCFAs was markedly increased whereas by reducing the transit time with cisapride, pH was elevated and the concentration of SCFAs was significantly reduced.

DISTRIBUTION OF MATERIALS IN THE COLON

Our early scintigraphic studies, in which Tc-99m pellets and In-111 labeled non-disintegrating tables were dosed together, suggested differential transit through the lower gut (9). This was confirmed in later studies in which small tablets and pellets labeled with In-111 and Tc-99m were dosed in colon-targeted dosage forms (46). The pellets appear to become trapped in the plaecal folds, whereas the solid units were propelled forward. This has been a consistent finding, which has great importance in terms of dosage form design to prolong release in the gut. Other workers using inert plastic flakes and granules have also investigated shape factors of non-nutrients on whole gut transit time (47). The plastic flakes showed a more rapid transit than the granules, supporting the scintigraphic evidence.

The anatomy of the distal colon, with its thick muscular walls, suggests a predominantly propulsive activity. Studies with single administrations of pellets or Pulsincap devices suggested that the distal part of the transverse colon area is difficult to treat since this area and the descending colon function as a conduit. Steady-state measurements confirm this assertion (48) and Weitschies' group have also reported data showing mass movements propel objects quickly through the distal transverse colon.

In order to look at the probable duration of treatment with topical agents for colonic drug delivery, we have conducted studies with normal subjects and patients with left-sided colitis. The subjects and patients were dosed daily with indium-111-labeled amberlite resin and imaged throughout the day. On the fourth, the division of activity in the colon was 67% in the proximal half and 33% in the distal half day for the control subjects, whereas for the patients with colitis the distribution was 90:10. These data emphasize the problem of treating left-sided colitis effectively during active periods of disease.

THE IMPORTANCE OF TIME OF DOSING

Time of dosing appears to be a further important factor in maximizing colonic contact, particularly in the ascending colon. Morning dosing without fasting is a common regimen in clinical trials, and patterns of motility under these conditions, at least in healthy volunteers, have been well established using scintigraphy. Following early morning dosing, a non-disintegrating unit clears the stomach in 1–2 hrs and has a SITT of 3.5–4.5 hr, although transit times as short as 2 hr or less have been noted in a few individuals. For most subjects dosed at 8 a.m., the unit will be expected to be at the ileocecal junction or to have entered the colon by around 1 p.m. Colonic transit through the proximal colon of intact objects such as non-disintegrating capsules is usually 5–7 hr, whereas transit of the dispersed particulate phase is longer, around 12 hr (49,50). For a non-disintegrating object dosed in the morning, the unit will have arrived at the hepatic flexure by 7–8 p.m. Thus, assuming the drug is absorbed in the colon, the maximum time window for absorption is 6–8 hr following morning dosing with a monolith and 12–15 hr with particulates.

Studies using the Pulsincap system (51) were carried out in our laboratories with the objective of targeting the distal colon with a pulsed delivery of a transcellular probe (quinine) and ^{51}CrEDTA, a paracellular probe. In these studies, subjects were dosed at 10 p.m. to ensure delivery to the descending colon by lunchtime the following day. The site of release was identified by incorporating ^{111}In -labeled resin into the unit and imaging the subjects with a gamma camera. A total of 39 subjects were investigated. Fifteen hours after nocturnal administration, the majority of the delivery systems were situated in the proximal colon at their predicted release time and had not advanced further than a similar set of systems viewed only 6 hr after dosing. This relative stagnation appears to reflect the lack of propulsive stimuli caused by the intake of food, and the effect of sleep in reducing colonic electrical and contractile activity (52–55). Delayed nocturnal gastric emptying (56) and reduced propagation velocity of the intestinal migrating motor complex (57) may also have contributed, as supported by the finding that in two individuals the delivery system did not enter the colon until 12.5 and 13.5 hr after ingestion.

If a delayed-release formulation is taken around 5 p.m., it will have progressed through to the ascending colon by the time the patient goes to bed. Quiescence of propulsive movements in the large bowel causes a relative stagnation, and units remain in the ascending colon overnight. Potentially, this can increase the time of contact to 11–13 hr even for a slowly dissolving matrix. On rising, the change in posture stimulates mass movements, felt by the subject as the urge to defecate, and contents move from the right to the left side of the colon.

From the studies conducted using gamma scintigraphy and MRI, it can be concluded that both temporal and dietary factors are important co-determinants of transit. For poorly soluble substances, the reserve time is an important determinant of bioavailability. Moving away from the current practice of dosing once-a-day formulations in the mornings might allow a reduction in the dosing frequency and increased efficacy of colon-targeted drugs and for formulations used to prevent acute disease episodes at night and in the early morning.

EFFECTS OF AGE, GENDER, AND OTHER FACTORS

Physiological functions naturally change with advancing age. However, there has always been great debate about the magnitude of age, gender and other non-meal-related factors, including posture and exercise, on GI transit (58). It is now generally accepted that gastric emptying and colonic transit are prolonged in women compared with men (59). However, there is still some debate about the effects of gender on SITT. Bennink et al. (60) concluded that SITT of a dual radionuclide-labeled test meal in healthy men and women are the same. Madsen's group has conducted studies on GI transit using a similar meal on various cohorts of healthy subjects utilizing gamma scintigraphy over a number of years. In a recent publication, the group concludes that age and gender do have an effect. Their measurements indicated that women have slower GI transit than men in all regions of the GI tract, particularly with regard to a slower mean colon transit in middle age. In contrast, aging was shown to accelerate the gastric emptying and intestinal transit significantly (61). A recent study showed that postprandial proximal gastric relaxation in women was prolonged, which is consistent with delayed gastric emptying (62). The differences in GI transit between the sexes have been attributed to the actions of the female sex hormones. A study by Hutson et al. (63) found that pre-menopausal women, and post-menopausal women taking hormone replacement therapy (HRT), showed slower gastric emptying of solids than post-menopausal women not taking HRT. Furthermore, those post-menopausal women not taking HRT showed similar gastric emptying times to men. That being the case, one would expect that the fluctuations of female sex hormones during the menstrual cycle would also have an effect. Again, studies on this topic have yielded conflicting results: some studies have shown that GI transit is delayed during the luteal phase of the menstrual cycle (64,65), while others have found no effect (55,66).

Quigley's group in Cork, Ireland, have concluded that normal aging is associated with changes in motility but the pattern is varied and no clear clinical consequence can be identified (67). More important in their view are the pathophysiological influences, including depression (and treatment with anti-cholinergics and opiates), hypothyroidism, and chronic renal failure.

CONCLUDING REMARKS

The relationship between GI transit and drug absorption is well established and investigative tools such as gamma scintigraphy; MRI, and magnetic moment imaging have greatly contributed to our understanding. In recent years, the Biopharmaceutics Classification Scheme has helped the industry contain costs in clinical development and by appropriate choice of in vitro methods, we have a reasonable level of assurance that, for certain classes of compounds, we can reasonably predict performance on the basis of laboratory tests. There is no doubt that the issues of dissolution, absorption, and transit are the key variables for simple tablet, pellet, and capsule formulations. For more sophisticated formulations, particularly delayed-release preparations, the situation is probably too complex to allow adoption of standard compendial dissolution tests irrespective of the choice of dissolution media. Our ability to progress in this area is dependent on arriving at a better understanding of the stirring and viscosity characteristics of the lower small intestine and large bowel. This will require more investment in the development of investigative methods and multi-modal imaging to ascertain the true conditions experienced by a formulation in the unprepared human bowel.

ACKNOWLEDGMENTS

The authors are grateful to Professor Weitschies for permission to use his data on MRI (Weitschies et al., Pharm Res 2003 In press.)

REFERENCES

1. Washington N, Washington C, Wilson CG. Physiological Pharmaceutics: Barriers to Drug Absorption. Taylor & Francis, 2001.
2. Perkins AC, Wilson CG, Frier M, Blackshaw PE, Danserau RJ, Vincent RM, Wenderoth D, Hathaway S, Li Z, Spiller RC. The use of scintigraphy to demonstrate the rapid esophageal transit of the oval film-coated placebo risedronate tablet compared to a round uncoated placebo tablet when administered with minimal volumes of water. Int J Pharm 2001; 222:295–303.
3. Ekeberg O, Feinberg MJ. Altered swallowing function in elderly patients without dysphagia: radiological findings in 56 cases. Am J Roentgenol 1991; 156:1181–1184.
4. Perkins AC, Wilson CG, Frier M, Blackshaw PE, Juan D, Danserau RJ, Hathaways S, Li Z, Long P, Spiller RC. Oesophageal transit, disintegration and gastric emptying of a film-coated risedronate placebo tablet in gastro-oesophageal reflux disease and normal control subjects. Aliment Pharmacol Ther 2001; 15:115–121.

5. Perkins AC, Wilson CG, Blackshaw PE, Vincent RM, Danserau RJ, Juhlin KD, Bekker PJ, Spiller RC. Impaired oesophageal transit of capsule versus tablet formulations in the elderly. Gut 1994; 35:1363–1367.
6. Perkins AC, Wilson CG, Frier M, Vincent RM, Blackshaw PE, Danserau RJ, Juhlin KD, Bekker PJ, Spiller RC. Esophageal transit of risedronate cellulose-coated tablet and gelatin capsule formulations. Int J Pharm 1999; 186:169–175.
7. Hardy JG, Wilson CG. Radionuclide imaging in pharmaceutical, physiological and pharmacological research. Clin Phys Physiol Meas 1981; 2:71–121.
8. Wilson CG, Washington N. Assessment of disintegration and dissolution of dosage forms in vivo using gamma scintigraphy. Drug Dev Ind Pharm 1988; 14:211–218.
9. Davis SS, Hardy JG, Taylor MJ, Whalley DR, Wilson CG. A comparative study of the gastrointestinal transit of a pellet and tablet formulation. Int J Pharm 1984; 21:167–177.
10. O'Reilly S, Wilson CG, Hardy JG. The influence of food on the gastric emptying of multiparticulate dosage forms. Int J Pharm 1987; 34:213–216.
11. Wilson CG, Washington N, Greaves JL, Kamali F, Rees JA, Sempik AK, Lampard JF. Bimodal release of ibuprofen in a sustained-release formulation—a scintigraphic and pharmacokinetic open study in healthy volunteers under different conditions of food-intake. Int J Pharm 1989; 50:155–161.
12. Collins PJ, Horowitz M, Maddox A, Myers JC, Chatterton BE. Effects of increasing solid component size of a mixed solid/liquid meal on solid and liquid gastric emptying. Am J Physiol 1996; 271:G549–G554.
13. Faas H, Steingoetter A, Feinle C, Rades T, Lengsfeld H, Boesiger P, Fried M, Schwizer W. Effects of meal consistency and ingested fluid volume on the intragastric distribution of a drug model in humans—a magnetic resonance imaging study. Aliment Pharmacol Ther 2002; 16:217–224.
14. Meyer JH, Lake R. Mismatch of duodenal deliveries of dietary fat and pancreatin from enterically coated microspheres. Pancreas 1997; 15:226–235.

15. Washington N, Moss HA, Washington C, Greaves JL, Steele RJC, Wilson CG. Non-invasive detection of gastro-oesophageal reflux using an ambulatory system. Gut 1993; 34:1482–1486.

16. Grattan T, Hickman R, Darby-Dowman A, Hayward M, Boyce M, Warrington S. A five way crossover human volunteer study to compare the pharmacokinetics of paracetamol following oral administration of two commercially available paracetamol tablets and three development tablets containing paracetamol in combination with sodium bicarbonate or calcium carbonate. Eur J Pharm Biopharm 2000; 49:225–229.

17. Rostami-Hodjegan A, Shiran MR, Ayesh R, Grattan TJ, Burnett I, Darby-Dowman A, Tucker GT. A new rapidly absorbed paracetamol tablet containing sodium bicarbonate. I. A four way crossover study to compare the concentration-time profile of paracetamol from the new paracetamol/sodium bicarbonate tablet and a conventional paracetamol tablet in fed and fasted volunteers. Drug Dev Ind Pharm 2002; 28:523–531.

18. Hunt JN, Pathak JD. The osmotic effects of some simple molecules and ions on gastric emptying. J Physiol 1960; 154:254–269.

19. Kelly K, O' Mahony B, Lindsay B, Jones T, Grattan TJ, Rostami-Hodjegan A, Stevens HNE, Wilson CG. Comparison of the rates of disintegration, gastric emptying, and drug absorption following administration of a new and a conventional paracetamol formulations, using Gamma scintigraphy. Pharm Res 2003; 20:1668 1673.

20. Hunt JN, Knox MT. A relation between the chain length of fatty acids and the slowing of gastric emptying. J Physiol 1968; 194:327–336.

21. Kumar D, Ritman EL, Malagelada JR. Three-dimensional imaging of the stomach: role of pylorus in the emptying of liquids. Am J Physiol 1987; 253:G79–G85.

22. Kleibeuker JHH, Beekhuis JNBJ, Jansen J, Piers DA, Lamers CBHW. Cholecystokinin is a physiological hormonal mediator of fat-induced inhibition of gastric emptying in man. Eur J Clin Invest 1988; 18:173–177.

23. Groning R, Heun G. Oral dosage forms with controlled gastrointestinal transit. Drug Dev Ind Pharm 1984; 10:527–539.

24. Groning R, Heun G. Dosage forms with controlled gastrointestinal passage—Studies on the absorption of nitrofurantoin. Int J Pharm 1989; 56:111–116.
25. Jamali F, Axelson JE. Influence of metoclopramide and propantheline on GI absorption of griseofulvin in rats. J Pharm Sci 1977; 66:1540–1543.
26. Marathe PH, Wen YD, Norton J, Greene DS, Barbhaiya RH, Wilding IR. Effect of altered gastric emptying and gastrointestinal motility on metformin absorption. Br J Clin Pharmacol 2000; 50:325–332.
27. Read NW, McFarlane A, Kinsman RI, Bates TE, Blackhall NW, Farrar GBJ, Hall JC, Moss G, Morris AP, O'Neill B, Welch I, Lee Y, Bloom SR. Effect of infusion of nutrient solutions into the ileum on gastrointestinal transit and plasma levels of neurotensin and enteroglucagon. Gastroenterology 1984; 86:274–280.
28. Spiller R. The ileal brake—inhibition of jejunal motility after ileal fat perfusion in man. Gut 1984; 25:365–374.
29. Dobson CL, Davis SS, Chauhan S, Sparrow RA, Wilding IR. The effect of oleic acid on the human ileal brake and its implications for small intestinal transit of tablet formulations. Pharm Res 1999; 16:92–96.
30. Dobson CL, Davis SS, Chauhan S, Sparrow RA, Wilding IR. Does the site of intestinal delivery of oleic acid alter the ileal brake response?. Int J Pharm 2000; 195:63–70.
31. Lin HC, Van Citters GW, Heimer F, Bonorris G. Slowing of gastrointestinal transit by oleic acid. Dig Dis Sci 2001; 46:223–229.
32. Kaus LC, Fell JT, Sharma H, Taylor DC. The intestinal transit of a single non-disintegrating unit. Int J Pharm 1984; 20:315–323.
33. Von Schonfeld J, Evans DF, Renzing K, Castillo FD, Wingate DL. Human small bowel motor activity in response to liquid meals of different caloric value and different chemical composition. Dig Dis Sci 1998; 43:265–269.

34. Von Schonfeld J, Evans DF, Wingate DL. Effect of viscous fiber (guar) on postprandial motor activity in human small bowel. Dig Dis Sci 1997; 42:1613–1617.

35. Bryan AJ, Kaur R, Robinson G, Thomas NW, Wilson CG. Histological and physiological studies on the intestine of the rat exposed to solutions of Myrj 52 and PEG 2000. Int J Pharm 1980; 7:145–156.

36. Basit AW, Podczeck F, Newton JM, Waddington WA, Eli PJ, Lacey LF. Influence of polyethylene glycol on the gastrointestinal absorption of ranitidine. Pharm Res 2002; 19:1368–1374.

37. Bourquin LD, Titgemeyer EC, Fahey GC Jr. Vegetable fiber fermentation by human fecal bacteria: cell wall polysaccharide disappearance and short-chain fatty acid production during in vitro fermentation and water-holding capacity of unfermented residues. J Nutr 1993; 123:860–869..

38. Anti M, Pignataro G, Armuzzi A, Valenti A, Iascone E, Marmo R, Lamazza A, Pretaroli AR, Pace V, Leo F, Castelli A, Gasbarrini G. Water supplementation enhances the effect of high fiber diet on stool frequency and laxative consumption in adult patients with functional constipation. Hepatogastroenterology 1998; 45:727–732.

39. Goni I, Martin-Carron N. In vitro fermentation and hydration properties of commercial dietary fiber-rich supplements. Nutr Res 1998; 18:1077–1089.

40. Kendrick ML, Zyromski NJ, Tanaka T, Duenes DA, Libsch K, Sarr MG. Postprandial absorptive augmentation of water and electrolytes in the colon requires intraluminal glucose. J Gastrointest Surg 2002; 6:310–315.

41. Pimentel H, Chow EJ, Lin HC. Eradication of small intestinal bacterial overgrowth reduces symptoms of irritable bowel syndrome. Am J Gastroenterol 2000; 95:3503–3506.

42. Campbell JM, Fahey GC. Psylium and methylcellulose properties in relation to insoluble and soluble fiber standards. Nutr Res 1997; 17:619–629.

43. Tsukahara T, Azuma Y, Ushida K. The effect of a mixture of live lactic acid bacteria on intestinal gas production in pigs. Micr Ecol Health Dis 2001; 13:105–110.

44. Nobaek S, Johansson ML, Molin G, Ahrme S, Jeppson B. Alteration of intestinal microflora is associated with reduction abdominal bloating and pain in patients with irritbale bowel syndrome. Am J Gastroenterol 2000; 95:1231–1238.

45. Oufir LE, Barry JL, Flourie B, Cherbut C, Cloarec D, Bornet F, Galmiche JP. Relationships between transit time in man and in vitro fermentation of dietary fiber by fecal bacteria. Eur J Clin Nutr 2000; 54:603–609.

46. Watts PJ, Barrow L, Steed KP, Wilson CG, Spiller RC, Melia CD, Davies MC. The transit rate of different sized model dosage forms through the human colon and the effects of a lactulose-induced catharsis. Int J Pharm 1992; 87:215–221.

47. Lewis SJ, Heaton KW. Roughage revisited (the effect on intestinal function of inert plastic particles of different sizes and shapes). Dig Dis Sci 1999; 44:744–748.

48. Hebden JM, Perkins AC, Frier M, Wilson CG, Spiller RC. Limited exposure of left colon to daily dosed oral formulation in active distal ulcerative colitis: explanation of poor response to treatment?. Gut 1997; 40:28A.

49. Hardy JG, Wilson CG, Wood E. Drug delivery to the proximal colon. J Pharm Pharmacol 1985; 37:874–877.

50. Barrow L, Spiller RC, Wilson CG. Pathological influences on colonic motility: implications for drug delivery. Adv Drug Del Rev 1991; 7:201–218.

51. Stevens HNE, Wilson CG, Welling PG, Bakhshaee M, Binns JS, Perkins AC, Frier M, Blackshaw EP, Frame MW, Nichols JD, Humphrey MJ, Wicks SR. Evaluation of Pulsincap to provide regional delivery of dofetilide to the human GI tract. Int J Pharm 2002; 236:27–34.

52. Frexinos J, Bueno L, Fioramonti J. Diurnal changes in myoelectric spiking activity of the human colon. Gastroenterology 1985; 88:1104–1110.

53. Narducci FG, Basotti G, Gaburri M, Morelli A. Twenty four hour manometric recording of colonic motor activity in healthy man. Gut 1987; 28:17–25.

54. Soffer EE, Scalabrini P, Wingate D. Prolonged ambulant monitoring of human colonic motility. Am J Physiol 1989; 257:G601-G606.

55. Basotti G, Betti C, Imbimbo BP, Pelli MA, Morelli A. Colonic high-amplitude propogated contractions (mass movements): repeated 24 h manometric studies in healthy volunteers. J Gastrointest Mot 1992; 4:187–191.

56. Goo RH, Moore JG, Greenberg E, Alazraki NP. Circadian variation in gastric emptying of meals in humans. Gastroenterology 1987; 93:515–518.

57. Kumar D, Wingate D, Ruckebusch Y. Circadian variation in the propagation velocity of the migrating motor complex. Gastroenterology 1986; 91:926–930.

58. Wilson CG, O'Mahony B, Lindsay B. Physiological factors affecting oral drug delivery. In: Swarbrick J, ed. Encyclopaedia of Pharmaceutical Technology. New York: Marcel Dekker, 2002:2214–2222.

59. Degen LP, Phillips SF. Variability of gastrointestinal transit in healthy women and men. Gut 1996; 39:299–305.

60. Bennink R, Peeters M, Van den Maegdenbergh V, Geypens B, Rutgeerts P, De Roo M, Mortelmans L. Evaluation of small-bowel transit for solid and liquid test meal in healthy men and women. Eur J Nucl Med 1999; 26:1560–1566.

61. Graff J, Brinch K, Madsen JL. Gastrointestinal mean transit times in young and middle-aged healthy subjects. Clin Physiol 2001; 21:253–259.

62. Mearadji B, Penning C, Vu MK, van der Schaar PJ, van Peterson AS, Kamerling IMC, Masclee AAM. Influence of gender on proximal gastric motor and sensory function. Am J Gastroenterol 2001; 96:2066–2073.

63. Hutson WR, Roehrkasse RL, Wald A. Influence of gender and menopause on gastric emptying and motility. Gastroenterology 1989; 96:11–17.

64. Wald A, Thiel DHV, Hoechstetter L, Gavaler JS, Egler KM, Verm R, Scott L, Lester R. Gastrointestinal transit: the effect of the menstrual cycle. Gastroenterology 1981; 80:1497–1500.

65. Gill RC, Murphy PD, Hooper HR, Bowes KL, Kingma YJ. Effect of the menstrual cycle on gastric emptying. Digestion 1987; 36:168–174.

66. Horowitz M, Maddern GJ, Chatterton BE, Collins PJ, Petrucco OM, Seamark R, Shearman DJ. The normal menstrual cycle has no effect on gastric emptying. Br J Obstet Gynaecol 1985; 92:743–746.

67. O'Mahony D, O'Leary P, Quigley EM. Aging and intestinal motility: A review of factors that affect intestinal motility in the aged. Drugs Aging 2002; 19:515–527.

6

Physiological Parameters Relevant to Dissolution Testing: Hydrodynamic Considerations

STEFFEN M. DIEBOLD
Leitstelle Arzneimittelüberwachung Baden–Württemberg, Regierungspräsidium Tübingen, Tübingen, Germany

HYDRODYNAMICS AND DISSOLUTION

Dissolution

Why Is Hydrodynamics Relevant to Dissolution Testing?

Release-related bioavailability problems have been encountered in the pharmaceutical development of formulations for a number of quite different chemical entities, including ciclosporin, digoxin, griseofulvin, and itraconazole, to name but a few. A thorough knowledge of hydrodynamics is useful in the course of dissolution method development and formulation development, as well as for the pharmaceutical industry's quality needs, e.g., batch-to-batch control. Occasionally, quality control specifications are not met due to "minor" variations involving hydrodynamics, such as the use of different volumes, or modified stirring devices or sampling procedures. The development of drug formulations is facilitated by the choice of an appropriate dissolution apparatus based on insight into its specific hydrodynamic performance. Using the right test might make it easier, for instance, to isolate the impact of different excipients and process parameters on drug release at an early stage of pharmaceutical formulation development. Furthermore, a sound knowledge of in vivo hydrodynamics may help to better understand and possibly to improve forecasting of in vivo dissolution and absorption of biopharmaceutical classification system (BCS) II compounds. Although gastrointestinal (GI) fluids are well-characterized and biorelevant dissolution media [e.g., Fasted State Simulated Intestinal Fluid (FaSSIF) and Fed State Simulated Intestinal Fluid (FeSSIF)] have been developed to simulate various states in the GI tract, knowledge of hydrodynamics appears to be relatively scant both in vitro and in vivo. This chapter gives a brief introduction of the basic hydrodynamics relevant to in vitro dissolution testing, including the convective diffusion theory. This section is followed by hydrodynamic considerations of in vitro dissolution testing and hydrodynamic problems inherent to in vivo bioavailability of solid oral dosage forms.

The Dissolution Process

Dissolution can be described as a mass transfer process. Although mass transfer processes commonly are under the combined influence of both thermodynamics and hydrodynamics, usually one of these prevails in terms of the overall dissolution process (1–3). Hydrodynamics is predominant for the overall dissolution rate if the mass transfer process is *mainly* controlled by convection and/or diffusion, as is usually the case for poorly soluble substances. This is of great practical relevance for pharmaceutical development, since new drug compounds often exhibit poor solubility in aqueous media.

The Dissolution Rate

The dissolution rate ($\mathrm{d}C/\mathrm{d}t$) of a pure drug compound is represented by an equation based on the work of Noyes, Whitney, Nernst, and Brunner (4–6), which is in turn based on earlier observations made by Schükarew in 1891 (7):

$$\frac{\mathrm{d}C}{\mathrm{d}t} = \frac{A \cdot D}{\delta_{\mathrm{HL}} \cdot V}(C_\mathrm{s} - C_\mathrm{t})$$

The proportionality constant k

$$k = \frac{A \cdot D}{\delta_{\mathrm{HL}} \cdot V}$$

is addressed as the "apparent dissolution rate constant." C_s represents the saturation solubility, C_t describes the bulk concentration of the dissolved drug at time t, D is the effective diffusion coefficient of the drug molecule, A stands for the surface area available for dissolution, and V represents the media volume employed in the test. According to the equations of Noyes, Whitney, Nernst, and Brunner, the dissolution rate depends on a small fluid "layer," called the hydrodynamic boundary layer (δ_{HL}), adhering closely to the surface of a solid particle that is to be dissolved (solvendum, solute). As can be seen from the combined equation, an inverse proportionality exists between the dissolution rate and the hydrodynamic boundary layer. If the latter is reduced, the dissolution rate increases.

Hydrodynamic Basics Relevant to Dissolution

Laminar and Turbulent Flow

Laminar flow is characterized by layers ("lamellae") of liquid moving at the same speed and in the same direction (Fig. 1). Little or no exchange of fluid mass and fluid particles occurs across these fluid layers. The closer the layers are to a given

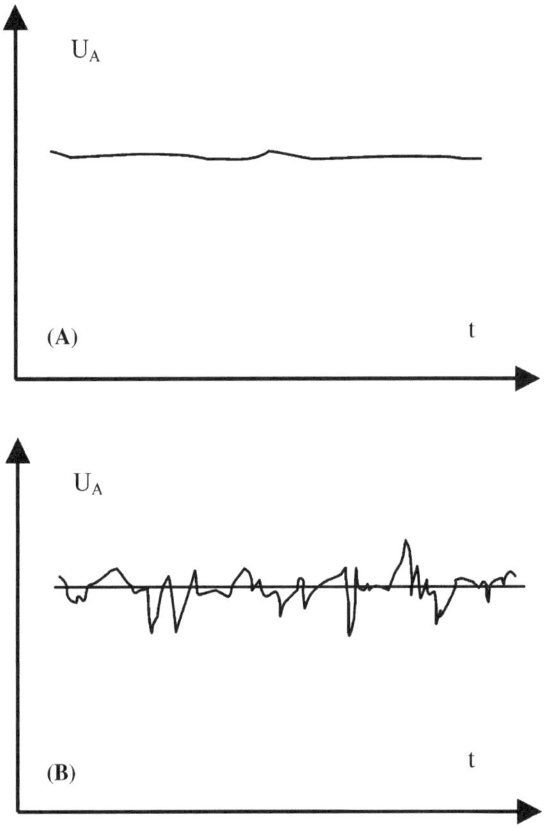

Figure 1 (**A**) Laminar and (**B**) turbulent flow: *t* describes the time scale, U_A represents the velocity component acting in the direction of the flow. *Source*: From Ref. 10.

surface, the slower they move. In an ideal fluid, the flow follows a curved surface smoothly, with the layers central in the flow moving fastest and those at the sides slowest. In turbulent flow, by contrast, the streamlines or flow patterns are disorganized and there is an exchange of fluid between these areas. Momentum is also exchanged such that slow-moving fluid particles speed up and fast-moving fluid particles give up their momentum to the slower-moving particles and slow down themselves. All, or nearly all, fluid flow displays some degree of turbulence. If the fluid velocity exceeds a crucial number, flow becomes turbulent rather than laminar since the frictional force can no longer compensate for other forces acting on the fluid particles. This event depends on the fluid viscosity, the fluid velocity, and the geometry of the hydrodynamic system and is described by the Reynolds number.

Reynolds Number

The dimensionless Reynolds number (Re) is used to characterize the laminar–turbulent transition and is commonly described as the ratio of momentum forces to viscous forces in a moving fluid. It can be written in the form

$$Re = \frac{\rho \cdot U_{\mathrm{A}} \cdot L}{\eta} = \frac{U_{\mathrm{A}} \cdot L}{\nu}$$

where ν represents the kinematic viscosity of the liquid (ρ and η are the density and dynamic viscosity, respectively). U_{A} describes the flow rate, and L represents a characteristic distance or length of the hydrodynamic system, for example, the diameter of a tube or pipe. Laminar flow patterns turn into turbulent flow if the Reynolds number of a particular hydrodynamic system exceeds a *critical* Reynolds number (Re_{crit}).

Particle–Liquid Reynolds Numbers

With respect to the hydrodynamics of particles in a stirred dissolution medium, the Reynolds numbers determined for the bulk flow have to be distinguished from the Reynolds numbers characterizing the particle–liquid system. The latter hydrodynamic subsystem consists of the dissolving particles and the surrounding fluid close to their surfaces. Thus, it is the relative velocity of the solid particle surface to the bulk flow (the "slip velocity") that counts. However, it is permissible to approximate the slip velocity to U_{A}, provided that the drug particles are suspended in the moving fluid and the density difference between particle and dissolution medium is at least $\sim 0.3\,\mathrm{g/cm^3}$ (8). In this case, L represents a characteristic length on the (average) particle surface and may arbitrarily be identified with the particle diameter. With respect to particle–liquid systems, the laminar–turbulent transition at the particle surface is decisive. Laminar flow turns turbulent if Re_{crit} for the flow *close to the particle surface* is exceeded. Thus, Re_{crit} (particle) is not necessarily identical with the Reynolds number of the bulk flow—although the latter may sometimes serve as a sufficient approximation (9,10).

"Eddies," Dissipation, and Energy Cascade

"Eddies" are turbulent instabilities within a flow region (Fig. 2). These vortices might already be present in a turbulent stream or can be generated downstream by an object presenting an obstacle to the flow. The latter turbulence is known as "Karman vortex streets." Eddies can contribute a considerable increase of mass transfer in the dissolution process under turbulent conditions and may occur in the GI tract as a result of short bursts of intense propagated motor activity and flow "gushes."

The mean velocity of eddies changes at a definitive distance called the "scale of motion" (SOM) (11). The bigger these eddies are, the longer is the SOM [(9), Sec. 4]. Apart from "large scale eddies," a number of "small scale eddies"

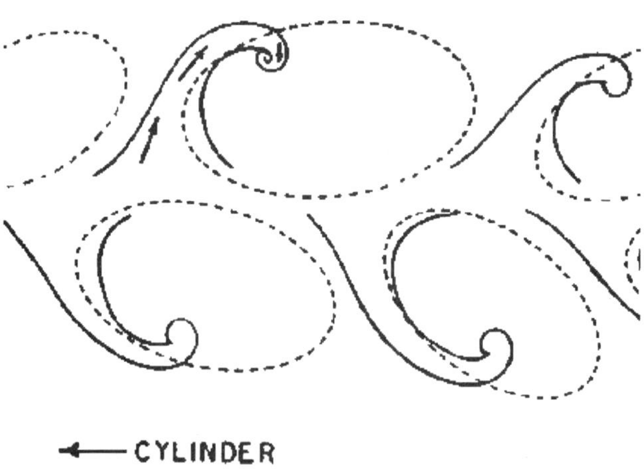

Figure 2 "Eddies" (large scale type) downstream of an object exposed to flow. *Source*: Adapted from Ref. 13, Sec. 21.4 (original by Grant HL. J Fluid Mech 1958; 4:149).

exist in turbulent flow. Under turbulent conditions, eddies transport the majority of the kinetic energy. Energy fed into the turbulence goes primarily into the larger eddies. From these, smaller eddies are generated, and then still smaller ones. The process continues until the length scale is small enough for viscous action to be important and dissipation to occur. This sequence is called the energy cascade. At high Reynolds numbers the cascade is long; i.e., there is a large difference in the eddy sizes at its ends. There is then little direct interaction between the large eddies governing the energy transfer and the small, dissipating eddies. In such cases, the dissipation is determined by the rate of supply of energy to the cascade by the large eddies and is independent of the dynamics of the small eddies in which the dissipation actually occurs. The rate of dissipation is independent of the magnitude of the viscosity. An increase in Reynolds number to a still higher value extends the cascade only at the small eddy end. Still, smaller eddies must be generated before dissipation can occur.

Energy Input ε

For closed dissolution systems, it can be hypothesized that the hydrodynamics depends on the input of energy in a general way. The energy input may be characterized by the power input per unit mass of fluid or the turbulent energy dissipation rate per unit mass of fluid (ε). Considering various *paddle* apparatus, the power input per unit mass of fluid (Fig. 3) can be calculated according to Plummer and Wigley [(12), Appendix B, nomenclature adapted]:

$$\varepsilon = \frac{p \cdot I^5 \cdot \omega^3}{V}$$

where ε has the dimension length^2/time^3. ω stands for the rotations per minute, I is the mean diameter of the paddle or impeller, p is a model constant dependent on the hydrodynamic flow pattern (laminar or turbulent), and V is the fluid volume. As expected, ε is influenced mainly by the diameter of the impeller and the rotation rate. Based on this equation,

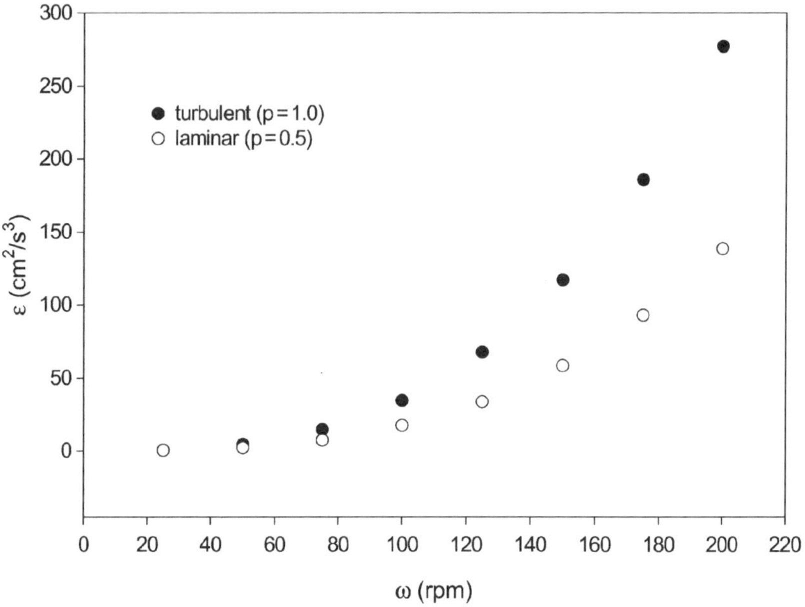

Figure 3 Power input per unit mass of fluid: paddle apparatus, 900 mL. Calculations shown for extremes of completely laminar and completely turbulent hydrodynamic conditions. The actual energy input lies in between the two curves, depending on the stirring rate. *Source*: From Ref. 10.

the power input per unit mass of fluid for the compendial paddle apparatus has been calculated [(10), Chapter 5.6.2]. The fluid mass specific energy input rises exponentially with paddle speed. The exponential form of the observed relationship suggests that there is a transition from laminar ($p = 0.5$) to turbulent flow ($p = 1.0$) within the system, and indicates that the energy input to the media and flow pattern in the vessels are related.

The power input per unit mass of fluid is greater for a dissolution volume of 500 mL than for 900 mL, at a given stirring rate. Remarkably, ε calculated for laminar conditions ($p = 0.5$) employing 500 mL of dissolution medium (not plotted) results in approximately the same hydrodynamic effectiveness as when turbulent conditions are assumed ($p = 1.0$) for a dissolution volume of 900 mL (10). This implies more effective hydrodynamics for the lower volume. Thus, it *cannot* be assumed that there are no hydrodynamic implications when volumes used for a specific dissolution test method are changed, but rather, that the change would require meticulous validation!

Hydrodynamic Boundary Layer Concept

Concept and Structure of the Boundary Layer

A boundary layer in fluid mechanics is defined as the layer of fluid in the immediate vicinity of a limiting surface where the layer and its breadth are affected by the viscosity of the fluid. The concept of the hydrodynamic boundary layer goes back to the work of the German physicist and mathematician Ludwig Prandtl (1875–1953) and was first presented at Göttingen and Heidelberg in 1904 (Fig. 4). According to the Prandtl concept, at high Reynolds numbers, the flow close to the surface of a body can be separated into two main regions. Within the bulk flow region viscosity is negligible ("frictionless flow"), whereas near the surface a small region exists that is called the

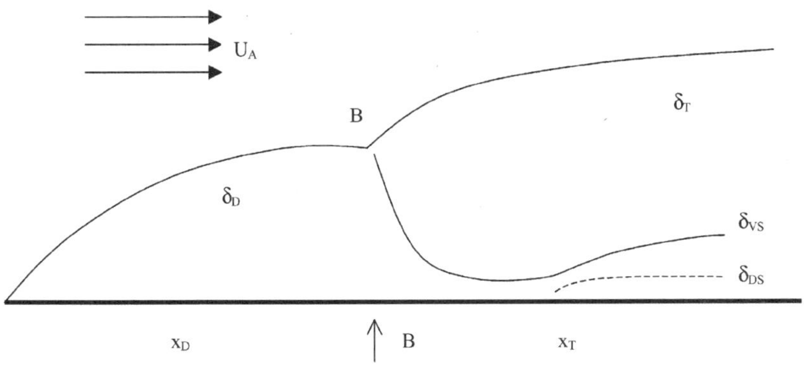

Figure 4 Hydrodynamic boundary layer development on the semi-infinite plate of Prandtl. δ_D = laminar boundary layer, δ_T = turbulent boundary layer, δ_{VS} = viscous turbulent sub-layer, δ_{DS} = diffusive sub-layer (no eddies are present; solute diffusion and mass transfer are controlled by molecular diffusion—the thickness is about 1/10 of δ_{VS}), B = point of laminar–turbulent transition. *Source*: From Ref. 10.

hydrodynamic boundary layer. In this region, adherence of molecules of the liquid to the surface of the solid body slows them down. The hydrodynamic boundary layer is dominated by pronounced velocity gradients within the fluid that are continuous, and does not, as is sometimes purported, consist of a "stagnant" layer. According to Newton's law of friction, pronounced velocity gradients lead to high friction forces near the surface of a solid particle. The hydrodynamic boundary layer grows further downstream of the surface since more and more fluid molecules are slowed down.

In terms of hydrodynamics, the boundary layer *thickness* is measured from the solid surface (in the direction perpendicular to a particle's surface, for instance) to an arbitrarily chosen point, e.g., where the velocity is 90–99% of the stream velocity or the bulk flow (δ_{90} or δ_{99}, respectively). Thus, the breadth of the boundary layer depends ad definitionem on the selection of the reference point and includes the laminar boundary layer as well as possibly a portion of a turbulent boundary layer.

Laminar and Turbulent Boundary Layer

Apart from the nature of the bulk flow, the hydrodynamic scenario close to the surfaces of drug particles has to be considered. The nature of the hydrodynamic boundary layer generated at a particle's surface may be laminar or turbulent regardless of the bulk flow characteristics. The turbulent boundary layer is considered to be thicker than the laminar layer. Nevertheless, mass transfer rates are usually increased with turbulence due to the presence of the "viscous turbulent sub-layer." This is the part of the (total) turbulent boundary layer that constitutes the main resistance to the overall mass transfer in the case of turbulence. The development of a viscous turbulent sub-layer reduces the overall resistance to mass transfer since this viscous sub-layer is much narrower than the (total) laminar boundary layer. Thus, mass transfer from turbulent boundary layers is greater than would be calculated according to the total boundary layer thickness.

Boundary Layer Separation

Both laminar and turbulent boundary layers can separate. Laminar layers usually require only a relatively short region of adverse pressure gradient to produce separation, whereas turbulent layers separate less readily. A few examples of turbulent boundary layer separation include golf ball design to stabilize trajectory, airfoil design to reduce aerodynamic resistance (Fig. 5), and, in nature, in sharkskin to improve the shark's ability to glide. The overall flow pattern, when separation occurs, depends greatly on the particular flow. The flow upstream of the separation point is fed by recirculation of some of the separating fluid. Sometimes the effect is quite localized, but more often it is not. The consequent post-separation pattern is affected by the fact that the separated flow becomes turbulent and so there is a highly fluctuating recirculation motion over the whole surface of the body. With respect to the dissolution of drug particles from oral solid formulations, recirculation flow is expected to increase mass transfer and can take place even at a low Reynolds numbers of $Re \sim 10$ (13).

As mentioned, a laminar boundary layer separates a greater distance from the surface of a curved body than a turbulent one. The laminar boundary layer in the upper photograph of Figure 5 is shown separating from the crest

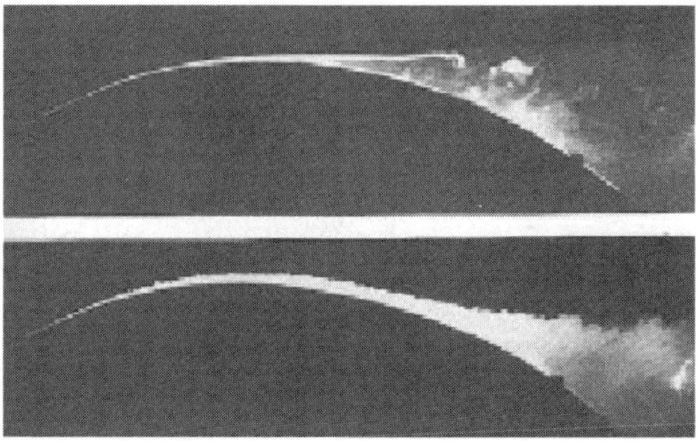

Figure 5 Boundary layer separation: Turbulent vs. laminar boundary flow close to an airfoil. *Source*: From Ref. 89.

of the convex surface, while the turbulent boundary layer in the second photograph remains attached longer, with the point of separation occurring further downstream. Turbulent layer separation occurs when the Reynolds stresses are much larger than the viscous stresses.

Prerequisites for the Hydrodynamic Boundary Layer Concept

Originally, the concept of the Prandtl boundary layer was developed for *hydraulically "even"* bodies. It is assumed that any characteristic length L on the particle surface is much greater than the thickness (δ_{HL}) of the boundary layer itself ($L > \delta_{HL}$). Provided this assumption is fulfilled, the concept can be adapted to curved bodies and spheres, including "real" drug particles. Furthermore, the classical ("macroscopic") concept of the hydrodynamic boundary layer is valid solely for high Reynolds numbers of $Re > 10^4$ (14,15). This constraint was overcome for the "microscopic" hydrodynamics of dissolving particles by the "convective diffusion theory" (9).

The "Convective Diffusion Theory"

The "convective diffusion theory" was developed by V.G. Levich to solve specific problems in electrochemistry encountered with the rotating disc electrode. Later, he applied the classical concept of the boundary layer to a variety of practical tasks and challenges, such as particle–liquid hydrodynamics and liquid–gas interfacial problems. The conceptual transfer of the hydrodynamic boundary layer is applicable to the hydrodynamics of dissolving particles if the Peclet number (Pe) is greater than unity ($Pe > 1$) (9). The dimensionless Peclet number describes the relationship between convection and diffusion-driven mass transfer:

$$Pe = \frac{U_A \cdot L}{D}$$

D represents the diffusion coefficient. For example, low Peclet numbers indicate that convection contributes less to the total mass transfer and the latter is mainly driven by diffusion. In contrast, at high Peclet numbers, mass transfer is dominated by convection. The quotient of *Pe* and *Re* is called the Prandtl number (*Pr*), or, if we are talking about diffusion processes, the Schmidt number (*Sc*):

$$Pr = \frac{Pe}{Re} = \frac{\nu}{D} = Sc$$

The Schmidt number is the ratio of kinematic viscosity to molecular diffusivity. Considering liquids in general and dissolution media in particular, the values for the kinematic viscosity usually exceed those for diffusion coefficients by a factor of 10^3 to 10^4. Thus, Prandtl or Schmidt numbers of about 10^3 are usually obtained. Subsequently, and in contrast to the classical concept of the boundary layer, *Re* numbers of magnitude of about $Re \geq 0.01$ are sufficient to generate Peclet numbers greater than 1 and to justify the hydrodynamic boundary layer concept for particle–liquid dissolution systems ($Re * Pr = Pe$). It can be shown that [(9), term 10.15, nomenclature adapted]

$$\delta \approx D^{1/3} \cdot \nu^{1/6} \cdot \sqrt{\frac{L}{U_{\mathrm{A}}}}$$

Note that the hydrodynamic boundary layer depends on the diffusion coefficient. Introducing the proportionality constant K_{e}° results in an equation valid for any desired hydrodynamic system based on relative fluid motion as proposed in Ref. 10:

$$\delta_{\mathrm{HL}} \approx K_{\mathrm{e}}^{\circ} \cdot D^{1/3} \cdot \nu^{1/6} \cdot \sqrt{\frac{L}{U_{\mathrm{A}}}}$$

K_{e}° consists of a combination of Prandtl's original proportionality constant used for the hydrodynamic boundary layer at a semi-infinitive plate, K_{e}, and a constant, K°, characterizing a particular hydrodynamic system that is under consideration. The latter constant has to be determined experimentally.

$$K_{\mathrm{e}}^{\circ} = K_{\mathrm{e}} \cdot K^{\circ}$$

Among other factors, K° is influenced by particle geometry and surface morphology (roughness, edges, corners, and defects). For instance, K° would equal 1 in the case of a smooth semi-infinite plate, and in this case K_e° is identical to K_e. Considering the "rotating disc system" in particular, Levich found K° to be 0.5. Given that a semi-infinite plate dissolves in a liquid stream and K_e equals 5.2 (which represents Prandtl's proportionality constant in the case of a semi-infinite plate; thus $K_e{}^{\circ} = 2.6$), we arrive at the following term for the thickness of Levich's *effective* hydrodynamic boundary layer (10):

$$\delta_{\mathrm{HL}} \approx 2.6 \cdot D^{1/3} \cdot \nu^{1/6} \cdot \sqrt{\frac{L}{U_{\mathrm{A}}}}$$

The Combination Model

A reciprocal proportionality exists between the square root of the characteristic flow rate, U_{A}, and the thickness of the effective hydrodynamic boundary layer, δ_{HL}. Moreover, δ_{HL} depends on the diffusion coefficient D, characteristic length L, and kinematic viscosity ν of the fluid. Based on Levich's convective diffusion theory the "combination model" ("Kombinations-Modell") was derived to describe the dissolution of particles and solid formulations exposed to agitated systems [(10), Chapter 5.2]. In contrast to the rotating disc method, the combination model is intended to serve as an approximation describing the dissolution in hydrodynamic systems where the solid solvendum is not necessarily fixed but is likely to move within the dissolution medium. Introducing the term

$$\delta_{\mathrm{HL}} \approx 2.6 \cdot D^{1/3} \cdot \nu^{1/6} \cdot \sqrt{\frac{L}{U_{\mathrm{A}}}}$$

into the well-known equation adapted from Noyes, Whitney, Nernst, and Brunner

$$\frac{\mathrm{d}C}{\mathrm{d}t} = \frac{A \cdot D}{\delta_{\mathrm{HL}} \cdot V}(C_{\mathrm{s}} - C_{\mathrm{t}})$$

and employing the proportionality constant k as the apparent dissolution rate constant:

$$k = \frac{A \cdot D}{\delta_{\mathrm{HL}} \cdot V}$$

results in the combination model according to Diebold (10):

$$\frac{\mathrm{d}C}{\mathrm{d}t} = 0.385 \cdot D^{2/3} \cdot \nu^{-1/6} \cdot \left(\frac{L}{U_{\mathrm{A}}}\right)^{-1/2} \cdot \frac{A}{V} \cdot (C_{\mathrm{s}} - C_{\mathrm{t}})$$

where C_{s} represents the saturation solubility of the drug, C_t describes the concentration of the dissolved drug in the bulk at time t, D stands for the effective diffusion coefficient of the dissolved compound, A represents the total surface area accessible for dissolution of the drug particles, and V is the volume of the dissolution medium employed in the test. Note that the apparent dissolution rate constant k is now a function of the flow rate that a particle surface "sees" (slip velocity) and also a function of L, the characteristic length on the particle surface: $k(U_{\mathrm{A}}; L)$. The proportionality constant k can be determined by appropriately performed dissolution experiments or calculated using the following equation:

$$\mathrm{In}(C_{\mathrm{s}} - C_0) - \mathrm{In}(C_{\mathrm{s}} - C_{\mathrm{t}}) = k \cdot t$$

where C_0 is the initial concentration of the drug at $t = 0$. Since δ_{HL} is related to k as demonstrated above, the combination model permits calculation of an overall average hydrodynamic boundary layer for a given particle size fraction. Thus, the proposed relationship provides a tool for a priori prediction of the average hydrodynamic boundary layer of non-micronized drugs and hence to roughly forecast (!) dissolution rate in vitro under well-defined circumstances, e.g., for the paddle apparatus [(10), Chapter 5.5, pp. 61–62, and Chapters 12.3.8 and 13.4.10].

Further Factors Affecting the Hydrodynamic Boundary Layer

Apart from the flow rate, of course, properties of the dissolution medium as well as the drug compound influence the effective hydrodynamic boundary layer and hence the intrinsic dissolution rate.

Saturation Solubility (C_s)

Although the saturation solubility (C_s) influences the *apparent* dissolution rate constant, it is an *intrinsic* property of a drug compound and can therefore affect the hydrodynamic boundary layer indirectly. High aqueous solubility, for example, leads to concentration-driven convection at the surface of the drug particles. Thus, forced and natural convection are mixed together, and it is challenging to separate/forecast their hydrodynamic effects on dissolution rate. In vivo dissolution, however, offers additional problems to the control of hydrodynamics. The saturation solubility of a drug in intestinal chyme may vary greatly within the course of dissolution in vivo, as has been demonstrated previously (10). The in vivo solubility of felodipine in jejunal chyme (37°C), for example, was determined to be about 10 μg/mL on average (median), but varied greatly with time at mid-jejunum, ranging from 1 to 25 μg/mL or even from 1 to 273 μg/mL, depending on the conditions of administration (10). Solubility variations within the course of an in vivo dissolution experiment may, in such cases, override hydrodynamic effects. Thus, the observed time dependency of intestinal drug solubility should be taken into account by dissolution models, which otherwise may describe dissolution rates in vitro well but fail to do so in vivo.

Diffusion Coefficient (D)

The diffusion coefficient is linked to the intrinsic dissolution rate constant (k_i) as expressed by the term

$$k_i = \frac{D}{\delta_{HL}}$$

Thus, the thickness of the effective hydrodynamic boundary layer δ_{HL} obviously depends on the diffusion coefficient. The diffusion coefficient D further correlates to the diameter of the particle or molecule as demonstrated by the relation of Stokes and Einstein:

$$D = \frac{\kappa_B \cdot T}{3 \cdot d \cdot \eta \cdot \pi}$$

where T is the temperature in Kelvin and κ_B represents the Boltzmann constant (1.381×10^{-23} J/K). The term reveals that the diffusion coefficient D itself is dependent on the dynamic viscosity (η). In the GI tract, diffusion coefficients of drugs may be reduced due to alterations in the fluid viscosity. Larhed et al. (16) reported that diffusion coefficients for testosterone were reduced by 58% in porcine intestinal mucus. It has also been observed in dissolution experiments that the reduction of diffusion coefficients can counteract effects of increased drug solubility due to mixed micellar solubilization (17).

Kinematic Viscosity (ν)

The viscosity of upper GI fluids can be increased by food intake. The extent of this effect depends on the food components and the composition and volume of co-administered fluids. Aqueous-soluble fibers such as pectin, guar, and some hemicelluloses are able to increase the viscosity of aqueous solutions. Increasing the kinematic viscosity of the dissolution medium generally leads to a reduction of the effective diffusion coefficient and hence results in decreased dissolution. For instance, Chang et al. increased the viscosity of their dissolution media using guar as the model macromolecule. Subsequently, dissolution rates of benzoic acid were reduced significantly. However, dissolution rates were not at all affected when adjusting the same viscosity using propylene glycols (18).

Temperature (T)

The temperature influences the drug's saturation solubility and also affects the kinematic viscosity (density of the liquid!) as well as the diffusion coefficient. Therefore, when performing dissolution experiments, temperature should be monitored carefully or preferably kept constant.

Particle Morphology and Surface Roughness

Faster initial dissolution rates obtained by grinding or milling the drug can often be attributed to both an increase in surface area and changes in surface morphology that lead to a higher surface free energy (19,20). However, an increase in edges, corners defects, and irregularities on the surfaces of *coarse* grade drug particles can also influence the effective hydrodynamic boundary layer δ_{HL} and hence dissolution rate (12,21–23). Depending on the surface roughness (R) of the drug particle, the liquid stream near the particle surface may be turbulent even though the bulk flow remains laminar (9,10). Irregularities, edges, and defects increase the mass transfer in different ways according to the different kinds of hydrodynamic boundary layers generated. In the case of a turbulent boundary layer, the overall surface roughness is assumed to behave in a hydraulically "indifferent" (i.e., does not increase mass transfer itself) manner if the protrusions and cavitations are fully located within the viscous sub-layer (δ_{VS}). The so-called *allowable* (= indifferent) dimension of such a surface roughness (R_{zul}) can be estimated using an equation originally developed for tubes and pipes [(24), Sec. 21 d]:

$$R_{\mathrm{zul}} = 100 \cdot \frac{\nu}{U_{\mathrm{A}}}$$

For $R < R_{\mathrm{zul}}$, the surface roughness does not cause perturbations that increase mass transfer.

In contrast, in the case of a laminar hydrodynamic boundary layer, the *critical* dimension of surface roughness (R_{crit}) can be determined using the following relation:

$$R_{\mathrm{crit}} = 15 \cdot \frac{\nu}{\sqrt{\tau/\rho}}$$

with

$$\sqrt{\frac{\tau}{\rho}} = 0.332 \cdot U_{\mathrm{A}}^{2}\sqrt{\frac{\nu}{U_{\mathrm{A}} \cdot L}}$$

where τ represents the shear stress, ρ is the fluid density, and ν stands for the kinematic viscosity. If $R > R_{\mathrm{crit}}$, the effective hydrodynamic boundary layer close to the particle wall becomes turbulent even though the bulk flow still may be laminar! In contrast to R_{zul}, R_{crit} depends on the characteristic length L of the particle surface and is about 10 times greater [(24), Sec. 21 d]. In the case of a laminar hydrodynamic boundary, Levich (9,25) estimated that R_{crit} could be exceeded for Reynolds numbers as low as $Re = 20$. This implicates that even very small irregularities or roughnesses on the surface of drug particles can have momentous effects on the hydrodynamic boundary layer δ_{HL} and hence on the dissolution rate.

Flow along a particle surface can be affected either by cavitations or by protrusions. In both cases, the flow pattern on the particle surface is changed and the dissolution rate may be altered due to local perturbations.

Figures 6 and 7 are derived from laboratory experiments and illustrate that flow can become turbulent close to particle walls even when the bulk flow remains laminar. The turbulent vortices bore into the particle surface, magnifying cavitations and abrading protrusions, and hence accelerating the dissolution process [(10), Chapter 4.3.5]. However, irregulari-

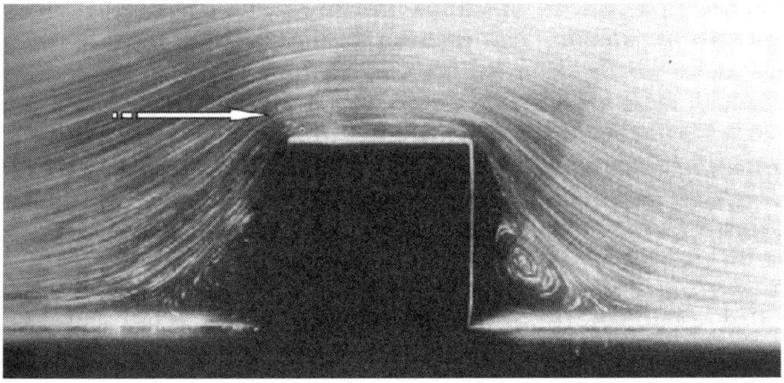

Figure 6 Flow along a simulated surface roughness (protrusion type) at $Re = 0.02$, visualized using aluminum powder. Note the vortex generated downstream of the cube. Flow is from left to right as indicated by the arrow (added by the author). *Source*: Adapted from Ref. 13, Sec. 12.1 (original by Taneda S. J Phys Soc Jpn 1979; 46:1935).

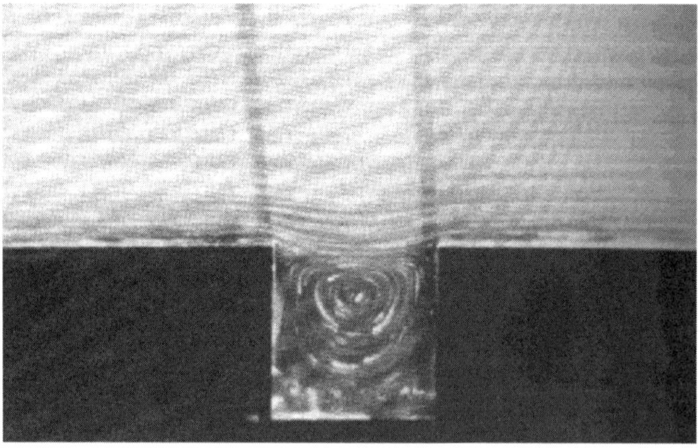

Figure 7 Flow along an artificial cavitation at low Reynolds number (visualized using aluminum powder). Flow is from left to right. *Source*: Adapted from Ref. 13, Sec. 12.4 (original by Taneda S. J Phys Soc Jpn 1979; 46:1935).

ties and roughnesses on the surface of drug particles are expected to influence the effective hydrodynamic boundary layer δ_{HL} of *coarse grade* drug particles only. For example, Mithani (26) investigated the dissolution of coarse dipyridamole (DPM) particles. The dissolution rate of single DPM crystals was increased with time due to a considerable increase in surface roughness, whereas the geometry of the crystals was maintained during dissolution. Particle geometry and morphology can be investigated using conventional scanning electron microscopy (Figs. 8 and 9) to predict these effects (10).

Figure 9 shows a magnification ($\times$7500) of the "smooth" and regular surface area indicated in Figure 8. The length of the edges of the cube was of the order of about 200–300 μm. The particle surface appeared to be smooth. Nevertheless, small "craters and hills" of the order of about 0.5–3 μm have to be taken into consideration. The observed cavitations and protrusion on the particle surfaces may cause perturbations, change the nature of the hydrodynamic boundary layer, and hence increase dissolution. Furthermore, as was confirmed by these microscopic observations, small

Figure 8 SEM picture of a single felodipine crystal (coarse grade). The regular cube shows an apparently smooth surface. The arrow indicates the point at which the next picture (Fig. 9) was taken. *Source*: From Ref. 10.

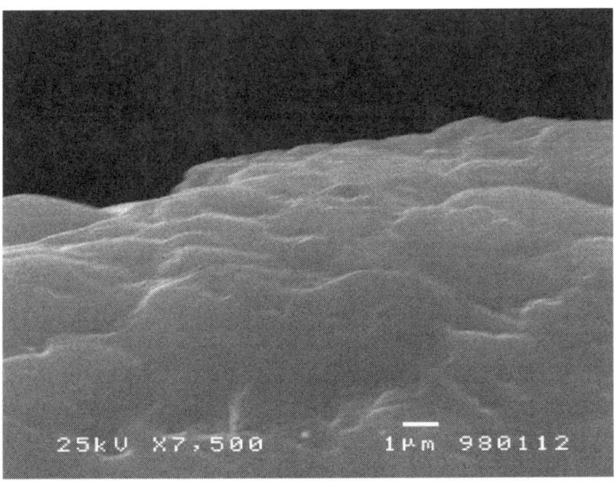

Figure 9 SEM picture of the surface of a smooth felodipine crystal apparently showing mounds, craters, and hills. *Source*: From Ref. 10.

particles adhere to the surfaces of the larger particles due to static charges (10). This occurs particularly if the powder fraction is obtained by sieving the bulk powder. In this case, dissolution might be biphasic. Subsequent to an initial "burst" phase, dissolution continues more slowly from the coarse grade "core" fraction (10,27). Thus, geometry and surface morphology appear to play a very important role in the dissolution of coarse grade drug particles.

Particle Size

The particle size of poorly soluble drugs is generally of major importance for dissolution and absorption. For example, in vitro investigations performed with sulfonamides showed that the initial dissolution rate increased with a decrease in particle size, other dissolution conditions remaining constant (27). As far back as in 1962, Atkinson and Kraml performed in vivo investigations and reported a two-fold enhancement in absorption of griseofulvin particles with a four-fold increased surface area (28,29). Similar results were obtained for the micronization of felodipine, particle size having a profound effect on its in vivo dissolution and absorption (30). Scholz et al. used a combination of infusion and oral administration of either normal saline or a 5% glucose solution to maintain and establish "fasted" and "fed" state motility patterns, respectively. The absorption characteristics of both a micronized and a coarse fraction of the drug were subsequently studied under these two motility patterns. The dissolution of the coarse grade fraction was improved by the "fed" state hydrodynamics, as reflected in the nearly doubled extent of absorption. In contrast, a micronized powder of the same chemical species showed less sensitivity to hydrodynamics, as was reported in former studies [(10), pp. 220 f, 235, and (31)].

Particle Size and Effective Hydrodynamic Boundary Layer

The mean hydrodynamic boundary layer generated on the surface of particles undergoing a dissolution process depends on the particle size and the particle size distribution. However, the thickness of the *effective* hydrodynamic boundary layer is contingent, in an interdependent manner, on the particle diameter and the flow rate at the particle surface. Considering particle sizes beyond 200 μm, mass transfer coefficients were found by Harriott (32) to be independent of particle size, provided that sufficient agitation was applied (stirring rates exceeded 300 rpm). Below particle sizes of about 200 μm, mass transfer coefficients and dissolution were considerably influenced by both stirring rate and particle sizes. The observed interdependency decreased gradually with decreasing particle sizes and was no longer measurable below 15 μm. Considering a combination of particle size and hydrodynamics, and further provided that the media viscosity is unaltered, it appears that three cases have to be distinguished [(10), Chapter 5.7]

- At a given stirring rate, the effective hydrodynamic boundary layer is expected to be independent of particle size beyond a maximal particle size range, since the particle surface cannot bind the surrounding fluid to an infinite distance into the bulk. As a matter of course, dissolution still depends on convection.
- Since the *absolute* thickness of the effective hydrodynamic boundary layer is very small, below a particular size range minimum, no hydrodynamic effects are perceived experimentally with varying agitation. This, however, does not mean, that there are no such influences! Further, the mechanisms of mass transfer and dissolution may change for very small particles depending on a number of factors, such as the fluid viscosity, the Sherwood number (the ratio of mass diffusivity to molecular diffusivity), and the power input per unit mass of fluid.
- In between these two extremes, the *effective* hydrodynamic boundary layer depends on the combined effects of particle size *and* hydrodynamics. Talking about "borderline particle sizes" is meaningful only if all other relevant data, such as the fluid viscosity,

the diffusivity, the temperature, and the saturation solubility of the compound, are additionally provided to characterize the hydrodynamic system.

Microparticles

Generally, micronized particles show less sensitivity to hydrodynamics compared to coarse grade material of the same chemical entity [(10), Chapters 5.7 and 12.3.5]. Armenante postulated a different mass transfer process for what he termed "microparticles" (33). The microparticle size range was defined in terms of the viscosity of the medium and the power input into the hydrodynamic system. The development of a boundary layer determines the mass transfer for macroparticles but contributes to a lesser extent to the dissolution of microparticles, since their behavior additionally depends on the hydrodynamics in micro-eddy regions. For very small particles (approximate diameters below about 5 μm in aqueous media), diffusion within the surface microclimate becomes predominant for mass transfer and particles behave more and more "like molecules" (34). Subsequently, the *relative* influence of the bulk flow, expressed by the Reynolds term, decreases gradually (10,35). Thus, local turbulences may occur at milder hydrodynamic conditions for the micro- than for the macroparticles, making them less sensitive to differences in the bulk hydrodynamics. Bisrat and Nyström (36) demonstrated that the thickness of the boundary layer increased with increase in mean volume diameter of the particles. This increase was found to be less pronounced above approximately 15 μm diameter. It was also shown that the intrinsic dissolution rates of digoxin and oxazepam of particles < 5 μm were not significantly affected by increased agitation intensities, while a sieve fraction of the same compounds in the range 25–35 μm was affected (31,37). Harriott (32) investigated the dependence of the boundary layer thickness upon the slip velocity for different particle sizes. The greater the slip velocity, the smaller the boundary layer generated at the surface of the particle. Harriott found that the slip velocity, the relative velocity of the solid to the fluid, was negligible for very small, suspended particles. Thus, bulk agitation should have relatively little influence on the dissolution rate of microparticles. However, at larger particle sizes, the slip velocity—and hence the boundary layer—becomes an important factor in the dissolution process.

HYDRODYNAMICS OF COMPENDIAL DISSOLUTION APPARATUS

Various dissolution test systems have been developed and several of them now enjoy compendial status in pharmacopeias, for example the reciprocating cylinder (United States Pharmacopeia Apparatus 3), the flow-through apparatus [European Pharmacopoeia (Pharm. Eur.) 2.9.3], or the apparatus for transdermal delivery systems, such as the paddle over disc. Hydrodynamic properties of these and other apparatus have been described only sparingly. The paucity of quantitative data related to hydrodynamics of pharmacopeial dissolution testers is lamentable, since well-controllable hydrodynamics are essential to both biopharmaceutical simulations and quality control. Here, we focus the discussion on the paddle and the basket apparatus, since these are the most important and widely used for oral solid dosage forms. A brief treatise on the hydrodynamics of the flow-through apparatus completes this section.

Methods Used for the Investigation of Flow Patterns and Flow Rates

Flow patterns of hydrodynamic systems like the compendial dissolution apparatus may be *qualitatively* characterized by means of dilute dye injection (e.g., methylene blue) or by techniques using particulate materials such as aluminum powders or polystyrene particles. Flow patterns may be also visualized by taking advantage of density or pH differences within the fluid stream. The "Schlieren" method, for instance, is based on refraction index measurement. Hot wire anemometry is an appropriate method to *quantitatively* characterize flow rates. The flow rate is proportional to the cooling rate of a thin hot wire presented to the stream. Using laser Doppler anemometry, flow rates as low as 1 μm/sec can be determined. This optic method is recognized as the gold standard since it is the most accurate available. However, the fact that the method can be used only for transparent media can be a disadvantage. Topics such as velocity measurement and flow visualization techniques are well covered by Tritton [(13), Sec. 25.2–4].

Flow Rate as a Function of Stirring Rate for Paddle and Basket

Recently, studies were performed to quantitatively examine the hydrodynamics of the two most common in vitro dissolution testers. Rotational (tangential) fluid velocities were corre-

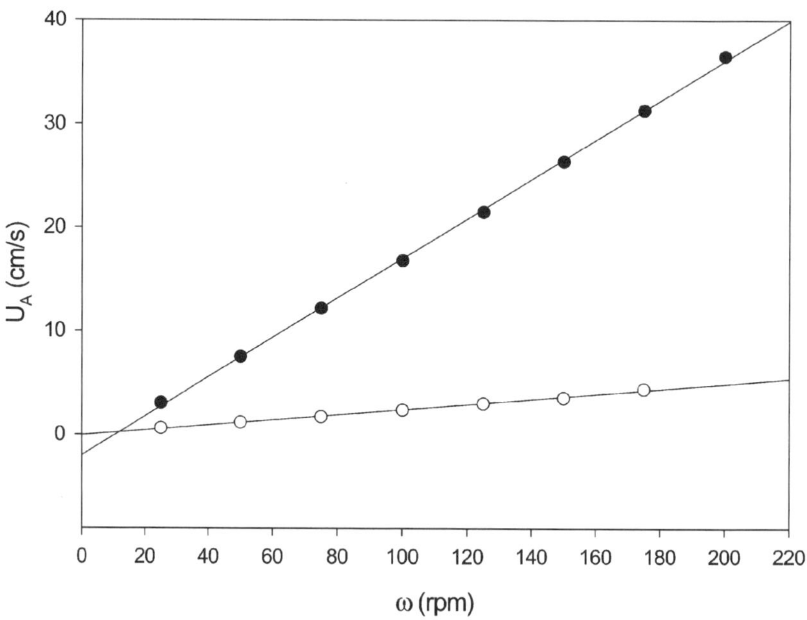

Figure 10 Rotational (tangential) flow (U_A) as a function of stirring rate (ω) for paddle (filled circles) and basket (open circles): Mean ± SD; position S2 approximately 1 cm above the paddle and midway between the paddle shaft and the wall of the dissolution vessel. (Please note that, in contrast to simulation techniques such as, for instance, computational fluid dynamics, these data are based on dissolution experiments.) *Source*: Data from Ref. 10, UPE method.

lated to stirring rates at various positions within the dissolution vessels of the paddle and the basket apparatus by means of an ultrasound pulse echo method [UPE method (10,38); Fig. 10]. This method permits direct characterization of hydrodynamics as opposed to indirect methods such as via the dissolution characteristics of dosage forms, the results of which are subject to varying properties from batch to batch (for example, USP calibrator tablets). Furthermore, the technique can be used for non-transparent media and suspensions, making it possible to study flow rate effects on excipient-loaded formulations. In general, fluid velocities (in cm/sec) for the paddle apparatus were determined to be about 8–10 times higher than those of the basket at a given stirring rate (rpm). At most positions, they correlated well and in a linear manner with the stirring rate for both the paddle and the basket.

Fluid velocities using the basket method were determined to range between 0.3 and 5 cm/sec [25–200 rpm], and for the paddle method, between 1.8 and 37 cm/sec [25–200 rpm]*. Possible applications of these fluid velocity data may include their use to forecast in vitro dissolution rates and profiles of pure drug compounds for the paddle test employing an appropriate mathematical scenario/formula like the combination model.

Flow Pattern in Paddle and Basket

Figures 11 and 12 illustrate the flow patterns for the basket and the paddle apparatus, respectively. An undertow can be observed visually in the paddle apparatus for stirring rates exceeding 125 rpm. The hydrodynamic region below the paddle, and, even more pronounced, below the basket, appears to be somehow "separated" from the region above the stirring device. Diffusion-driven exchange of dissolved mass between these two regions is unhampered, but little

* Detailed sets of fluid velocity data for the paddle and the basket apparatus, including various positions in the vessels and different volumes (500, 900, and 1000 mL) of dissolution medium, can be found in Ref. 10 (Chapter 11.3).

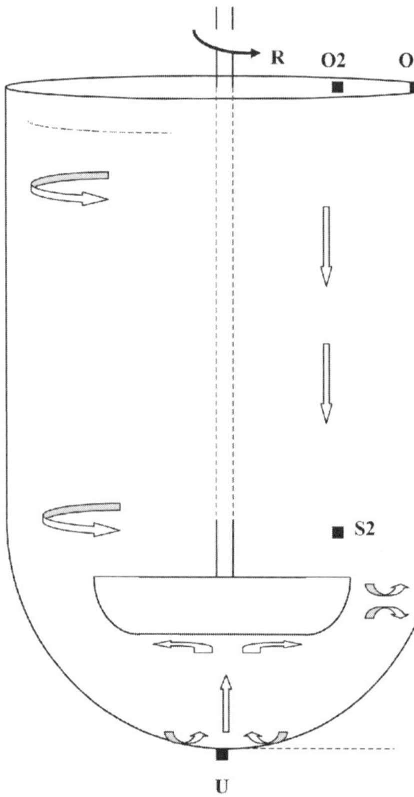

Figure 11 Schematic flow pattern for the paddle apparatus, based on *quantitative* experimental data (see also Fig. 12). *Source*: From Ref. 10.

(convective-driven) exchange of particulate material takes place. Flow rates given for the basket apparatus, however, are valid for the bulk flow only and likely do not reflect the influence of hydrodynamics on dissolution *inside the basket*. Nevertheless, vessel hydrodynamics of regions *outside the basket* may be relevant for dissolution of solid formulations with respect to fractions of particulate material that have fallen though the basket screen. Further, hydrodynamics inside the basket may also be influenced by the "outside" bulk hydrodynamics and the stirring rate in such a way that, starting with a rotational speed of about 100 rpm or more, contact between the bulk fluid and the formulation inside the basket

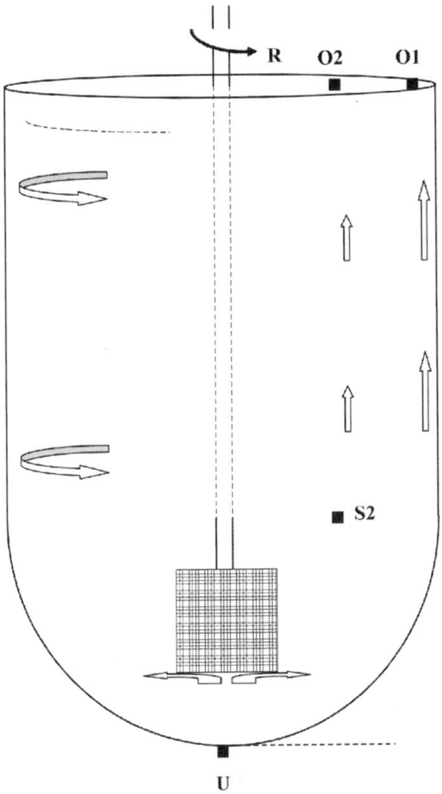

Figure 12 Schematic flow pattern for the basket apparatus. Because of the hemispheric symmetry of the dissolution vessel, it is sufficient to draw the flow just for one-half of the vessel. The arrows indicate flow direction. All designated flow patterns are based on *quantitative* experimental data. *Source*: From Ref. 10.

becomes restricted. At these rates, the basket may be regarded as a "closed container," with limited access to "fresh" dissolution medium and less turbulent flow conditions. For some specific purposes, the basket could even be used to serve as a "rotating cylinder," with the formulation placed outside the basket at the bottom of the vessel. Such a modified apparatus could be advantageous when mild but reproducible hydrodynamic conditions are desired.

Fluid Velocities at Various Positions and Volumes

Rotational Flow Below the Stirring Device

Fluid velocities for rotational (tangential) flow below the stirring device employing 900 mL of medium were determined to be 8.5 cm/sec at 50 rpm and 16 cm/sec at 100 rpm, the most widely used agitation rates in the paddle apparatus (10). The fluid velocities for the rotational flow measured at various (lateral) positions of the dissolution vessels do *not* differ significantly. This is true for the basket as well and indicates that the fluid is homogeneously accelerated within the vessel (10).

Vertical Flow Below the Stirring Device

Hydrodynamics at the bottom of the vessel (position U) is of particular interest since many non-floating tablet and (soft gelatin, primarily) capsule formulations remain there after disintegration and throughout the dissolution test and are therefore primarily exposed to this hydrodynamic flow regime. "Coning effects" are sometimes observed at low stirring rates in the paddle apparatus at about 50 rpm at the bottom of the hemispheric vessel. This undesired phenomenon generally occurs when disintegrating type tablets with high loads of insoluble, dense excipients are employed. There is no simple linear correlation between the stirring rate and the *vertical* (axial) flow rate (upward stream) at the bottom of the vessel (Fig. 13). The vertical flow rates at the bottom region of the vessel are very low (<1.5 cm/sec). An insufficient upward stream in combination with a far stronger rotational (horizontal, tangential) flow might explain the coning effects observed.

Vertical Flow Above the Stirring Device

Close to the wall of the dissolution vessel (position O1), the flow is directed upwards, creeping along the wall as indicated by a negative algebraic sign (figure not shown). For the basket, this is also true in position O2, indicating an upward directed stream for the bulk flow above the basket, whereas for the paddle an undertow is recorded at position O2 (positive algebraic sign) (Fig. 14).

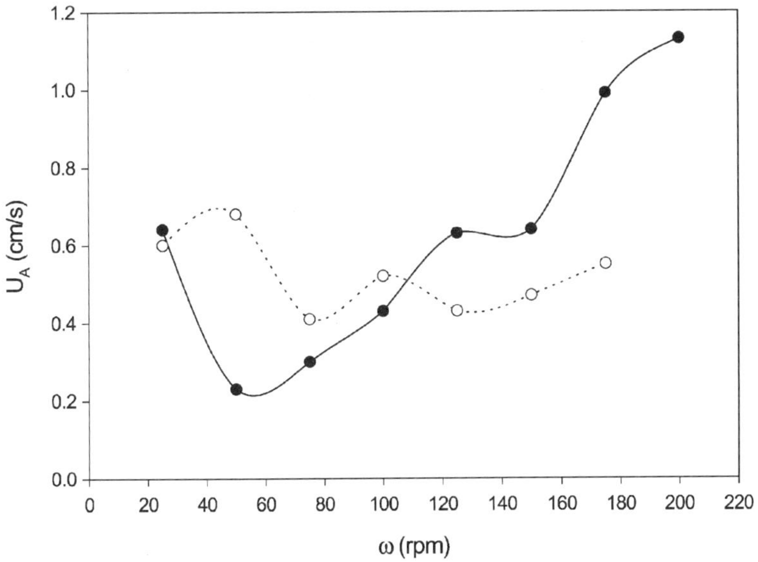

Figure 13 Vertical (axial) flow (U_A) below the stirring device as a function of stirring rate (ω) for paddle (filled circles) and basket (open circles) at the bottom of the hemispheric dissolution vessel filled with 900 mL. *Source*: From Ref. 10.

Fluid Velocities Employing Different Volumes

The lower the volume of medium employed in a dissolution test, the higher are the flow rates, ceteris paribus. A test volume of 500 mL results in a considerable increase in the fluid velocities at any given stirring rate compared to 900 mL of dissolution medium (Fig. 15). A significant mass/volume effect on hydrodynamics appears to exist. Up to the level of the paddle, for example, the rotational (tangential) fluid velocity at 100 rpm was determined to be 16.8 cm/sec using 900 mL of dissolution medium compared with 20.5 cm/sec employing a volume of just 500 mL (10). The undertow generated at the bottom of the dissolution vessel, where the formulations are often located during the tests, was also found to be higher using 500 than 900 mL. Thus, the volume used in the dissolution tests cannot be ignored and has an influence not only in terms of the concentration driving force

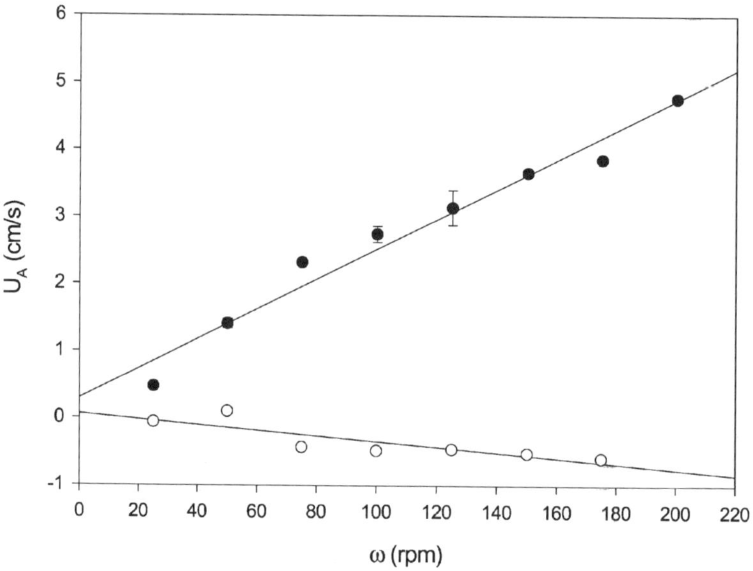

Figure 14 Vertical (axial) flow (U_A) above the stirring device as a function of stirring rate (ω) for paddle (filled circles) and basket (open circles). Mean ± SD; vertical position O_2. *Source*: From Ref. 10.

for dissolution but also from a hydrodynamic point of view [(10), Chapter 11.3.3 and Fig. 11.10, p. 185]. Therefore, special care has to be taken in method validation for quality control purposes when the volumes are changed, e.g., when the method is adapted for a higher strength dosage form. This statement also holds for the basket.

Prediction of Fluid Velocities for the Paddle and the Basket

The empirically gained knowledge of the fluid velocities in the dissolution vessels at rotational speeds from 25 to 200 rpm resulted in a number of parameters that find application in developing equations to correlate stirring rates and flow rates (tangential fluid velocities) at specific regions within the vessel. Flow rates (U_A) in the paddle and the basket apparatus can be calculated for any desired stirring rate (ω) by means of a simple linear relationship using the data for the

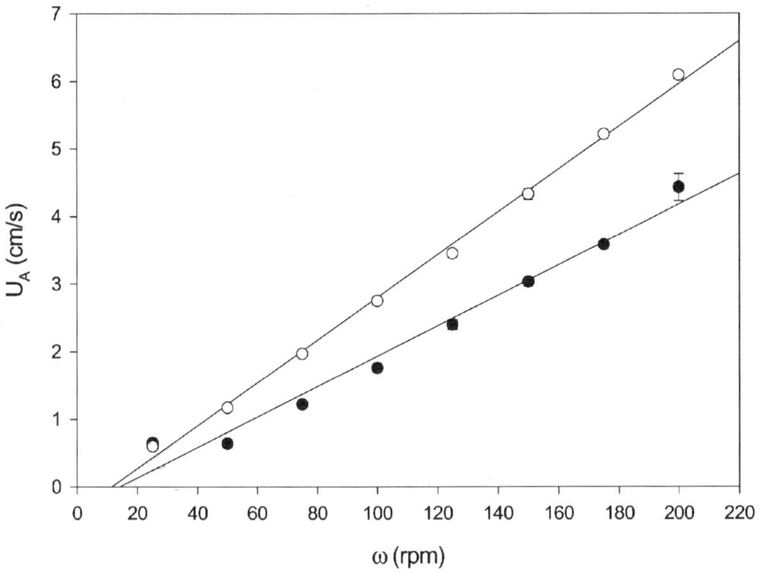

Figure 15 Rotational (tangential) flow (U_A) as a function of stirring rate (ω) for the basket using 900 mL (filled circles) and 500 mL (open circles). Mean ± SD, $n = 6$, $P < 0.001$, paired t-test; lateral position S2. *Source*: From Ref. 10.

parameters $b[1]$ and $b[0]$ reported in Refs. 10 and 38:

$$U_A = b[1] * (\omega) + b[0]$$

U_A is given in cm/sec and ω in min^{-1}. Two examples illustrate the applicability of this relationship:

1. The fluid velocity approximately 1 cm below the paddle (position S1)* at a stirring rate of 110 rpm employing 900 mL of dissolution medium was calculated to be 17.98 cm/sec. Indeed, at 100 rpm, the flow rate was determined to be 16.01 cm/sec using the UPE method, and at 125 rpm the flow rate was measured to be 20.29 cm/sec, both of which give some plausibility to the calculated value.

---

* Position S1 is not indicated in Figures 11 & 12; for exact *graphical* location (see Ref. 10, Page 162, Fig. 11.3).

2. The bulk flow rate up to the mark of the basket (position S2) employing 60 rpm and 500 mL of dissolution medium, for instance, was calculated to be 1.5 cm/sec. Comparison to experimental data verified the concept: 1.17 cm/sec was obtained for 50 rpm and 1.97 cm/sec for 75 rpm (10).

Rotational fluid velocities are calculated since horizontal (rotational) flow prevails in the hydrodynamic regime within the dissolution vessels. Thus, the overall hydrodynamics and hence dissolution is dominated by the substantially higher rotational (tangential) fluid velocities.

Reynolds Numbers In Vitro

Bulk Reynolds Numbers

In the paddle method, *bulk* Reynolds numbers range from $Re = 2292$ (25 rpm, 900 mL) up to $Re = 31025$ (200 rpm, 500 mL). In contrast, Reynolds numbers employing the basket apparatus range from $Re = 231$ to $Re = 4541$. These Reynolds numbers are derived from dissolution experiments in which oxygen was the solute [(10), Chapter 13.4.8] and illustrate that turbulent flow patterns may occur within the *bulk* medium, namely for flow close to the liquid surface of the dissolution medium. The numbers are valid provided that the whole liquid surface rotates. According to Levich (9), the onset of turbulent bulk flow under these conditions can then be assumed at $Re \sim 1500$.

Particle–Liquid Reynolds Numbers

As mentioned earlier, Reynolds numbers determined for the bulk flow have to be discerned from Reynolds numbers characterizing a particle–liquid dissolution system. The latter were calculated for drug particles of different sizes using the Reynolds term according to the combination model. The kinematic viscosity of the dissolution medium at 37°C is about 7×10^{-03} cm^2/sec. The fluid velocities (U_A) employing the paddle method at stirring rates of 50–150 rpm can be taken from the literature and may arbitrarily be used as the slip velocities at the particle surfaces.

Based on these data, particle–liquid Reynolds numbers were calculated to range from $Re = 25$ (50 rpm) to $Re = 90$ (150 rpm) for coarse grade particles with a median diameter of 236 μm. In contrast, Reynolds numbers for a batch of micronized powder of the same chemical entity with a median diameter of 3 μm were calculated to be significantly lower ($Re < 1$), indicating less sensitivity towards convective hydrodynamics [(10), Chapter 12.3.8]. Based on the aforementioned considerations for spheres, bulk Reynolds numbers of about $Re > 50$ appear to be sufficient to produce the laminar–turbulent transition around a rough drug particle of coarse grade dimensions.

Hydrodynamics of the Flow-Through Apparatus

The flow-through cell system (USP Apparatus 4) is described under monograph <724> dealing with drug release and is becoming more important for the dissolution of solid oral dosage forms. Standard flow rates of 4, 8, and 16 mL/min are prescribed and a sinusoidal flow profile is provided having a pulsation rate of 120 ± 10 pulses per minute. Cammarn and Sakr (39) used an alternate approach to describe hydrodynamics and dissolution performance of the flow-through cell system involving dimensionless analysis. Volumetric flow rates up to 53 mL/min were employed in these tests. These values corresponded to linear fluid velocities of less than 2.3 cm/sec. Reynolds numbers were calculated under these conditions to range from 7 to 292, indicating that bulk flow is laminar. For example, a $Re = 16.3$ was determined for a flow rate of 10.4 mL/min (12 mm cell, single vertical). Dissolution rates were determined to be a function of media linear velocity (in cm/sec) rather than being described by volumetric flow rate. Tablet diameter, shape, and surface were found to be critical to dissolution rate of, e.g., non-disintegrating tablets.

IN VIVO HYDRODYNAMICS, DISSOLUTION, AND DRUG ABSORPTION

Absorption of orally administered drugs depends mainly on dissolution if the compound is poorly soluble but highly permeable. A variety of factors can influence in vivo dissolution, such as the properties of the drug itself (polymorphism, pK_a, complexation behavior, diffusivity), the formulation variables (capsule shell, tablet hardness, particle size distribution of excipients), the composition of the GI fluids (pH, buffer capacity, solubilization and wettability properties), and—last but not least—the hydrodynamics of the GI tract. Many poorly soluble drugs fail to be completely bioavailable after oral dosing. In the case of dissolution rate limited absorption, the thickness of the boundary layer can influence the dissolution. The thickness of the boundary layer is, in turn, dependent upon the (in vivo) hydrodynamics. In vivo hydrodynamics, however, depend on GI *motility*. Although much is known about motility patterns, little is known about the relationship of motility patterns and GI *hydrodynamics*. To the best of our knowledge, it is not yet clear in which way exactly and to what extent GI motility correlates with intestinal flow rates, how fast the liquids progress, and what flow rates are produced in the gut by the different motility patterns. According to Johnson et al. (40), the *velocity* of propulsive contractions in the upper small intestine seems to be the major determinant of intestinal transit. Nevertheless, two important issues remain partially unresolved:

1. So far, we are not able to define or predict intestinal flow rates solely based on the knowledge of motility data.
2. It is still challenging to isolate hydrodynamic influences on drug dissolution in vivo from other factors that can play a role in absorption.

GI Motility

In the GI tract, different hydrodynamic conditions are present, depending on the fasted or the fed state. Contraction patterns are controlled in terms of electromechanical impulses (myoelectric activity) as well as by various hormones (cholecystokinin, secretin, glucagon, motilin, and insulin, for example). In the fasted state, the motility pattern is regulated by the (interdigestive) migrating myoelectric complex [(I)

MMC], a cyclic pattern consisting of mainly three phases (I, II, and III in that order) with a duration of approximately 90–120 min. IMMC starts at the proximal GI tract (lower esophagus, stomach, and proximal duodenum). During phase I (approximately 45–60 min), residence times are long but there is barely any fluid movement since there are no contractions. In phase III, lasting about 10 min and followed by a "quiescent phase" of about 0–5 min, all "slow waves" (rhythmic fluctuations of the cellular membrane potential) are associated with "spikes." As a result, about half of the contractions propagate the GI contents up to 30–40 cm aborally, and fluid movement is so rapid that often there might be insufficient time for dissolution to occur prior to reaching the absorptive sites. In contrast, phase II conditions, with a duration of 30–45 min, are most likely to favor drug dissolution. This IMMC phase is most similar to post-prandial status in terms of the percentage of slow waves associated with spikes, distribution between segmental and propagated contractions, and distances over which peristaltic waves are propagated.

The motility pattern of the fed state is more regular. Sixty-five percent of propagated contractions travel only 3–9 cm. There is sufficient chyme present in the gut lumen to serve as the dissolution medium, and the chyme is more or less in continuous movement. Due to the rhythmic segmentation contractions, a more frequent local acceleration of the chyme can be assumed. It is likely that the rate and the frequency (but not necessarily the type) of the bulk flow is different in the fed than in the fasted state and that this could lead to changes in dissolution, dependent on the sensitivity of the formulation. Taking these physiological variations into consideration, the dissolution of poorly soluble drugs and release from formulations sensitive to hydrodynamic changes are expected to be more effective in the fed than the fasted state.

GI Hydrodynamics

Hydrodynamics of the upper GI tract are characterized by: 1) the kinetics of gastric emptying, and 2) the small intestinal transit and the flow rate of intestinal fluid (chyme). Gastric emptying becomes important for the overall absorption of certain drugs because it can act as the "gatekeeper" controlling delivery of drugs to the absorptive sites in the intestines. This is of particular importance for drugs that are highly soluble in gastric juice, such as furosemide, acetaminophen, aspirin, lidocaine, or amoxicillin, to name but a few examples. Bioavailability of these compounds is limited by the time required for them to reach the absorptive sites in the duodenum, jejunum, and ileum, a time that is primarily controlled by gastric emptying. In the case of poorly soluble but highly permeable drugs, both the flow rate *and* the composition and volume of chyme available for dissolution are the predominant factors. Flow rate and volume are both of importance since they can influence intestinal transit and the time available for in vivo dissolution as well as the time available for contact of the dissolved drug with the absorptive sites.

Gastric Emptying

GI transit of formulations including solid pharmaceuticals and multi-particulate dosage forms is covered by Wilson and Kelly (Chapter 5). Therefore, the focus of the discussion here is on the hydrodynamics of gastric emptying and small intestinal transit of liquids. The volume, the temperature, and the composition (caloric content, osmolality, pH, viscosity) of gastric contents influence gastric emptying. Of these factors, caloric content is most important for the regulation of gastric emptying kinetics of liquids.

Non-caloric Liquids

The emptying of isotonic non-caloric fluids is proportional to the initial volume and the distension of the stomach. Quantities of about 600 mL most likely activate barostatic receptors. Gastric emptying of small volumes of non-caloric (non-nutrient) fluids correlates with the corresponding phase of the antral interdigestive migrating myoelectric complex (IMMC) in humans. During phase I gastric emptying is negligible, whereas it reaches maximum during phase III. Although gastric emptying of volumes < 50 mL is highly dependent on the motility phase, this is not so true for larger volumes ($>$ 200 mL), as demonstrated by Oberle et al. (41). First order kinetics tend to apply for volumes of about 200 mL or larger. Using the canine model, it has been shown that volumes larger than 300 mL establish fed state-like conditions. However, if the viscosity of the liquid is elevated, this induction can happen at lower volumes. Further, the emptying of viscous liquids is considerably slower compared to non-viscous liquids of the same volume (42,43). The half-life of gastric emptying ($GE_{50\%}$) of non-nutrient liquids ranges from 12 min (200 mL administered) to 22 min (50 mL administered)*. In general, gastric emptying of non-caloric liquids is much faster than that of caloric fluids.

Caloric Liquids

The rate of delivery of calories to the duodenum is kept within a very narrow range, regardless of whether the calories are presented as carbohydrate, protein, fat, or a mixed meal. Caloric liquids of volumes greater than 200 mL empty slower than non-nutrient liquids of identical volume. The energy content of the liquid is the most important determinant of the rate of gastric emptying and $GE_{50\%}$, and this determinant is regulated mainly in the duodenum. Glucose solutions (400 mL, orally administered) have been found to obey linear release kinetics and to empty at an average rate of 2.1 kcal/min regardless of concentration at which provided (44). McHugh et al. (45,46) were the first to report calorie-driven, linear emptying of orally administered glucose solutions with a constant rate of 0.4 kcal/min for *Macaca mulatta*. The authors demonstrated that $GE_{50\%}$ doubles for a given volume if the caloric density of the fluid administered is doubled. Thus, caloric fluids are emptied in a manner that presents a constant caloric delivery to the duodenum regardless of the glucose concentration. This rate is, however, species dependent. Neither motility phase I nor motility phase II of the

---

*See Ref. 10 (Chapter 15.1.2) for a detailed synopsis including original references.

IMMC has any significant impact on the gastric emptying rate of the glucose solutions (47).

Non-linear Initial Release Kinetics for Caloric Fluids

The larger the load of glucose delivered to the duodenum, the longer and more complete is the inhibition of gastric emptying. However, gastric emptying is not a continuous process. Rather, the stomach *initially* empties even a nutrient solution rapidly as though it were saline. Hunt et al. (48) administered 1134 polycose meals of different energy contents (0.5–2.0 kcal/ml) and various volumes (300, 400, and 600 mL) to 21 subjects. The mean rate at which the calories were delivered to the duodenum was found to be 2.5 kcal/min, confirming the previous results of Brener et al. (44). However, for the greater volumes (400 and 600 mL, respectively), the rate of calorie emptying was increased during the initial 30 min up to 3.3 and 4.0 kcal/min, revealing non-linear initial kinetics. Calbet and MacLean (49) described exponential release kinetics characterizing the *initial* phase of gastric emptying of 600 mL of glucose solution 2.5%. Schirra et al. (47) additionally reported non-linear kinetics for human gastric emptying of concentrated glucose solutions [400 mL, 12.5% and 25% (w/v)]. Thus, gastric emptying of caloric fluids is obviously of a biphasic nature. The short initial phase is dominated by first order kinetics and followed by a linear, steady-state release of the remaining fluid. Gastric contents have to reach the duodenal (and ileal) glucose receptors before feedback mechanisms are fully activated. The time gap between the administration of the caloric fluid and the subsequent activation of GI feedback mechanisms plays a role in this behavior. Half-lives ($GE_{50\%}$) of gastric emptying were found to range from 49 min (500 mL glucose 10%) to 118 min (500 mL glucose 25%) and from 23 min (200 mL glucose 25%) to 94 min (400 mL glucose 25%). A detailed synopsis of human gastric emptying data including kinetics and release rates of various nutrient solutions has been summarized by Diebold [(10), Chapter 15.1.2]. The delay in gastric emptying resulting from ingestion of proteins, lipids, or carbohydrates is similar to those summarized here, provided that the energy content is the same, with an emptying rate of about 2 kcal/min.

Interspecies Differences

A rank order of gastric emptying ($GE_{50\%}$) exists among species. Gastric emptying rates for monkeys (*M. mulatta*) and dog, which are considered comparable, are slowest. Corresponding values for humans are slightly higher, whereas porcine gastric emptying is much faster ((10), Table 15.6).

Osmolality

The influence of osmolality on gastric emptying appears to be of minor importance for liquids (49,50). However, employing hyperosmotic saline solutions (500 mL), $GE_{50\%}$ was demonstrated to increase from 4.9–13.8 min (iso-osmotic) up to 53.1 min (hyperosmotic) (51). The further the liquid deviates from iso-osmotic, the slower is its rate of emptying. Thus, hypotonic and hypertonic fluids empty more slowly than do isotonic fluids. It has been shown that the "osmoreceptor" for the feedback signal resides in the duodenum. So long as duodenal contents are kept isotonic, gastric emptying of non-caloric fluids is rapid. There is no negative feedback to slow gastric emptying when hypertonic fluids are placed directly in the jejunum. The nature of this feedback mechanism for inhibiting gastric emptying has not been elucidated but presumably is both neural and humoral in nature. The caloric load of ingested meals and liquids predominates the influence of osmolality on gastric emptying in the fed state (50).

pH

The lower the pH, the slower is gastric emptying. Secretin presumably modulates this effect since acid in the duodenum is the prime stimulus for its release, and it has been shown to delay gastric emptying. In addition, neural receptors that respond to acid are present in the duodenum.

Liquid–Solid Meals

If the per os meal consists of liquid *and* solid components, gastric emptying exhibits a biphasic mechanism. With the exception of emptying of solid particles in MMC phase III, gastric emptying of solids into the duodenum takes place only if these particles are smaller than 1–3 mm in diameter (43,52). These particles are emptied, after a short lag phase, according to linear kinetics, whereas the liquid fraction often exhibits exponential or biphasic-(exponential) release kinetics (53–55).

Variability of Gastric Emptying

GI flow rates in the upper small intestine were demonstrated to be highly variable following oral administration of both saline 0.9% and glucose solution 20% (Fig. 16) (10).

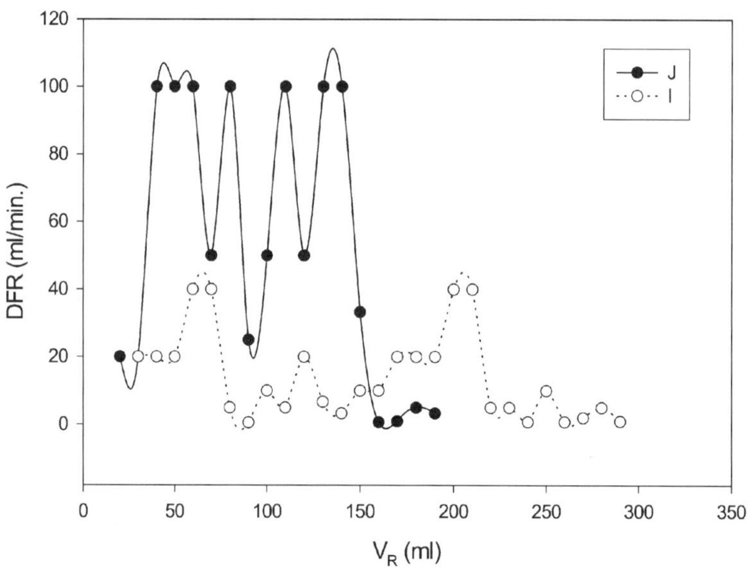

Figure 16 Variability (time dependency) of differential GI flow rates (DFR) in the small intestine of Labradors. V_R represents the cumulative volume of chyme collected at midgut following oral administration of 200 mL glucose solution 20% (I) and 200 mL NaCl 0.9% (J). *Source*: From Ref. 10.

The observed variability was more pronounced for the saline than for the glucose solution and was attributed mainly to the influence of gastric emptying rather than to MMC-driven transit variations (10). Variability of gastric emptying due to antral motility (typical of phase III contractions) and subsequent non-uniform gastric emptying can cause double peaks in the absorptive phase of concentration vs. time plots and can be seen with solids, suspensions, and solutions. This was demonstrated, e.g., for the absorption of cimetidine following oral administration in the fasted state in humans (56).

Intestinal Transit

Small intestinal transit time represents 10–25% of the total GI residence time and usually takes between 2 and 5 hr. Compared to transit through the large intestine, the overall small intestinal transit is shorter, varies less, and is more important for the absorption of both nutrients and drugs. The intestinal transit rate of fluids within a particular segment of the upper small intestine depends on fasted vs. fed state and, in the fasted states, on the phase of the MMC in the particular segment at the time of observation. Under physiological conditions, the chyme moves aborally, but short periods of retropulsion and gushes can occur intermittently. Propulsion of chyme is fastest in the duodenum and slowest in the ileum. It can be influenced by age, pregnancy, gender, or certain diseases, although small intestinal transit is generally less sensitive to these influences than large intestinal transit. Small intestinal transit can be accelerated artificially by co-administration of certain prokinetic drugs such as metoclopramide, bromopride, or domperidone and slowed down by inhibitors such as loperamide and opioids or by anticholinergics, such as ipratropium bromide, tropicamide, or trihexyphenidyl. The increase of the transit time is linked to an increase in time available for dissolution. On the other hand, motility-inducing agents, such as cisapride, which affects the small intestine as well as the colon, increase propagative contractions and hence may favor drug dissolution although limiting contact time of the dissolved drug with the absorptive sites.

Table 1 Mean Flow Rates (MFRs) in Various Intestinal Segments Are Related to the Phase of the MMC in Humans

MMC Phase	MFR (mL/min; mean ± SD) Jejunum	Ileum	Terminal ileum
I–II	0.58 ± 0.12	0.17 ± 0.03	0.33 ± 0.01
III	1.28 ± 0.18	0.50 ± 0.13	0.65 ± 0.01
Mean phase (I–III)	0.73 ± 0.11	0.33 ± 0.09	0.43 ± 0.06
Fed state (400 mL)	3.00 ± 0.67	2.35 ± 0.28	2.09 ± 0.16

Source: From Ref. 10. Calculated according to Ref. 65.

Transit Rates and Flow Rates in the Human Small Intestine

Mean and median transit rates of liquids passing through the upper small intestine employing different techniques and various liquid meals were determined to range between 1 and 4.8 cm/min (Table 1) see Ref. 10, Table 15.14 for a synopsis). Investigations on this subject were performed by the authors of Refs. 57–63. Jejunal and ileal flow rates in the human midgut range between 1 and 4.5 mL/min (see Ref. 10, Table 15.13 for a synopsis). Dillard et al. (64) reported 15 mL/min. However, these authors employed high perfusion rates of about 14 mL/min. Kerlin et al. (65) performed flow rate measurements on intestinal segments of about 20 cm. They used an aspiration method employing phenol red (PSP) at a perfusion rate of 1 mL/min. However, it seems questionable if such short distances are representative for the hydrodynamics of the small intestine in general. Jejunal flow rates are found to be greater than ileal flow rates, as was confirmed by Johnson et al. (40) for the relationship of jejunal and ileal transit rates in the canine upper intestine.

Jejunal and ileal flow rates are somewhat higher in the fed state than in the fasted state, as demonstrated by several authors (65–67).

Influence of Osmolality on Intestinal Transit and on Chyme Volume Available for Dissolution

There is clear evidence that in vivo hydrodynamics, namely mean intestinal fluid transit, depends on the osmotic conditions within the small intestine. Trendelenburg was the first author to perform systematic research on this subject, in 1917 (68). Holgate and Read (69) found that the intestinal transit rate was increased by hyperosmotic magnesium sulfate solutions despite the retardation of gastric emptying. Miller and co-workers reported oro-cecal transit times of intestinal chyme being significantly reduced from 205 to 35 min (median, $P < 0.01$) by co-administered lactulose [10 g per 300 mL standard meal (70)]. The authors concluded that intestinal transit was accelerated due to massive secretion of water into the lumen of the small intestine. Sellin and Hart (71) administered 250 mL of glucose solution 20%. Mean oro-cecal transit times were significantly decreased due to the hyperosmolality of the fluids. Similar observations have been reported using the canine model. Transit rates in the canine upper small intestine were significantly different after oral administration of hyperosmotic glucose solution (20%, 200 mL) compared to the same volume of 0.9% sodium chloride solution (2.7 cm/min vs. 1.1 cm/min, $n = 8$, $P < 0.001$, bifactorial ANOVA) (10). Ingestion of hypertonic liquids stimulated net water efflux across the intestinal wall into the GI lumen, possibly increased intestinal peristalsis, and accelerated the fluid transit *even though* gastric emptying was retarded. Apart from an acceleration of fluid transit, the increase of volume in the small intestine causes a considerable increase of in vivo dissolution of poorly soluble drugs, as was demonstrated with the use of an invasive aspiration method [(10), Chapter 16]. The (*c*umulative) *d*issolved (*n*ot *a*bsorbed) fraction of felodipine (F_{CDNA}) correlated well with the recovered volume at mid-jejunum of Labradors ($R = 0.972$, Pearson and Bravais, $P < 0.001$) (Fig. 17). The more liquid/chyme was available in the gut lumen, the faster was the in vivo dissolution. This result is in compliance with the equations adapted from Noyes, Whitney, Nernst, and Brunner.

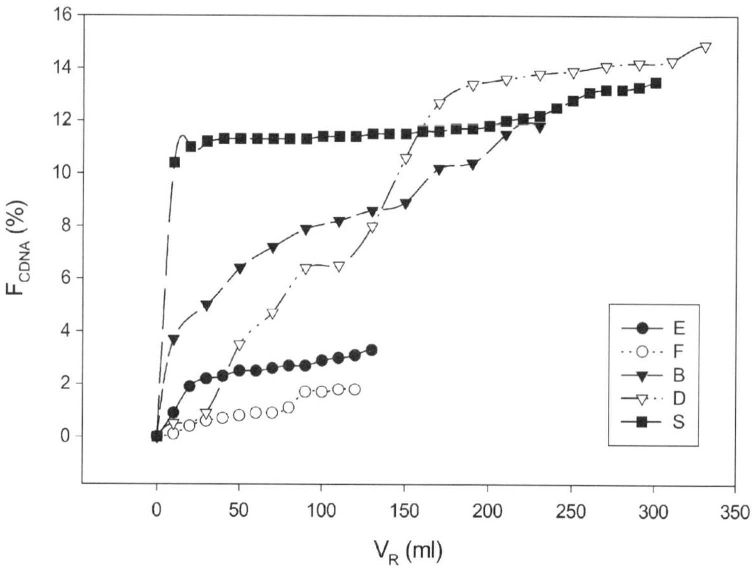

Figure 17 Volume dependent in vivo dissolution of micronized felodipine: F_{CDNA} indicates the dissolved fraction of felodipine aspirated at mid-jejunum of Labradors. The orally administered dose of 10 mg was suspended in 200 mL saline 0.9% (Experiments # E and F) or glucose 20% (Experiments # B, D, and S). V_R represents the recovered fluid volume. *Source*: From Ref. 10, Figure 16.12.

Transit Rates and Flow Rates in Canine Small Intestine

Due to the paucity of data for humans, it might be helpful to look at the canine model. In general, mean intestinal transit and flow rates of the dog correspond well to analogous data from humans. Flow rates in the canine jejunum after administration of 200–600 mL of various liquid meals ranged between 1 and 4 mL/min and sometimes up to 7 mL/min (72–76). Further, intestinal flow rates are highest in phase II/III of the MMC, followed by post-prandial flow rates. Flow rates in the canine duodenum and the proximal jejunum after administration of various liquids range between 2 and 13 mL/min (30,43,77). For instance, median duodeno-jejunal flow rates were determined to be 8.3 mL/min after oral administration of 200 mL glucose solution 20% (10). These flow rates obtained following the administration of glucose solutions are in good agreement with previous data of Brener et al. (44) for humans. They reported a gastric emptying rate of 2.13 kcal/min, which corresponds to a theoretical flow rate of about 10 mL/min. However, mean flow rates in the human upper small intestine often appear to be somewhat lower than those in the canine small intestine (41).

Variability of Intestinal Transit and GI Flow Rates

Considering the limited bioavailability of many poorly soluble drugs, any variability of GI flow or transit in the small intestine could have a pronounced influence on in vivo dissolution and absorption. Intestinal transit of liquids was shown to be variable both inter- and intra-individually. Caride et al. (61) compared a scintigraphic method to determine gastro-cecal transit times with the "hydrogen breath technique." Nineteen study participants received isotonic lactulose solution and ^{99m}Tc-DTPA-Diethylentriamine-N,N,N′,N″, N″-Penta acetic acid. Mean gastro-cecal transit times (MTTs) were found to be comparable for both experimental techniques (mean about 75 $\pm$ 8 min). However, individual transit times exhibited a relatively broad range, from 31 to 139 min. Cobden et al. (60) found inter-individual transit times to range from 25 to 150 min in a study with 21 participants. The authors employed the hydrogen breath technique and administered 200 mL of 10% lactulose orally as the test solution. Gushes, anterograde and retrograde directed fluid propulsions in the upper small intestine, constitute another prominent source of variability. These produce extremely high flow rates, particularly close to the pylorus, but these "flow peaks" are of short distance and duration (57,78). Therefore, they are unlikely to favor intestinal dissolution. The same is true for the transpyloric flow of non-caloric liquids from the stomach, which is not a continuous process but rather is linked to pyloric contractions and occurs in short episodes of 1–3 sec about three times a minute (79).

Techniques Used for the Investigation of GI Hydrodynamics

There are a number of experimental methods and techniques used for the investigation of GI hydrodynamics in humans. An introduction to this subject, including the intubation method, gamma scintigraphy, radiotelemetry, and the hydrogen breath technique, can be found in Macheras et al. [(80), Chapter 5.3.6]. Aspiration techniques and gamma scintigraphy are the most common methods used for the investigation of in vivo hydrodynamics of liquids. Of these two, scintigraphic experiments are less invasive. The dosage form (or a liquid carrier) is labeled with a gamma emitter (usually ^{99m}Tc or ^{111m}In). The transit is then followed by a gamma sensitive sensor or camera. Gastric emptying times and small intestinal transit rates can be selectively investigated within the course of the same experiment. This permits separation of any interdependencies of intestinal transit and gastric emptying (10). In contrast to most aspiration methods, the phases of gastric motility are not interrupted, e.g., by frequent intubation, since no fluid must be aspirated. Thus, duodeno-jejunal and ileal feedback mechanisms remain intact and can influence gastric emptying in a physiological manner. On the other hand, comparability to flow rate data already in literature is often limited—a common disadvantage of most scintigraphic methods. Moreover, Beckers et al. (81,82) found that scintigraphic techniques generate gastric emptying data that are up to 70% higher than those from aspiration experiments for methodical reasons. The authors found human gastric emptying half-lives ranging from 150 to 200 min (600 mL, 444 kcal). Another disadvantage of this method is that the drug itself cannot usually be labeled because carbon, nitrogen, and oxygen radionuclides are positron emitters with very short half-lives and high radiation burdens. A further limitation to this technique is that it cannot distinguish between a radionuclide present as a solid from one in solution.

Reynolds Numbers in the Upper Small Intestine

The overall situation in vivo is far more complicated than the hydrodynamics in dissolution apparatus. Moreover, only a few data are available to exactly characterize the flow rate and the transit rate for the different segments, motility patterns, and prandial states of the human small intestine. Therefore, it is a challenge to calculate meaningful and valid Reynolds numbers for the hydrodynamics of the small intestine.

Reynolds Number for Bulk Flow

The Reynolds number characterizing laminar–turbulent transition for *bulk* flow in a pipe is about $Re \sim 2300$ provided that the fluid moves unidirectionally, the pipe walls are even and behave in a hydraulically smooth manner, and the internal diameter remains constant. However, intestinal walls do not fulfill these hydraulic criteria due to the presence of curvatures, villi, and folds of mucous membrane, which are up to 8 mm in the duodenum, for instance (Fig. 18). Furthermore, the internal diameter of the small intestine is estimated to

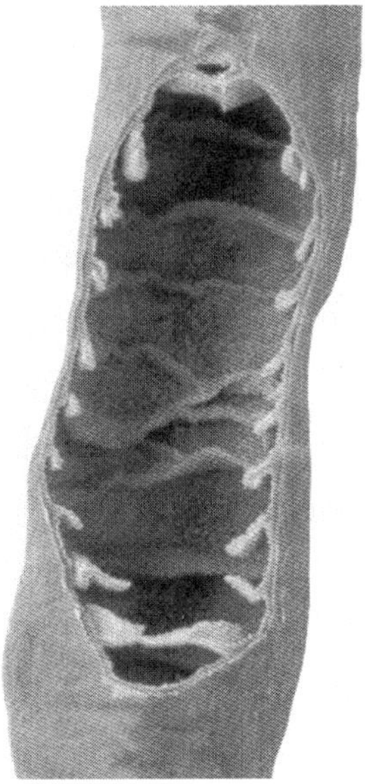

Figure 18 Segment of the human small intestine with folds of mucous membrane (prepared by plastination). The total length of the human small intestine is estimated to be about 3.5–3.8 m. *Source*: From Ref. 90.

be about 3–4 cm and does not remain constant. Not only does the diameter decrease with increasing distance from the pylorus, but the gut wall contracts, leading to momentary fluctuations in diameter.

Nevertheless, *approximate* bulk Reynolds numbers may be calculated using a kinematic viscosity of $\nu = 7 \times 10^{-3}$ cm^2/sec (water, 37°C) for intestinal chyme and an internal diameter of the small intestine of 3 cm. Employing jejunal flow rates of 0.5–4.5 mL/min, bulk Reynolds numbers of $Re \sim 0.5$ to $Re \sim 4.5$ are then obtained. As previously demonstrated, median flow rates of 35 mL/min, including (short period) spike flows beyond 100 mL/min,* can occur at midgut after administration of non-nutrient liquids (10). But even taking into account such extremely high flow rates, bulk Reynolds numbers of $35 < Re < 100$–125 are obtained. Thus, *bulk* flow at midgut is unlikely to be turbulent for considerable periods of time. This can be chiefly attributed to the relatively low flow rates and the somewhat elevated viscosity of the intestinal fluids. It would take consistently higher flow rates in both the fed and the fasted state to permanently induce turbulence in the chyme flow of the human small intestine. However, perturbations may occasionally occur close to the intestinal wall due to the folds, villi, and curvatures.

Particle–Liquid Reynolds Number

The diameter of drug particles and hence the surface specific length L is much smaller than the pipe diameter. For this reason, particle–liquid Reynolds numbers characterizing the flow at the particle surface are considerably lower than the corresponding bulk Reynolds numbers. Particle–liquid Reynolds numbers for particle sizes below 250 μm were calculated to be below $Re \sim 1$ for flow rates up to 100 mL/min. However, this circumstance does not limit the applicability of the boundary layer concept, since in aqueous hydrodynamic

---

*This apparently high flow rate may be an artefact of the canine experiments, in which removal of the fluids at mid-jejunum through a fistula may have eliminated *long-range* feedback inhibition of flow.

systems the Peclet number is still greater than 1 [(9,10), Chapters 5.1 and 12.3.8]. Furthermore, the surface of a drug particle is far from being smooth and even. Craters and protrusions may cause perturbations at the particle surface and elevate the corresponding Reynolds numbers so that the particle surface may experience turbulent conditions *even though* the bulk flow is laminar. Moreover, the shape of the particles differs more or less according to the origin of the fraction (ground, sieved, precipitated). Above all, the Stokes law of creeping (bulk) flow can be used for smooth spheres only if $Re < 0.5$! Thus, in the case of "rough" drug particles, $Re \sim 0.5$ might be an appropriate magnitude to characterize the laminar–turbulent transition for flow around a sphere. Ground or milled drug particles, with more defects, protrusions, and rough surfaces, can be reasonably expected to produce laminar–turbulent transition at much lower Reynolds numbers, e.g., in the range of $10^{-2} < Re < 1$. Thus, although neither fed state nor fasted state flows are likely to provoke a laminar–turbulent transition for the *bulk flow*, the drug *particle* potentially "sees" a turbulent flow pattern at physiological flow rates, since the crucial particle–liquid Reynolds number for the laminar–turbulent transition at a rough, edged, and spherical particle surface is about $Re_{\mathrm{crit}} \leq 0.5$.

In Vitro–In Vivo Comparison of Reynolds Numbers

Reynolds numbers calculated for the in vivo hydrodynamics are considerably lower than those of the corresponding in vitro numbers, both for bulk and particle–liquid Reynolds numbers. Remarkably, *bulk* Reynolds numbers in vivo appear to have about the same magnitude as *particle–liquid* Reynolds numbers characterizing the flow at the particle surface in vitro using the *paddle* apparatus. In other words, it appears that hydrodynamics per se play a relatively minor role in vivo compared to the in vitro dissolution. This can be attributed to physiological co-factors that greatly affect the overall dissolution in vivo but are not important in vitro (e.g., absorption and secretion processes, change of MMC phases, complex composition of chyme, bile acids, mucus, and further components). These influences may sometimes overrule hydrodynamic effects in vivo and make it difficult to selectively measure any hydrodynamic effects on in vivo dissolution.

Intestinal Hydrodynamics Can Influence Absorption

Intestinal Transit and Absorption of Nutrients

The purpose of the fasting motor pattern is to keep the small intestine swept clean of bacteria, indigestible meal residua, desquamated cells, and secretions. In contrast, the purpose of the fed pattern is to produce thorough mixing of the chyme with the digestive enzymes and provide maximal contact between the absorbing cells and the intestinal chyme. Thus, absorption is greatest during the fed motor pattern even though the motility is lower in terms of transit rate than in MMC phase III. For example, glucose, water, and electrolytes are considerably better absorbed from isolated canine gut in the fed than in the fasted state motility pattern, owing to a significant reduction of the small intestinal transit (83). Segmental contractions over distances of 1–4 cm encourage mixing of the lumenal contents in the fed state, leading, for example, to better digestion of 0.5 and 2 mm liver particles in the fed state (84). Apart from the fed state composition of chyme, the transit rate, and segmental contractions associated with an increase in mixing efficiency, absorption depends on the volume of chyme available for dissolution. Not only do the ingested food and fluids directly influence the volume in the upper GI tract, they also stimulate secretion of gastric acid, bile, and pancreatic juice.

Intestinal Transit and Drug Absorption

GI absorption of many poorly soluble drugs depends on small intestinal transit, as demonstrated for ketoprofen, nifedipine, haloperidol, miconazole, and others. Small intestinal transit rate and transit time become important factors in drug absorption, particularly when the ratio of dose to solubility is high and dissolution rate is very slow or when the drug is taken up selectively at a specific location of the intestine ("absorption window"). In this case, the extent of absorption is limited by the residence time at the uptake sites, as in the case of lithium carbonate, which is taken up by the small intestine but not by the colon. For drugs that are highly soluble in gastric juice, like atenolol, for instance, no influence on the absorption was observed when intestinal transit rate was reduced about 50% by co-administration of codeine phosphate (91). In contrast, depending on particle size, hydrodynamics can influence drug absorption of *poorly soluble* drugs, as demonstrated in pharmacokinetic studies of felodipine with fistulated Labradors (30). The hydrodynamic influence on the bioavailability of felodipine (aqueous solubility: 1.2 μg/mL at 37°C, log *P* 4.5 for toluol/water) was selectively investigated and revealed a dependency on the particle size in vivo (Fig. 19). A two-fold higher bioavailability after administration of a felodipine suspension under hydrodynamic conditions representative of the fed state compared to the fasted state was observed for the coarse grade compound. In contrast, no change in the

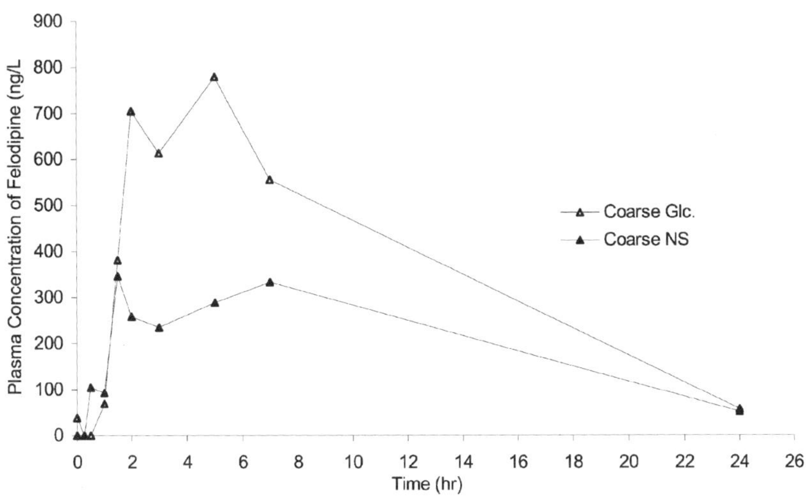

Figure 19 Mean plasma concentrations following the administration of felodipine suspension to Labradors. Median particle size: 125 μm ($n = 6$); dose: 10 mg, in either 0.9% saline (NS) or 5% glucose (Glc.) solution. *Source*: From Ref. 30.

bioavailability with hydrodynamic conditions was observed for micronized drug. The coarse grade particles appeared to be more sensitive to hydrodynamics than the micronized ones (10,31,36). In vivo, however, the particle size itself appears to have a more important influence on bioavailability than the hydrodynamics per se. Subsequently, improved absorption attributed to the reduced particle size often overrules the influence of altered hydrodynamics, although the latter affects dissolution, too.

"Leveling" of In Vivo Hydrodynamics?

Often, no overt influence of GI hydrodynamics on the absorption of drugs is observable in vivo. Therefore, one may ask, what role do GI hydrodynamics play in relation to other physiological factors relevant to the absorption of drugs? Arguing in a more teleologic and speculative way, one must point out that the GI tract of mammalians was surely not designed for the GI absorption of drugs but primarily optimized for food uptake and exploitation of nutritional components. Evolution had to take care of an efficient transport, digestion, and absorption system for nutritional substrates of all kinds and provenience. Thus, it might have been advantageous if a species had been able to efficiently absorb *small quantities* of food, exploit *different sources* of food (various plants and animals), and cope with varying nutritional components (fats, carbohydrates, peptides, etc.), regardless of their availability and relative proportions. Adapted omnivores like primates may have had some benefit compared to specialists like carnivores or herbivores, since good times can change for animals in nature over short time spans as well as on an evolutionary time scale. Intestinal hydrodynamics that are extremely sensitive to different "input variables" would also have been vulnerable to environmental changes. Of course, this would not have been conducive to efficient absorption or nutritional supply and might have been a permanent source of malabsorption, leading to crucial negative selection. These considerations may perhaps explain the leveling of GI hydrodynamics in the light of evolution.

Representation of GI Motility Patterns and Flow Rates by In Vitro Hydrodynamic Conditions

Abrahamsson demonstrated that human intestinal hydrodynamics were reflected in vitro using the paddle method at stirring rates of about 140 rpm [(85), Paper V]. The author used erosion sensitive HPMC-Hydroxypropylmethylcellulose matrix tablets containing a poorly soluble, neutral, and lipophilic ingredient. The formulations were susceptible to mechanical stress. However, human studies to establish such correlations are expensive and time consuming. As the anatomy and the physiology of the GI tract of Labradors resemble those of the human GI tract, this canine breed can serve as a model to simulate human intestinal hydrodynamics. Preliminary results indicate that, following oral dosing of micronized felodipine powder under hydrodynamic conditions representative of the fed state, canine intestinal hydrodynamics were reflected in vitro employing the paddle method at stirring rates of 100–150 rpm [(10), Chapter 16.3.4]. Recently, Scholz et al. (86) studied the dissolution performance of micronized and coarse grade felodipine in a biorelevant medium using the USP paddle apparatus at various paddle speeds. Ratios of percentage dissolved were calculated pairwise for slower as well as for faster stirring rates. These ratios were then compared to AUC-Area under the curve ratios obtained in a corresponding pharmacokinetic study in Labradors, in which the absorption of both the micronized and coarse grade felodipine had been compared under two GI hydrodynamic conditions (86). The authors proposed to use a paddle speed combination of 75 and 125 rpm to represent the motility patterns in response to administration of normal saline and 5% glucose, respectively. In vitro AUC-Area under the curve ratios of this particular experimental setup showed best agreement with the pharmacokinetic data (30). It seems that the compendial paddle apparatus can be used both to simulate intestinal hydrodynamics as well as to reflect *variations* in hydrodynamic conditions in the upper GI tract.

Recommendations on the Choice of an Appropriate Dissolution Test Apparatus

The following considerations may support the choice of an appropriate dissolution test apparatus based on different hydrodynamic scenarios in vivo. Constant flow rates, such as those that may occur in MMC phase I–II, in the regular fed state, or at distal segments of the small intestine, are best simulated by the paddle method (Pharm. Eur. 2.9.3.–1, USP Apparatus 2). Dissolution is mainly driven by convection and the hydrodynamics of the paddle are easy to select and standardize. Thus, provided an appropriate composition, volume, and particle size range are chosen for the dissolution test, the paddle apparatus can be used to reflect hydrodynamic conditions in the upper GI tract under certain dosing conditions (86). However, if the flow rates to be reflected in vitro vary with time (e.g., pulsatile flow rates of MMC phase III or transpyloric flow), the flow-through tester may be the more suitable apparatus since the flow rates in vitro can be varied with time using appropriate pumps and control software. At an early developmental stage, it might sometimes be desirable to produce mechanical stress acting on the drug formulation in vitro. This could be required to simulate the effects of the "antral mill" (on the formulation) or of grinding by the intestinal wall (on particle agglomerates). In this case, drug release and particle dissolution are furthered by erosion and thus increased by abrasive processes [(87,10), with additional references]. The best choice for this kind of application might be the Biodis™ apparatus. Alternatively, the paddle method could be appropriate, provided the vessels are filled with glass beads (88). However, mechanical forces are only relevant for the dissolution of particle agglomerates and drug release from formulations that are *susceptible* to mechanical stress, such as HPMC-Hydroxypropylmethylcellulose matrix tablets. In contrast, erosion and abrasion play a minor role for smaller units such as single drug particles or microparticles, which are primarily subject to convective diffusion hydrodynamics.

CONCLUSION

Hydrodynamics in the upper GI tract contribute to in vivo dissolution. Our ability to forecast dissolution of poorly soluble drugs in vitro depends on our knowledge of and ability to control hydrodynamics as well as other factors influencing dissolution. Provided suitable conditions (apparatus, hydrodynamics, media) are chosen for the dissolution test, it seems possible to predict dissolution limitations to the oral absorption of drugs and to reflect variations in hydrodynamic conditions in the upper GI tract. The fluid volume available for dissolution in the gut lumen, the contact time of the dissolved compound with the absorptive sites, and particle size have been identified as the main hydrodynamic determinants for the absorption of poorly soluble drugs in vivo. The influence of these factors is usually more pronounced than that of the motility pattern or the GI flow rates per se.

REFERENCES

1. Ramtoola Z, Corrigan OI. Effect of agitation intensity on the dissolution rate of indomethacin and indomethacin–citric acid compressed discs. Drug Dev Ind Pharm 1988; 14(15–17):2241–2253.
2. Raines MA, Dewers TA. Mixed transport/reaction control of gypsum dissolution kinetics in aqueous solutions and initiation of gypsum karst. Chem Geol 1997; 140(1–2):29–48.
3. Liu Z, Dreybrodt W. Dissolution kinetics of calcium carbonate minerals in H_2O–CO_2 solutions in turbulent flow—the role of the diffusion boundary layer and the slow reaction $H_2O + CO_2 <\!-\!> H^+ + HCO_3^-$. Geochim Cosmochim Acta 1997; 61(14):2879–2889.
4. Brunner E. Reaktionsgeschwindigkeit in heterogenen Systemen. Z Phys Chem 1904; 47:56–102.
5. Nernst W. Theorie der Reaktionsgeschwindigkeit in heterogenen Systemen. Z Phys Chem 1904; 47:52–55.

6. Noyes AA, Whitney WR. Über die Auflösungsgeschwindigkeit von festen Stoffen in ihren eigenen Lösungen. Z Phys Chem 1897; 23:689–692.

7. Schükarew A. Reaktionsgeschwindigkeiten zwischen Metallen und Haloiden. Z Phys Chem 1891; 8(76):76–82.

8. Levins DM, Glastonbury JR. Particle–liquid hydrodynamics and mass transfer in a stirred vessel. Trans Inst Chem Eng 1972; 50:132–146.

9. Levich VG. Physicochemical Hydrodynamics. 2nd ed. Englewood Cliffs, NJ: Prentice Hall, 1962.

10. Diebold SM. Hydrodynamik and Lösungsgeschwindigkeit—Untersuchungen zum Einfluß der Hydrodynamik auf die Lösungsgeschwindigkeit schwer wasserlöslicher Arzneistoffe (Hydrodynamics and Dissolution—Influence of Hydrodynamics on Dissolution Rate of Poorly Soluble Drugs). 1st ed. Aachen, Germany: Shaker Verlag, 2000.

11. Sherwood TK, Woertz BB. The role of eddy diffusion in mass transfer between phases. Trans Am Inst Chem Eng 1939; 35:517–540.

12. Plummer LN, Wigley TML. The dissolution of calcite in CO_2-saturated solutions at 25°C and 1 atmosphere total pressure. Geochim Cosmochim Acta 1976; 40:191–202.

13. Tritton DJ. Physical Fluid Dynamics. 2nd ed. Oxford: Oxford University Press, 1995.

14. Blasius H. Grenzschichten in Flüssigkeiten mit kleiner Reibung. Z Math Phys 1908; 56:1–37.

15. Levich VG. Theory of concentration polarization. J Phys Chem 1944; 18(9):335–355.

16. Larhed AW, Artursson P, Grasjo J, Bjork E. Diffusion of drugs in native and purified gastrointestinal mucus. J Pharm Sci 1997; 86(6):660–665.

17. De Smidt JH, Grit M, Crommelin DJA. Dissolution kinetics of griseofulvin in mixed micellar solutions. J Pharm Sci 1994; 83(9):1209–1212.

18. Chang SH, Parrot EL. Hydrodynamics of dissolution in a newtonian and a non-newtonian solution. Drug Dev Ind Pharm 1991; 17(13):1731–1751.

19. Dundon ML, Mack E. The solubility and surface free energy of calcium sulfate. J Am Chem Soc 1923; 45(11):2479–2485.

20. Roller PS. Chemical activity and particle size. II. The rate of solution at slow stirring of anhydrite and gypsum. J Phys Chem 1932; 36:1202–1231.

21. Grijseels H, de Blaey CJ. Dissolution at porous interfaces. Int J Pharm 1981; 9:337–347.

22. Compton RG, Daly PJ, House WA. The dissolution of Iceland spar crystals; the effect of surface morphology. J Colloid Interface Sci 1986; 113:12–20.

23. Dreybrodt W, Buhmann D. A mass transfer model for dissolution and precipitation of calcite from solutions in turbulent motion. Chem Geol 1991:107–122.

24. Schlichting H. Grenzschicht-Theorie. Karlsruhe: G. Braun Verlag, 1951.

25. Levich VG. The theory of concentration polarisation. Acta Physicochim URSS 1942; 17(5–6):257–307.

26. Mithani SD. Dissolution and Precipitation of Dipyridamole: Effect of pH and Bile Salt Concentration. Thesis: The University of Michigan, Michigan, 1998.

27. Kaneniwa N, Watari N. Dissolution of slightly soluble drugs. I. Influence of particle size on dissolution behaviour. Chem Pharm Bull 1974; 22(8):1699–1705.

28. Atkinson RM, Bedford C, Child KJ, Tomich EG. Effect of particle size on blood griseofulvin levels in man. Nature 1962; 193:588–589.

29. Kraml M, Dubuc J, Gaudry R. Gastrointestinal absorption of griseofulvin: 2. Influence of particle size in man. Antibiot Chemother 1962; 12:239–242.

30. Scholz A, Abrahamsson B, Diebold SM, Kostewicz E, Polentarutti BI, Ungell AL, Dressman JB. Influence of hydrodynamics and particle size on the absorption of felodipine in Labradors. Pharm Res 2002; 19(1):42–46.

31. Anderberg EK, Bisrat M, Nyström C. Physicochemical aspects of drug release. VII. The effect of surfactant concentration and drug particle size on solubility and dissolution rate of felodipine, a sparingly soluble drug. Int J Pharm 1988; 6:67–77.

32. Harriott P. Mass transfer to particles. I. Suspended in agitated tanks. AIChE J 1962; 8:193–101.

33. Armenante PM, Kirwan DJ. Mass transfer to microparticles in agitated systems. Chem Eng Sci 1989; 44(12):2781–2796.

34. Batchelor GK. Mass transfer from small particles suspended in turbulent fluid. J Fluid Mech 1980; 98(3):609–623.

35. Diebold SM, Dressman JB. Dissolution of microparticles—a hydrodynamically based contribution to an unresolved pharmaceutical issue. Pharm Res. In preparation.

36. Bisrat MC. Nyström, Physicochemical aspects of drug release. VIII. The relation between particle size and surface specific dissolution rate in agitated suspensions. Int J Pharm 1988; 47:223–231.

37. Bisrat M, Anderberg EK, Barnett MI, Nyström C. Physicochemical aspects of drug release: XV. Investigation of diffusional transport in dissolution of suspended, sparingly soluble drugs. Int J Pharm 1992; 80:191–201.

38. Diebold SM, Dressman JB. Hydrodynamik kompendialer Lösungsgeschwindigkeits-Testapparaturen: Paddle und Basket. Pharm Ind 2001; 63(1):94–104.

39. Cammarn S, Sakr A. Predicting dissolution via hydrodynamics: salicylic acid tablets in flow through cell dissolution. Int J Pharm 2000; 201:199–209.

40. Johnson C, Sarna SK, Baytiyeh R, Cowles V, Zhu YR, Telford GL, Roza AM, Adams MB. Postprandial motor activity and its relationship to transit in the canine ileum. Surgery 1997; 121(2):182–189.

41. Oberle R, Chen TS, Lloyd C, Barnett JL, Owyang C, Meyer J, Amidon GL. The influence of the interdigestive migrating myoelectric complex on the gastric emptying of liquids. Gastroenterology 1990; 99:1275–1282.

42. Keinke O, Schemann M, Ehrlein HJ. Mechanical factors regulating gastric emptying of viscous nutrient meals in dogs. Q J Exp Physiol 1984; 69:781–795.

43. Sirois PJ. Size and Density Discrimination of Nondigestible Solids During Emptying from the Canine Stomach. A Hydrodynamic Correlation. Thesis: The University of Michigan, Ann Arbor, 1989.

44. Brener W, Hendrix TR, McHugh PR. Regulation of the gastric emptying of glucose. Gastroenterology 1983; 85:76–82.

45. McHugh PR, Moran TH. Calories and gastric emptying: a regulatory capacity with implications for feeding. Am J Physiol 1979; 236:R254–R260.

46. McHugh PR, Moran TH, Wirth JB. Postpyloric regulation of gastric emptying in rhesus monkeys. Am J Physiol 1982; 243:R408–R415.

47. Schirra J, Katschinski M, Weidmann C, Schafer T, Wank U, Arnold R, Goke B. Gastric emptying and release of incretin hormones after glucose ingestion in humans. J Clin Invest 1996; 97(1):92–103.

48. Hunt JN, Smith JL, Jiang CL. Effect of meal volume and energy density on the gastric emptying of carbohydrates. Gastroenterology 1985; 89:1326–1339.

49. Calbet JAL, MacLean DA. Role of caloric content on gastric emptying in humans. J Physiol 1997; 498(2):553–559.

50. Vist GE, Maughan RJ. The effect of osmolality and carbohydrate content on the rate of gastric emptying of liquids in man. J Physiol 1995; 486(2):523–531.

51. Meeroff JC, Go VLW, Phillips SF. Control of gastric emptying by osmolality of duodenal contents in man. Gastroenterology 1975; 68:1144–1151.

52. Meyer JH, Dressman JB, Fink A, Amidon G. Effect of size and density on canine gastric emptying of nondigestible solids. Gastroenterology 1985; 89:805–813.

53. Notivol R, Carrio I, Cano L, Estorch M, Vilardell F. Gastric emptying of solid and liquid meals in healthy young subjects. Scand J Gastroenterol 1984; 19:1107–1113.

54. Carbonnel F, Rambaud JC, Mundler O, Jian R. Effect of energy density of a solid–liquid meal on gastric emptying and satiety. Am J Clin Nutr 1994; 60:307–311.

55. Ziessman HA, Fahey FH, Collen MJ. Biphasic solid and liquid gastric emptying in normal controls and diabetics using continuous acquisition in LAO view. Dig Dis Sci 1992; 37(5): 744–750.

56. Takamatsu N, Welage LS, Hayashi Y, Yamamoto R, Barnett JL, Shah VP, Lesko LJ, Ramachandran C, Amidon GL. Variability in cimetidine absorption and plasma double peaks following oral administration in the fasted state in humans: correlation with antral gastric motility. Eur J Pharm Biopharm 2002; 53:37–47.

57. Barreiro MA, McKenna RD, Beck IT. Determination of transit time in the human jejunum by the single-injection indicator-dilution technique. Am J Dig Dis 1968; 13(3):222–233.

58. Wald A, Back C, Bayless TM. Effect of caffeine on the human small intestine. Gastroenterology 1976; 71(5):738–742.

59. Matshese JW, Malagelada JR, Phillips SF. Intestinal fluid flow and pancreatic enzyme output after mixed solid–liquid meals of different composition. Gastroenterology 1978; 74:1135.

60. Cobden I, Barker MCJ, Axon ATR. Gastrointestinal transit of liquids. Ann Clin Res 1983; 15:119–122.

61. Caride VJ, Prokop EK, Troncale FJ, Buddoura W, Winchenbach K, McCallum RW. Scintigraphic determination of small intestinal transit time: comparison with the hydrogen breath technique. Gastroenterology 1984; 86:714–720.

62. Wisen O, Hellström PM, Johansson C. Meal energy density as a determinant of postprandial gastrointestinal adaptation in man. Scand J Gastroenterol 1993; 28:737–743.

63. Beaugerie L, Flourie B, Marteau P, Pellier P, Franchisseur C, Rambaud JC. Digestion and absorption in the human intestine of three sugar alcohols. Gastroenterology 1990; 99:717–723.

64. Dillard RL, Eastman H, Fordtran JS. Volume–flow relationship during the transport of fluid through the human small intestine. Gastroenterology 1965; 49(1):58–66.

65. Kerlin P, Zinsmeister A, Phillips S. Relationship of motility to flow of contents in the human small intestine. Gastroenterology 1982; 82:701–706.

66. Stephen AM, Haddad AC, Phillips SF. Passage of carbohydrate into the colon—direct measurements in humans. Gastroenterology 1983; 85(5):89–95.

67. Soergel KH. In: Demling L, Ottenjann R, eds. Gastrointestinal Motility. 1971:81–92.

68. Trendelenburg P. Physiologische und pharmakologische Versuche über die Dünndarmperistaltik. Arch Exp Pathol Pharmakol 1917; 81:55–129.

69. Holgate AM, Read NW. Relationship between small bowel transit time and absorption of a solid meal. Dig Dis Sci 1983; 28(9):812–819.

70. Miller A, Parkman HP, Urbain JC, Brown KL, Donahue DJ, Knight LC, Maurer AH, Fisher RS. Comparison of scintigraphy and lactulose breath hydrogen test for assessment of orocecal transit. Dig Dis Sci 1997; 42(1):10–18.

71. Sellin JH, Hart R. Glucose malabsorption associated with rapid intestinal transit. Am J Gastroenterol 1992; 87(5):584–589.

72. Schemann M, Ehrlein HJ. Postprandial patterns of canine jejunal motility and transit of luminal content. Gastroenterology 1986; 90:991–1000.

73. Sarr MG, Kelly KA. Patterns of movement of liquids and solids through canine jejunum. Am J Physiol 1980; 239:G497–G503.

74. Soper NJ, Geisler KL, Sarr MG, Kelly KA, Zinsmeister AR. Regulation of canine jejunal transit. Am J Physiol 1990; 259:G928–G933.

75. Johnson C, Sarna S, Cowles V, Baytiyeh R, Zhu YR, Buchmann E, Bonham L, Roza AM, Adams MB. Effects of transection and reanastomosis on postprandial jejunal transit and contractile activity. Surgery 1995; 117:531–537.

76. Bueno L, Fioramonti J, Ruckebusch Y. Rate of flow of digesta and electrical activity of the small intestine in dogs and sheep. J Physiol 1975; 249:69–85.

77. Hinder RA, Kelly KA. Canine gastric emptying of solids and liquids. Am J Physiol 1977; 233:E335–E340.

78. Malbert CH, Ruckebusch Y. In: Nueten JM, Schuurkes JAJ, Akkermans LMA, eds. Gastro-Pyloro-Duodenal Coordination. 1st ed. Petersfield: Wrightson Biomedical Publishing, 1990.

79. Heading RC, King PM. 1st ed. Nueten JM, Schuurkes JAJ, Akkermans LMA, eds. Gastro-Pyloro-Duodenal Coordination. 1st ed. Petersfield: Wrightson Biomedical Publishing, 1990.

80. Macheras P, Reppas C, Dressman JB. Biopharmaceutics of Orally Administered Drugs. 1st ed. Hemel Hempstead, Hertfordshire, UK: Ellis Horwood Ltd., 1995.

81. Beckers EJ, Rehrer NJ, Saris WHM, Brouns F, Hoor FT, Kester ADM. Daily variation in gastric emptying when using the double sampling technique. Med Sci Sports Exerc 1991; 23(10):1210–1212.

82. Beckers EJ, Leiper JB, Davidson J. Comparison of aspiration and scintigraphic techniques for the measurement of gastric emptying rates of liquids in humans. Gut 1992; 33:115–117.

83. Sarr MG, Kelly KA, Phillips SF. Canine jejunal absorption and transit during interdigestive and digestive motor states. Am J Physiol 1980; 239:G167–G172.

84. Williams NS, Meyer JH, Jehn D, Miller J, Fink AS. Canine intestinal transit and digestion of radiolabeled liver particles. Gastroenterology 1984; 86:1451–1459.

85. Abrahamsson B. Biopharmaceutical Aspects of Extended Release Tablets based on the Hydrophilic Matrix Principle. Thesis, University of Uppsala, Uppsala, Sweden, 1997.

86. Scholz A, Kostewicz E, Abrahamsson B, Dressman J. Can the USP paddle method be used to represent in vivo hydrodynamics? J Pharmacol 2003; 55:443–453.

87. Kamba M, Seta Y, Kusai A, Nishimura K. Comparison of the mechanical destructive force in the small intestine of dog and human. Int J Pharm 2002; 237:139–149.

88. Aoki S, Ando H, Tatsuishi K, Uesugi K, Ozawa H. Determination of the mechanical impact force in the in vitro dissolution

test and evaluation of the correlation between in vivo and in vitro release. Int J Pharm 1993; 95:67–75.

89. Milton VD. An Album of Fluid Motion. Stanford Parabolic: Parabolic, 1982:91.

90. Institut für Plastination. Körperwelten, Einblicke in den menschlichen Körper (Ausstellungskatalog). Vol. 3. Heidelberg: Institut für Technik und Arbeit, 1997.

91. Riley SA, Sutcliffe F, Kim M, Kapas M, Rowland M, Turnberg LA. The influence of gastrointestinal transit on drug absorption in healthy volunteers. Br J Clin Pharm 1992; 34:32–39.

7

Development of Dissolution Tests on the Basis of Gastrointestinal Physiology

SANDRA KLEIN, MARTIN WUNDERLICH, and JENNIFER DRESSMAN

Institute of Pharmaceutical Technology, Biocenter, Johann Wolfgang Goethe University, Frankfurt, Germany

ERIKA STIPPLER

Phast GmbH, Biomedizinisches Zentrum, Homburg/Saars, Germany

INTRODUCTION

Almost half a century after the first attempts at dissolution testing, we are still grappling with the question of "which media to use to run which dissolution tests." This is not a trivial question, since the outcome of a test can be greatly dependent on the dissolution medium, especially if the drug itself and/or key excipients are poorly soluble and/or ionizable. In addition, dissolution tests are run for different reasons at different points in the product life cycle. In pre-clinical development, dissolution of the pure drug is often studied under biorelevant conditions to assess whether dissolution is likely to be a rate-limiting factor in the oral absorption of a drug. Later, various formulations will be compared, again under biorelevant conditions, to determine which are most suitable for taking into clinical studies. During the progression through phase II and III clinical trials, batch sizes are increased and the formulation is often optimized. At this stage, it may well be desirable to develop an in vitro–in vivo correlation (IVIVC) so that the biopharmaceutical properties after further scale-up and minor formulation changes in the product can be assessed with in vitro studies instead of having to perform a pharmacokinetic bioequivalence study. At this time, dissolution tests for routine quality control (QC) of the drug product are also being developed. These QC procedures should also reflect insofar as possible the gastrointestinal (GI) conditions under which the product has to perform. At times, this can be quite a challenge with today's standard apparatus due to the parallel need to confirm that the product can release 100% (or near to) of the drug.

Even after the drug product has been approved, research on formulation and dissolution testing does not stop. Quite the contrary: often new dosage strengths and modified release (MR) products are brought onto the market to provide the medical practitioner with more prescribing flexibility. Last but not least, as the patent protection for the drug substance runs out, other manufacturers may desire to bring competitor products onto the market. Approval of these multisource products may under certain circumstances be contingent on the ability to pass an array of specially designed dissolution tests according to the so-called bioavailability–bioequivalence (BABE) guidance (1) rather than having to show bioequivalence in a pharmacokinetic study.

To assist the reader with the question of "which dissolution test to apply when?" the first part of this chapter is divided into two primary sections—one dealing with drugs that have few or no solubility problems, in which case developing dissolution tests at all stages of the product life cycle is a relatively straightforward process, and the other dealing with compounds where the dissolution test design may have to undergo a transition as the compound moves from early development into clinical trials and later to an approved product. The second part of the chapter deals more specifically with the question of developing dissolution tests that can predict in vivo performance for MR products.

GETTING STARTED: SOLUBILITY AND THE DOSE: SOLUBILITY RATIO

First and foremost, it is important to arrive at a thorough understanding of the compound's solubility behavior over the usual pH range encountered in the GI tract. Table 1 summarizes typical pH values in the GI tract in young, healthy individuals, as well as approximates residence times for pellets and (non-disintegrating) tablets in the various GI segments.

Table 1 Typical Values [Average (Range)] of pH and Mean Residence Times (MRT) in Various Segments of the GI Tract of Young, Healthy Volunteers

Segment	pH	MRT (pellets)	MRT (tablets)[a]
A. Pre-prandial			
Stomach	1.8 (1–3)	30 min	60 min
Duodenum	6.0 (4–7)	< 10 min	< 10 min
Upper jejunum	6.5 (5.5–7)	60 min	30 min
Lower jejunum	6.8 (6–7.2)	60 min	30 min
Upper ileum	7.2 (6.5–7.5)	60 min	60 min
Lower ileum	7.5 (7–8)	60 min	120 min
Proximal colon	5.5–6.5	4–12 hr	4–12 hr
B. Post-prandial			
Stomach	4 (3–6)	2–4 hr	2–10 hr
Duodenum	5.0 (4–7)	< 10 min	< 10 min
Upper jejunum	5.5 (5.5–7)	60 min	60 min
Lower jejunum	6.5 (6–7.2)	60 min	60 min
Upper ileum	7.2 (6.5–7.5)	60 min	60 min
Lower ileum	7.5 (7–8)	60 min	60 min
Proximal colon	5.5–6.5	4–12 hr	4–12 hr

[a]Non-disintegrating tablets.

The solubility should be measured at all of these pH values with a suitable, validated method such as shake-flask or pSol (2) at 37°C to determine whether the (envisaged) dose of the drug can be completely dissolved at all points of interest in the GI tract (see Chapter 11 for more discussion of solubility determination). Typically, this would be the upper GI pH (stomach and proximal small intestine) for immediate release (IR) products, the pH in the small intestine for enteric-coated products and, additionally for MR dosage forms intended to release over a period of six hours or more, the pH in the proximal colon.

With these data on hand, some rules of thumb can now be applied to steer dissolution efforts. A dose:solubility ratio (*D*:*R*) of less than 250 mL at all pH values of interest indicates that dissolution is very unlikely to limit drug absorption. For these highly soluble compounds, a simplified dissolution program can be followed, as outlined in the section "Development of Dissolution Tests for Products Containing Drugs with Good Solubility." If the *D*:*R* lies between 250 and 1000 mL in simple buffers across the pH range of interest, the compound is still unlikely to exhibit dissolution rate-limited absorption, but this should be confirmed by studying the dissolution of the pure compound in so-called biorelevant media (see section "Development of Dissolution Tests for Less Soluble Drugs"). At most, the compound is likely to require micronization, use of an appropriate salt form and/or addition of a small amount of surfactant to the formulation to achieve acceptable dissolution in simple buffer solutions. Further development of dissolution tests then follows the procedures outlined in the section "Development of Dissolution Tests for Products Containing Drugs with Good Solubility." Finally, if the *D*:*R* for the compound is greater than 1000 mL even in biorelevant media, it should be recognized that development of an oral dosage form is going to "require allocation of considerable resources." These three general solubility categories are depicted in Figure 1 along with the accompanying degree of dissolution-related challenge in product development.

Of course, the dose of a new drug is often not well defined early in the development process, so at this stage calculating

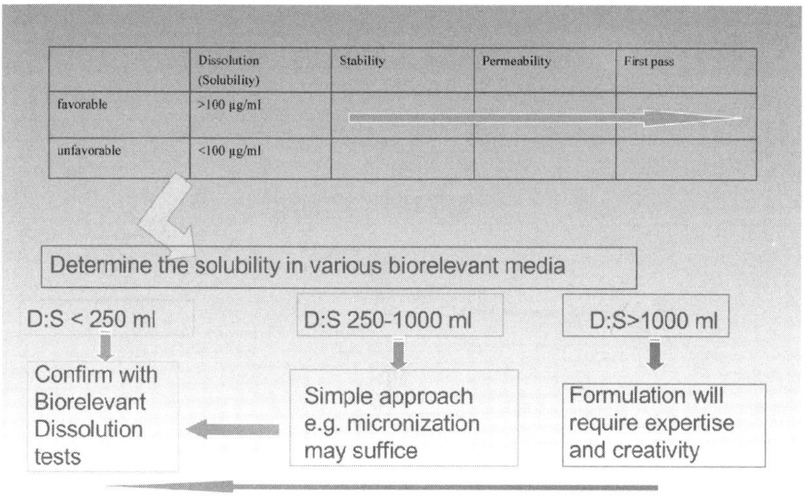

Figure 1 Using the dose:solubility ratio and solubility as a guide to assessing the level of formulation challenge.

D:S involves a lot of guesswork. An alternative to *D:S* as a yardstick for compounds still early in development is to use a solubility of 100 µg/mL as a criterion. In our experience, few compounds with aqueous solubilities >100 µg/mL across the pH range of interest exhibit dissolution problems in vivo.

As an example, data for solubility characteristics of phenoxymethylpenicillin potasasium are shown in Table 2. The solubility of phenoxymethylpenicillin is well over 100 µg/mL. However, the drug is dosed at very high levels; market products with 980.4 mg of the potassium salt are common on the European market. At this high dose, the drug just fails to meet the Biophamaceutical Classification System (BCS) specification for a highly soluble drug. However, all seven market products tested in our laboratories released > 85% of the label claim within 20 min (data for seven formulations at the 980.4 mg dose, Ref. (3)) indicating that drug dissolution is unlikely to pose a problem for either for formulation development or for bioavailability. Indeed, at a 250 mg dose (which corresponds to the WHO recommended dose) the drug would be classed as "highly soluble" according to the BCS and can be considered to belong to Class I (4).

Table 2 BCS-relevant Characteristics of Potassium Phenoxymethylpenicillin

	mg/mL	D:S ratio	BCS classification
Solubility			
SGFsp (USP 27)	1.16	~900 at $D = 980.4$ mg	High at $D = 250$ mg (WHO recommended dose), low at available market dose (980.4 mg)
Water	>10	<250	
SIFsp (USP 27)	>10	<250	
Permeability			High

A couple of words of warning about solubility experiments (see also Chapter 11):

(1) For ionizable compounds, especially salts, it is very important to check the pH of the medium before, during, and at the end of the solubility experiment when using the shake-flask method. The buffer capacity of water, often used for solubility determination, is essentially zero, so dissolution of the salt moiety can result in a huge change in the pH of the medium. Many buffers that are used in solubility experiments also have insufficient buffer capacity to withstand pH changes due to dissolution of a salt. For this reason, it is important to check the pH of the medium not only prior to adding the solute but also during and at the end of the experiment. If necessary, the pH can be adjusted to the desired value by adding NaOH or HCl, respectively. An alternative is to use the pSol approach (5) which has been shown to generate results concordant with the shake-flask method for poorly soluble compounds (2).

(2) Use of DMSO or other organic solvents to pre-dissolve the compound is to be strongly discouraged as this may lead to a supersaturated solution or crystallization of the drug in a high-energy polymorph, both of which can lead to a crass over estimate of the true solubility and thus generate unanticipated

problems further along in the development of the compound. At some point, it will become obvious that the drug is exhibiting typical problems associated with poor solubility/dissolution such as e.g., inability to generate adequate exposure in animal-toxicity studies, difficulties to formulate parenteral solutions and problems with oral bioavailability.

Development of Dissolution Tests for Products Containing Drugs with Good Solubility

For formulation development purpose, drugs can be defined as drugs as having good solubility characteristics (i.e., dissolution is unlikely to be rate-limiting to absorption) when $D{:}S < 1000\,\text{mL}$ across a pH range of approximately 1–7 in simple buffer solutions and $D{:}S < 250\,\text{mL}$ in biorelevant media. *For these compounds, it is often possible to use the same dissolution test procedure throughout the product life cycle*. Exceptions to this rule of thumb would include development of a completely different type of dosage form such as an orally disintegrating dosage form ("flash tab"), enteric-coated dosage form, MR product etc. The most appropriate dissolution apparatus for IR products of compounds with good solubility is the paddle tester (USP Type 2).

Dissolution of the Pure Compound

After establishing that the solubility is appropriately high over a pH range of approximately 1–7 in simple buffer media, the next step is to verify that the dissolution of the pure drug powder is rapid at a pH values of about 2 and 6.5, typical of the gastric and small intestinal pH, respectively, in young, healthy subjects (i.e., those with the same GI characteristics as the subjects who will be later enrolled in bioavailability/bioequivalence studies). This test can be simply performed by sprinkling the (envisaged) dose on 500 mL of pre-warmed medium in the paddle apparatus and starting the test. A suitable set of test conditions is given in Table 3.

If dissolution of the pure drug powder is complete in 10–15 min in both media, this is an indication that any well-designed IR formulations (powder, granule, tablet, capsule

Table 3 Suitable Dissolution Test Methods for Compounds with Good Solubility[a]

Parameter	Setting
Apparatus	Paddle
Volume of dissolution media	500 mL
Degassing	Degassing if needed
Dissolution media	(1) Phosphate standard buffer pH 6.8 TS (3rd Ph Int Vol. 1:196) or simulated intestinal fluid, pH 6.8 without pancreatin (USP 27) (2) 0.01 N HCl plus sodium chloride 0.2%
Agitation	75 rpm
Temperature	37°C
Sampling times	10, 15, 20, 30, 45, 60 min (also 90 and 120 min if necessary to complete release)

[a]Defined in the section "Getting Started: Solubility and the Dose: Solubility Ratio" for formulation development purposes.

etc.) should be able to achieve 85% release of the labeled content within 30 min under similar test conditions. Failure of the pure powder to completely dissolve within 15 min or great variability among samples in the % dissolved at 15 min may indicate that the drug has some wetting problems that should be addressed during formulation (see the section "Getting started: Solubility and the Dose:Solubility Ratio" for one or two suggestions).

Choice of Dissolution Tests to Compare Formulations During Development

The same test conditions used for the pure drug powder can now be used to compare formulations. The dissolution characteristics of potassium phenoxymethylpenicillin and several IR formulations of this drug that are available on the German market were compared, along with the dissolution of the pure drug powder (Fig. 2) at both low and almost neutral pH. The results show that dissolution is formulation-dependent. For the formulations tested, dissolution from some was virtually

(A)

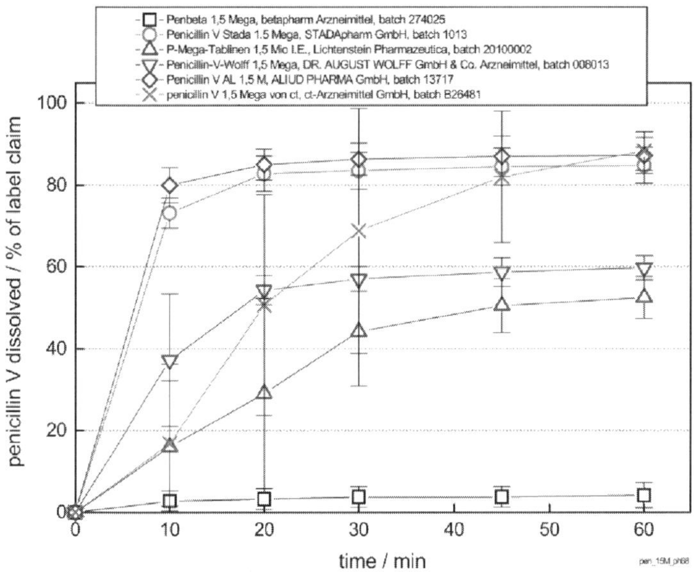

(B)

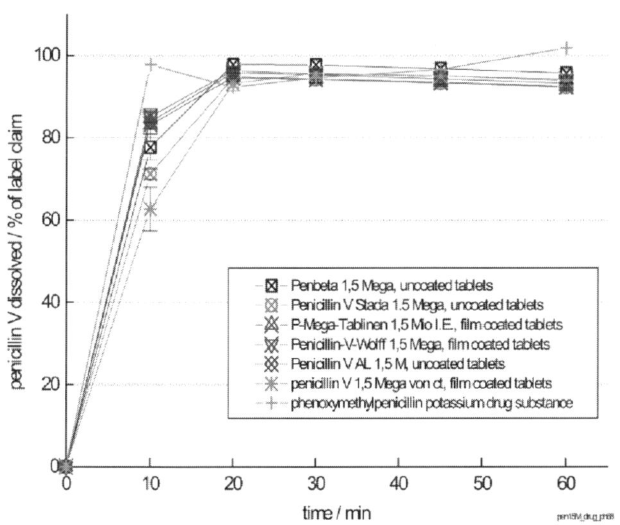

Figure 2 Dissolution characteristics of potassium phenoxymethylpenicillin pure drug and several formulations available on the German market according to the test conditions given in Table 3: (A) At acid pH (the pH used to generate the data shown here was pH 1.2 rather than pH 2 as indicated in the Table) and (B) at pH 6.8.

identical to dissolution of the pure drug at both pH values, indicating that the excipients and processing have no negative impact on dissolution. In other cases, dissolution from the product was very slow at low pH. Comparison with the profile for the pure drug indicates that slow release can be definitively attributed to the formulation rather than the drug itself.

In one case, the formulation barely released any drug under the pH 1.2 condition. This could be traced back to the disintegration behavior, as little or no disintegration was observed at the low pH. Subsequently, a full-change method was used to determine whether exposure to low pH would harm release at pH 6.8 (Fig. 3). As can be seen from the graph, release was almost as complete when tested after exposure to pH 1.2 for an hour as when the tablet was placed in a pH 6.8 medium from the outset. These results underscore the

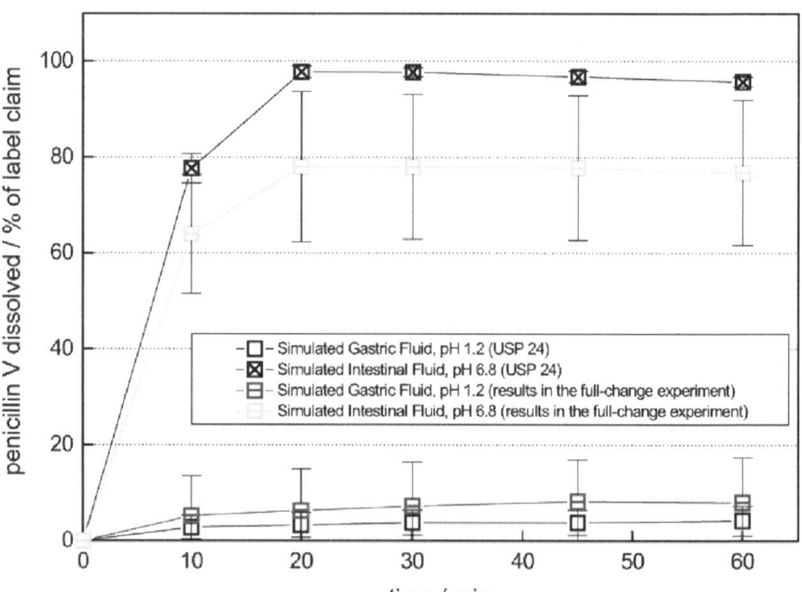

Figure 3 Full-change method to determine whether poor disintegration at pH 1.2 would adversely affect subsequent dissolution behavior at pH 6.8.

need to observe the dissolution process closely during development, as recommended in Chapter 2.

In general, it is preferable to choose excipients and processes for IR dosage forms that do not result in a formulation that requires a particular pH to function well. In the general population, the pH in the stomach is quite variable (see the subsection "Choice of Dissolution Test Conditions for Quality Control") and there is no guarantee that the dosage form will be exposed to acid, so dosage forms that require acid to facilitate release are unlikely to perform robustly in the clinical practice setting.

Another reason to avoid highly acidic conditions for QC purposes is that many drugs show poorer stability in this range than at near neutral pH, due to acid catalysis of the decomposition reaction (e.g., acid-catalyzed hydrolysis). An exception might be compounds that undergo oxidation: these compounds are usually stable at acid pH but start to decompose more quickly in the near neutral to basic region.

Choice of Dissolution Test Conditions for Quality Control

As a quality control test, a test at near-neutral pH (e.g., either of the pH 6.8 test media described in Table 3) is generally to be preferred over a test under low pH conditions. As alluded to in the previous section, gastric pH is elevated in several significant subpopulations. Examples include patients receiving H_2-receptor antagonist or proton pump inhibitor therapy, a subgroup of the elderly (variously estimated as 10–20% in the Western countries, with an incidence of over 50% in the Japanese elderly) as a result of an asympomatic decrease in gastric acid secretion with aging, and also in some pathological conditions e.g., in advanced AIDS patients. So it is unlikely that the drug product would experience a low pH environment in all those who receive the medication. Further, since gastric emptying time is highly variable (gastric emptying time in the fasted state is highly dependent on the so-called IMMC (interdigestive migrating motility cycle) and can vary from just a few minutes to over an hour depending on the motility pattern at the time of ingestion and the volume of fluid ingested with the dosage form, (6), adequate contact time to dissolve the drug product in the stomach cannot be guaranteed. By contrast, the vast majority of humans have a small intestinal pH in the range of 6–7 (see Table 1) and the residence time in the small intestine is consistently in the range 2–5 hr, providing a reliable environment for dissolution of the drug from the IR dosage form.

Scale-up and Formulation Changes, Generic Formulations

For IR dosage forms of highly soluble drugs, it is likely to be difficult to produce batches with widely enough varying dissolution characteristics to be able to establish an IVIVC (see Chapter 10). This is because of the need to have side-batches whose dissolution and absorption rates vary by at least 10% (each side of) the batch of interest, typically the pivotal batch or the marketed product.

However, in many cases a biowaiver, based purely on a comparison of the dissolution characteristics of the product, can be achieved for IR products containing highly soluble drugs. The reader is referred to the Food and Drug Administration (FDA) guidances (1,7,8) for more details about the role of dissolution testing in scale-up and postapproval changes on the one hand and approval of generic drug products (multisource products) on the other hand. It should be also noted that the WHO is in the process of updating its guidelines on registration requirements to establish interchangeability of multisource products and the new guidelines, which are considerably more flexible in terms of biowaivers (product approval without need for a pharmacokinetic determination of bioequivalence), should be available in 2005 (9).

According to the FDA guidances, if the drug is sufficiently highly soluble and permeable, and dissolution of the drug from the reference and test products occurs to an extent of 85% of label strength or better within 30 min in three media (pH 1.2, 4.5, and 6.8 are currently recommended), this is viewed as adequate proof of bioequivalence, provided the products are also pharmaceutically equivalent: same drug (i.e., active pharmaceutical ingredient), same dose, same dosage form type.

Note that a choice of pH 6.8 test conditions for quality control assures that at least one of these three criteria will be met by the product, thus harmonizing quality control measures with biopharmaceutical tests for bioequivalence.

Development of Dissolution Tests for Less Soluble Drugs

Less soluble drugs are defined for the purposes of this chapter as those for which the D:S is > 250 mL at some pH between 1 and 7, even in biorelevant media. However, it would be unwise to simply lump all less soluble drugs together: features of the molecule such as lipophilicity, ionization at physiological pH, and crystal lattice energy (melting point) can all significantly affect the magnitude of the solubility/dissolution problem and the ease with which appropriate dissolution methods can be developed. That said, this section is arranged in subsections which reflect the physicochemical properties of the compounds, in increasing degree of difficulty from the point of view of developing both formulations for oral delivery and appropriate dissolution tests for these formulations.

Solubility and Dissolution of the Pure Compound

The first step is to assess the solubility and dissolution characteristics of the pure drug in biorelevant media which cover the usual pH range in the GI tract. Some useful compositions are shown in Table 4.

The composition of fasted state simulated gastric fluid (FaSSGF) is similar to that of simulated gastric fluid without pepsin (SGFsp) (USP 27), the composition of which is provided in the table as a reference. However, the pH of FaSSGF is closer to average values of gastric pH observed in the literature (according to a survey of over 20 studies published on the subject) in the fasted state and a minor amount of a non-ionic surfactant (Triton X 100) has been, added to lower the surface tension to that observed in aspirated human gastric juice

Table 4 Some Useful Media For Preparation and Use as Biorelevant Media

FaSSGF pH 1.8	
Sodium chloride	2 g
Hydrochloric acid conc.	3g
Triton X 100	1 g
Deionized water *qs ad*	1 L
Blank FaSSIF pH 6.5	
$NaH_2PO_4 \times H_2O$	3.438 g
NaCl	6.186 g
NaOH	0.348 g
Deionized water *qs ad*	1 L
Blank FeSSIF pH 5.0	
Glacial acetic acid	8.65 g
NaCl	11.874 g
NaOH pellets	4.04 g
Deionized water *qs ad*	1 L
SCoF pH 5.8	
1 M Acetic acid	170 mL
1 M NaOH	157 mL
Deionized water *qs ad*	1 L
SGFsp pH 1.2	
Sodium chloride	2 g
Hydrochloric acid conc.	7 g
Deionized water *qs ad*	1 L
FaSSIF	
Sodium taurocholate	1.65g
Lecithin	0.591g
Blank FaSSIF *qs ad*	1 L
FeSSIF	
Sodium taurocholate	8.25 g
Lecithin	2.954 g
Blank FeSSIF *qs ad*	1 L

(35–50 mN/m e.g. (10). Alternatively, Vertzoni et al. (11) have proposed that the surface tension could be lowered appropriately with a combination of pepsin and very low concentrations of bile salts (11).

A composition for the upper small intestine in the fasted state (FaSSIF) is presented, as well as the buffer ($FaSSIF_{blank}$) solution which forms the basis of this medium. In order to precisely assess the effect of bile salts on solubility and dissolution of the drug substance, results in $FaSSIF_{blank}$ and FaSSIF should be compared. Analogous compositions are also presented for the fed state in the upper small intestine ($FeSSIF_{blank}$ and FeSSIF). The preparation of these media has been described in the literature, most recently by Marques (12).

To simulate conditions lower in the small intestine, media are buffered at higher pH values and contain progressively lower concentrations of bile salts (see also the section "Dissolution Test Design for MR Products"). These compositions reflect the active re-absorption of bile salts from the ileum, a process which is about 95% efficient, and the trend to higher pH as one moves further away from the pylorus.

Due to fermentation of hitherto undigested carbohydrates by the cecal and colonic bacteria (the large bowel contains concentrations of bacteria of up to 10^{10}–10^{12} bacteria/mL), the pH in the proximal colon is usually lower than that of the ileum. This is reflected in the composition of SCoF, which is essentially an acetate buffer. The use of acetate is appropriate as it is known that the products of carbohydrate fermentation include very short chain acids (acetate, propionate, and butyrate are typical).

To challenge the ability of MR dosage forms to resist exposure to high ionic strength, the ionic strength of any of the above-mentioned media can be increased, typically with sodium chloride in the first instance. However, it must be said that the osmolarity in the GI tract rarely falls outside the range 50–600 mOsm/Nm and that if this range is exceeded an artefactual discrimination may result.

Dissolution Tests for Weak Acids with Borderline Solubility Characteristics

In addition to potassium phenoxymethylpenicillin (aqueous solubility >10 mg/mL except at low pH), which just fails to meet the BCS criteria for "highly soluble" at higher doses, there are numerous other examples of compounds which are unable to meet the criteria at low pH but which fall well within the required *D:S* range at typical pH in the small intestine. Notable examples include ibuprofen and indomethacin, both carboxylic acids used orally as anti-inflammatory agents, furosemide, nitrofurantoin, and hydrochlorothiazide. These have all been classified as class II or IV drugs according to the current FDA guidance criteria (4).

Although the pure drug form of compounds such as these may dissolve more slowly than their "true Class I" counterparts, it is relatively easy to formulate products from which they can dissolve quickly at pH values typical of the small intestine by using standard formulation techniques such as micronization or addition of small amounts of surfactants (sodium lauryl sulfate is a popular choice) to the formulation.

A typical example is ibuprofen. The BCS-relevant characteristics of the drug are given in Table 5. Obviously, there will be little or no dissolution of ibuprofen under typical gastric conditions in the fasted state. However, the *D:S* falls almost within the BCS limit of <250 mL at pH 6.8, so it can be assumed that dissolution into a standard volume of medium (e.g., 500 mL, as recommended in Table 3) can be completed. This assumption is borne out by the results for dissolution of the pure drug and several IR oral drug products available on the European market as shown in Figure 4.

Whereas the pure drug goes into solution slowly over a period of about one hour, all of the formulations release the drug quickly. This phenomenon is likely due to the fact that poor wetting characteristics of the substance are overcome by the use of surfactants or hydrophilic excipients in the formulation. Since the high permeability of ibuprofen in the small intestine reduces any bioavailability risks associated with a slightly slower rate of release, and since gastric emptying is

Table 5 BCS-Relevant Characteristics of Ibuprofen

	mg/mL	*D:S* ratio	BCS classification
Solubility			
SGFsp (USP 27)	0.037	~21,600	Low
Purified water	0.089	~8,900	
SIFsp (USP 27)	2.472	323 mL	
Permeability			High

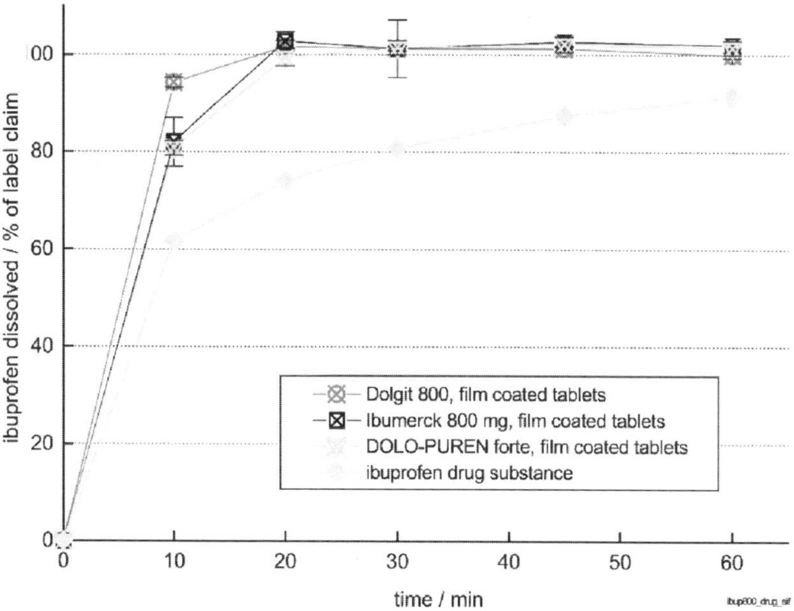

Figure 4 Dissolution of ibuprofen from the pure drug and several formulations available on the European market under the pH 6.8 test conditions shown in Table 3.

likely a further factor which influences the pharmacokinetic profile, it is unlikely that small variations in release rate would be expressed as changes in the bioavailability of the drug product. So it could be reasonably argued that to allow biowaivers (see Chapter 11) for IR products of poorly soluble, acidic compounds, they would need to exhibit similar pH/solubility and pH/dissolution behavior to that of penicillin V or ibuprofen.

Dissolution Tests for Neutral Compounds and Weak Acids with Very Poor Solubility Characteristics

For even less soluble, weak acid drugs, the situation is not so simple, because the solubility even in biorelevant media is very low. A typical example is troglitazone, an antidiabetic drug previously marketed by GlaxoWellcome. The BCS-relevant characteristics of troglitazone are shown in Table 6.

In this case, the solubility is extremely poor, even at pH 7, which is considerably above the pKa of troglitazone and corresponds to pH values commonly found in the mid section of the small intestine. Other well-known compounds with analogous behavior are mefenamic acid, glyburide, and phenytoin. For troglitazone, the presence of bile salts improves the solubility quite dramatically and lipophilic constituents in the dissolution medium (e.g., in full-fat milk) lead to better dissolution, and in turn better absorption when troglitazone is administered in the fed than the fasted state, as reported by Nicolaides (13). Use of biorelevant dissolution testing permitted these authors not only to qualitatively predict the food effect, but also to predict relative bioavailability of three test formulations.

When administered in the fasted state, poorly soluble, neutral drugs actually behave quite similarly to very poorly soluble, weakly acid drugs, in that the main site of dissolution is often the small intestine—due to the longer residence time there compared with the residence time in the stomach (Table 1). Only when they are highly lipophilic and can be incorporated in the lipid part of the meal and/or solubilized by mixed micelles in the small intestine are these compounds likely to dissolve quickly enough in the upper GI tract to effect good oral bioavailability. As a result of longer gastric residence, presence of lipids and their digestive products as well as high bile concentrations, these compounds often show positive food effects i.e., the bioavailability increases when they are

Table 6 BCS-Relevant Characteristics of Troglitazone (pKa 6.1)

	μg/mL	*D:S* ratio	BCS classification
Solubility			
pH 7	1.7	~117 L	Low
FaSSIF	70	~2.85 L	
FeSSIF	300	670 mL	
Permeability			High

administered with food. A typical case is danazol, used in the therapy of endometriosis, the bioavailability of which increases three-fold when administered with a meal (Fig. 5). These results can be simulated by dissolution in biorelevant media simulating the fasted and fed states (14).

In such cases, it is obviously advantageous to use biorelevant dissolution tests to characterize the drug substance, to compare formulations and to make a preliminary assessment of possible food effects. However, for routine quality control work, the manufacture of media containing bile components is not only rather time-consuming but may also present difficulties in terms of quality assurance and validation of the raw materials, as is the case with many chemicals obtained from natural sources.

A reasonable way to proceed is to determine the concentration at which a well-defined surfactant (e.g., sodium lauryl sulfate or Tween 80) produces the same $D:S$ ratio as the physiological concentration of bile components. Dissolution

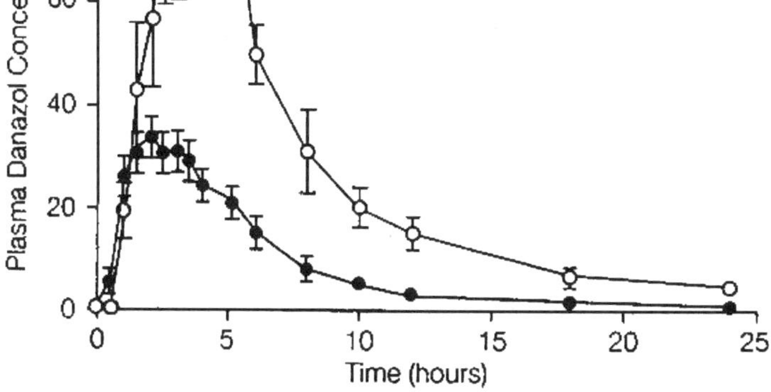

Figure 5 Bioavailability of danazol in the fasted and fed state. Open circles represent fasted state administration and closed circles fed state administration. *Source*: From Ref. 16.

is then performed in a buffer containing this concentration of the surfactant to assess whether the dissolution profile can be matched to that in the bile component-containing medium in terms of rate and extent of dissolution and the form of the dissolution profile (this can be determined by the application of the Weibull function to the results, see Chapter 8). If this is the case, dissolution is run again at a surfactant concentration which corresponds to, but does not exceed, sink conditions for the compound (defined as the conditions in which the final concentration of the drug, when the given dose has been completely dissolved, corresponds to one-third of the solubility of the drug in that medium). If the dissolution curve is still homomorphic (has the same general shape characteristics) to that in the medium containing physiological concentrations of bile components, use of this medium for quality control purposes can be justified. Especially useful would be the development of an IVIVC in this medium (see Chapters 8–10).

It should be noted that this procedure needs to be carried out on a case-by-case basis—there is no indication that the relative solubilization capacity (ability of bile components or surfactants to enhance solubility/dissolution of a drug) is consistent from drug to drug. Therefore, use of a "standard" medium containing a synthetic surfactant to correspond to either FaSSIF or FeSSIF results is not possible.

Dissolution Tests for Poorly Soluble Weak Bases

The dissolution of poorly soluble, weakly basic drugs in the GI tract is somewhat more complicated to simulate owing to the variability in gastric conditions. The pH is likely to be the greatest influence on solubility since the influence of the pH on solubility is exponential whereas the effects of bile components on solubility are linear. Therefore, even a modest change in pH can create an orders of magnitude change in solubility whereas it takes a substantial increase in bile output to have a pronounced effect on solubility. The influence of pH on solubility is exemplified by the data shown in Figure 6.

Now, theoretically, since the gastric pH tends to be low in the fasted state, one might be tempted to assume that the

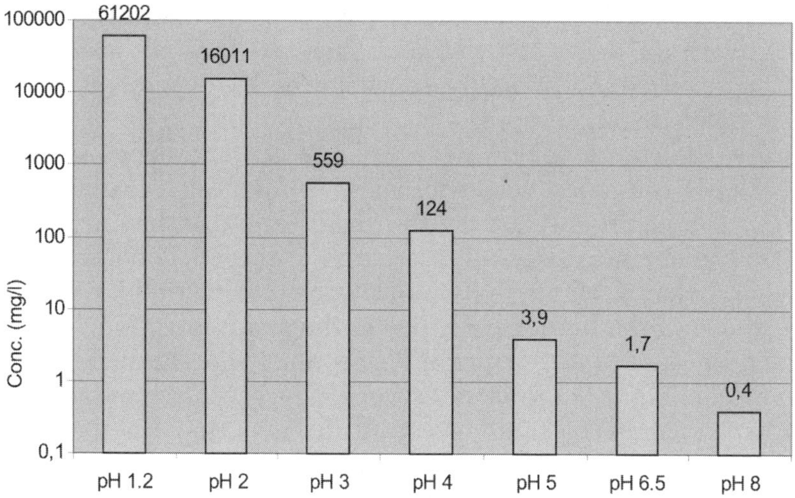

Figure 6 Typical solubility behavior for a poorly soluble weak base as a function of pH. The intrinsic solubility is 0.4 μg/mL. At pH values typical of the small intestine, solubility is minimally better than the intrinsic solubility (solubility of the free base form) but at gastric pH (~2) the solubility is about 16 mg/mL.

drug will go quickly into solution at this pH and be readily absorbed from the GI tract. The flaws in this argument are the following:

(a) First, as mentioned earlier in this chapter, gastric residence time in the stomach in the fasted state is quite variable, so an adequate residence time cannot be guaranteed for the dissolution of a poorly soluble weak base.

(b) Second, not all poorly soluble weak bases are soluble enough in gastric juice to effect complete dissolution, even if the gastric residence time is on the order of a half- to one hour. An example is itraconazole, with a solubility of 1.8 μg/mL even at pH values as low as pH 1.2.

(c) Third, gastric pH is not always as acidic in patient populations as in young, healthy volunteers. *Helicobacter pylori* infection is widespread and often leads to elevations in gastric pH. Certain populations tend towards hypo- or even achlorhydria with aging—this is well documented in the Japanese population with more than half of elderly Japanese hypo-to achlorhydric, but less prevalent in the Western countries with incidence calculated at about 10–20% of those over 60 years of age. Additionally, sales figures for the major gastric acid-blockers (H_2-receptor antagonists and proton pump inhibitors) indicate a very widespread use of these drugs in the developed countries, with subsequent influence on gastric pH.

(d) Fourth, only a very few drugs are absorbed directly from the stomach (ethanol being one of these). Thus, for the great majority of poorly soluble weak bases, there will be exposure to the higher pH fluids of the small intestine before the drug arrives at the site of absorption. The solubility data shown in Figure 6 illustrates very clearly the potential for precipitation in the small intestine and consequent non-availability for uptake across the mucosa.

As a result, if dissolution from formulations is studied exclusively under low pH conditions, the formulators are likely to be in for a rude shock when the results come back from the pharmacokinetic studies—poor and highly variable absorption is the order of the day for drugs that have been formulated without an eye to robustness of the release from the dosage form as a function of pH. Instead, it is recommended that a formulation be sought that can release the drug even when there is not enough acid in the stomach to provide a sufficient boost to the solubility or when the gastric residence time is short.

The Hypoacidic Stomach Model

To test the robustness of the formulation to variations in gastric pH, dissolution results should be obtained in both the pH 2 medium described in Table 3 and a model which reflects the conditions in the hypochlohydric stomach. A good choice would be acetate buffer adjusted to pH 5 and having a very low buffer capacity, since hypochlorhydria is generated by a *reduction in HCl secretion* rather than the *addition of buffer species*.

Results for the release of the drug whose solubility is depicted in Figure 6 from two differently constituted formulations in SGFsp at pH 1.2 and an acetate buffer at pH 5 are shown in Figure 7. The discrimination with respect to robustness of release using the higher pH medium is clearly illustrated, with formulation B exhibiting a release profile that is virtually independent of the pH of the dissolution medium.

Comparing only the results at low pH, one would expect both formulations to perform equally in the clinic. However, as would be expected from the dissolution profiles at *both* pH values, formulation B produced far less variability of absorption in the clinical studies and was also better absorbed than formulation A. This example illustrates clearly the value of the hypochlorhydic model for screening formulations prior to taking them into the clinic.

The Transfer Model

The transfer model (15) can be used to answer the question of whether the drug is successfully released in the stomach, only to precipitate when it moves into the higher pH environment of the small intestine. As depicted in Figure 8, the pure drug (or formulation) is added to a gastric simulating medium at time zero, after which it is allowed to dissolve and simultaneously transferred into a second vessel containing FaSSIF or other suitable biorelevant medium.

Figure 9 shows results from a typical run in which precipitation occurs after a certain concentration is reached in the receptor medium. The solid line shows how the concentration would climb in the receptor medium in the absence of precipitation. The curve with the error bars shows the actual concentrations measured by taking samples and analysing them for dissolved drug. The discrepancy between the two curves can be attributed to precipitation, which also becomes visually obvious after some time. Especially interesting for the prediction of the likelihood of precipitation in vivo is the horizontal dotted line. This corresponds to the solubility of the compound in the receptor medium (in this case FaSSIF), clearly indicating that a substantial supersaturation can be reached in the presence of even rather low concentrations of bile salts and lecithin. It is hypothesized that the bile components serve as nucleation inhibitors thus facilitating high concentrations of drug in the small intestine which, of course, is very favorable for drug absorption.

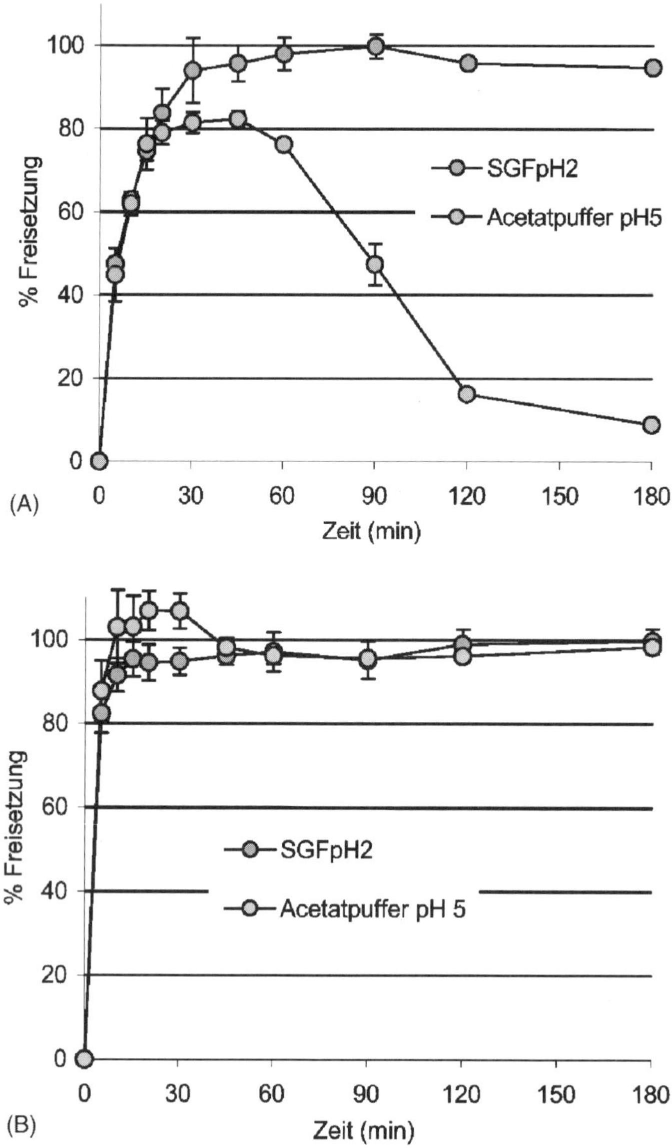

Figure 7 Behavior of two formulations of a poorly soluble, weakly basic drug (solubility characteristics shown in Figure 6) in media composed at two pHs—one to represent acidic conditions in the stomach, the other to represent the hypochlorhydric stomach. (A) Formulation with non-robust dissolution characteristics and (B) Formulation with robust dissolution characteristics.

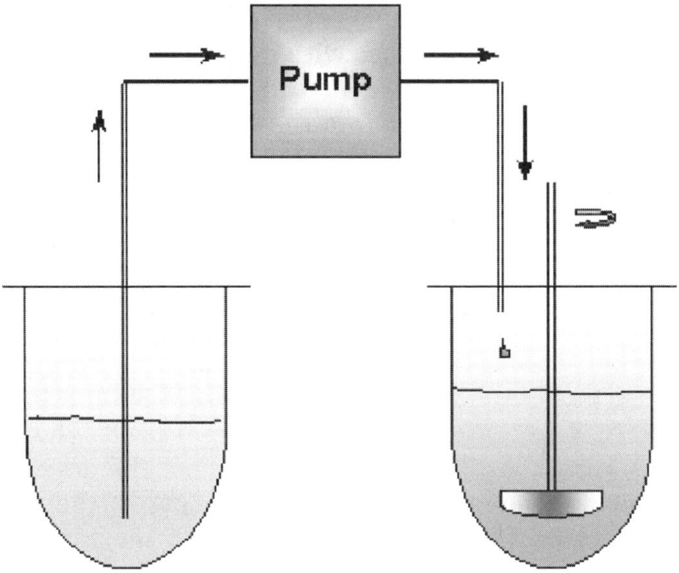

Figure 8 Transfer model for poorly soluble, weakly basic drugs.

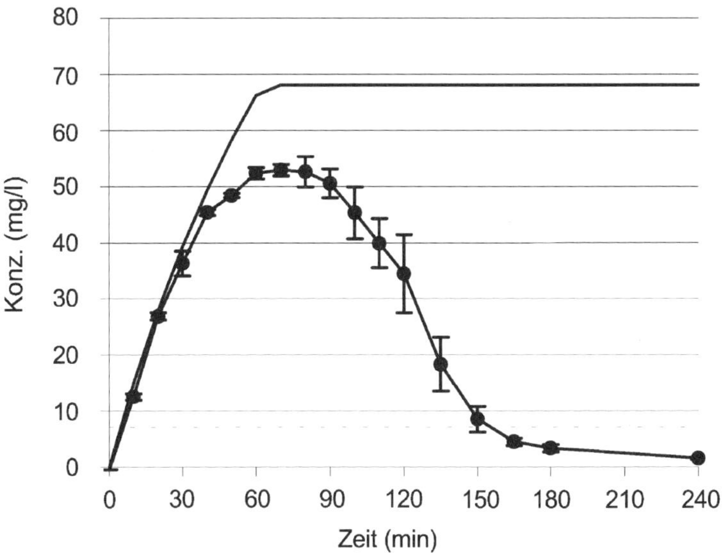

Figure 9 Typical results observed during the transfer of a poorly soluble, weak base from an acidic medium to FaSSIF.

In summary, use of biorelevant media to determine solubility in the upper gut combined with assessment of formulations with respect to robustness and ability to protect the drug from precipitation are key to an efficient development process for compounds that are poorly soluble and weakly basic.

Dissolution Test Design for MR Products

A quick look through the standard USP dissolution tests for dosage forms with modified release suggests they have been developed primarily with a view to facilitate quality control procedures and little attention has been given to simulating GI conditions. In many cases, just one medium is used, which is in quite stunning contrast to the experience of the dosage form as it moves through the different segments of the GI tract. In these tests, the most commonly used medium is (inexplicably) dilute acid (e.g., SGFsp or simple dilutions of HCl), others use water. These media can hardly be accused of simulating the lumenal environment throughout the passage of the dosage form through the GI tract. The use of single media to attempt IVIVC for MR dosage forms probably explains why many attempts at IVIVC have been unsuccessful. In fact, single media are only likely to predict in vivo release from an MR dosage form when the mechanism governing the release is extremely robust to the changing physiological GI environment and the drug itself is highly soluble over the complete GI pH range. Although many osmotic pump formulations can meet these requirements, for most other mechanisms of release the single medium approach is likely to at best result in a correlation with poor robustness to variations in formulation and may lead to no correlation at all.

For some products, e.g., propanolol extended release formulations (USP 27), a modification of the standard method for enteric-coated dosage forms have been introduced to reflect the change from conditions in the stomach to those in the small intestine. This is a step in the right direction, but to achieve dissolution testing that can differentiate between formulations which are robust and those which are not, and especially to be able to predict food effects on the release from

MR products, it is necessary to simulate the passage through the GI tract somewhat more physiologically.

General Considerations: Using Compendial Dissolution Apparatus to Model GI Passage of MR Dosage Forms

To illustrate how misleading single medium tests can be with respect to release from MR products, commercially available mesalazine products were compared at pH 6.8 and 7.5. These two pH values are of interest because they represent performance at mid-jejunum and in the ileum, respectively. Since mesalazine products are intended for local action in the small intestine to treat chronic inflammatory conditions like Crohn's disease and ulcerative colitis, knowing whether the dosage form can release the drug at the site of inflammation is *necessary* to guide the development of the formulation. However, testing at just the pH of the segment targeted for release may not be *sufficient*: what if the drug is actually released at sites proximal to the targeted segment and therefore prematurely absorbed to the systemic circulation and no longer locally available to exert its anti-inflammatory effect?

Figure 10 shows the release of four commercially available products at pH 6.8 and at pH 7.5. The two formulations with the slow-release coatings tend to release mesalazine more quickly at the higher pH. The two enteric-coated products, Claversal® and Salofalk® release mesalazine abruptly after a certain lag time. At pH 6.8, this lag time is much longer for Claversal® than for Salofalk® even though the coating material is the same Eudragit type. At pH 7.5, the lag time is shorter and the same for both formulations. The single media experiments are thus able to pick up formulations differences among various formulations but it is still not evident whether the drug is released appropriately at the sites of inflammation.

Table 7 shows the "pH-gradient" sequence of media which can be used to simulate passage through the GI tract in the BioDis (USP Type 3) apparatus to help identify the sites of release of mesalazine from the various formulations.

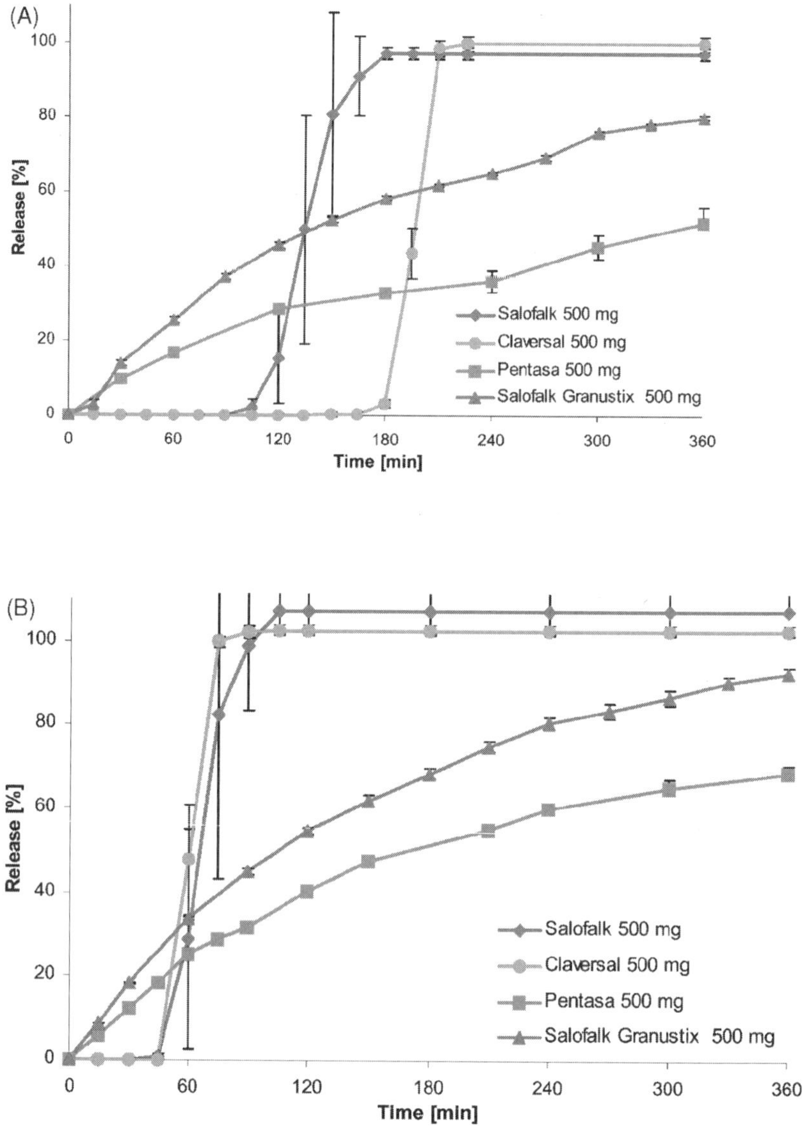

Figure 10 Release from four commercially available mesalazine products in single media. (A) pH 6.8 and (B) pH 7.5.

Table 7 The "pH-gradient" Method used to Compare Mesalazine Formulations in the BioDis (USP Type 3) Dissolution Tester

	pH	Medium	Residence time (min)	
			Tablets	Pellets
Stomach	1.80	SGFsp (mod).	60	20
Proximal jejunum	6.50	Phosphate buffer (Ph. Eur)	15	45
Distal jejunum	6.80	SIFsp (USP 25)	15	45
Proximal ileum	7.20	Phosphate buffer (Ph. Eur)	30	45
Distal ileum	7.50	SIFsp (USP 23)	120	45
Ascending colon	6.50	Phosphate buffer (Ph. Eur)	360[a]	360[a]
Transverse colon	6.50/6.80	Phosphate buffer (Ph. Eur)	240/240[a]	240/360[a]
Descending colon	6.80	Phosphate buffer (Ph. Eur)	360[a]	360[a]

[a]Residence time in the colon varies greatly.

Release results with this method are shown in Figure 11. The BioDis method enables the release pattern to be interpreted in terms of release at sites of inflammation. In Crohn's disease, the inflammation often starts at the ileocecal junction and spreads from there in the proximal and/or distal direction and may affect the entire GI tract in severe cases, whereas in colitis the inflammation is restricted to the large bowel. The release patterns in Figure 11 can be used in combination with a knowledge of the sites of inflammation in a given patient to choose the most suitable dosage form available on the market for that patient (Klein, 18).

Fed vs. Fasted State Testing—Can Meal-Related Failures of the MR Mechanism Be Detected In Vitro?

The example in the preceding section illustrates the utility of the Type 3 tester and use of sequential media to simulate

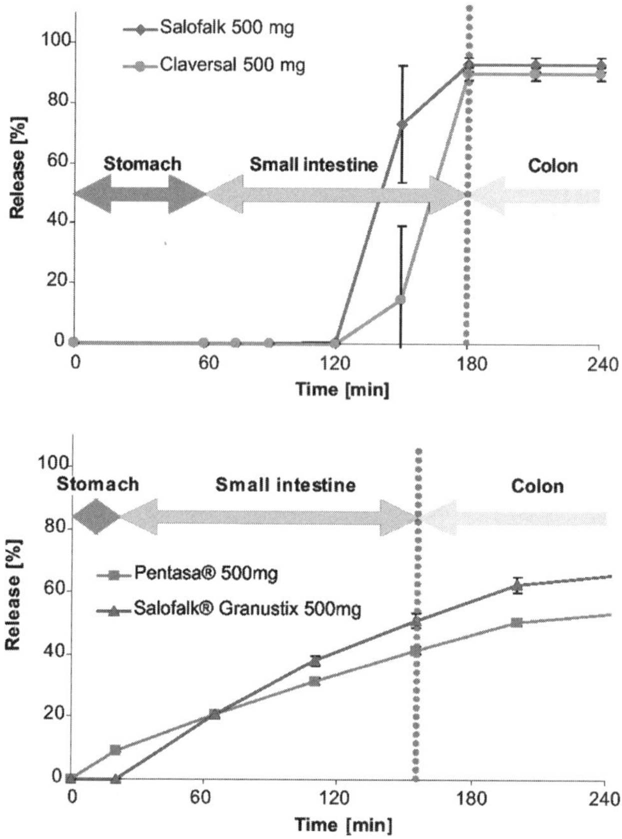

Figure 11 Comparison of release from the mesalazine formulations shown in Figure 10 using sequential media in the USP Type 3 "BioDis" apparatus.

release from enteric-coated dosage forms during passage along the GI tract. Another key question for enteric coated as well as other types of MR dosage forms is their ability to perform robustly, irrespective of whether they are administered in the fed or fasted state. Factors such as interactions with meal components, increases in gastric, bile, and pancreatic secretions and changes in the motility pattern can all play a role here.

For these purposes, one needs to be able to simulate, at least in a general way, the stomach in the fed state and also to take into account the longer upper GI passage time of non-disintegrating formulations in the fed state due to the switch in the gastric and small intestinal motility pattern from the fasted to the fed state. Possible combinations are shown in Table 8 for experiments with the Type 3 tester. The media can be combined with the passage times shown in Table 1 to arrive at a reasonable test set-up.

Data have been obtained with this set-up for several different types of MR products and the ability to predict food effects, at least on a qualitative basis, appears to be very promising. An example of a known food effect which can be simulated in vitro is that of a salbutamol MR formulation. The in vitro results are shown in Figure 12.

The release is somewhat slower under simulated fed state than under simulated fasted state conditions, which

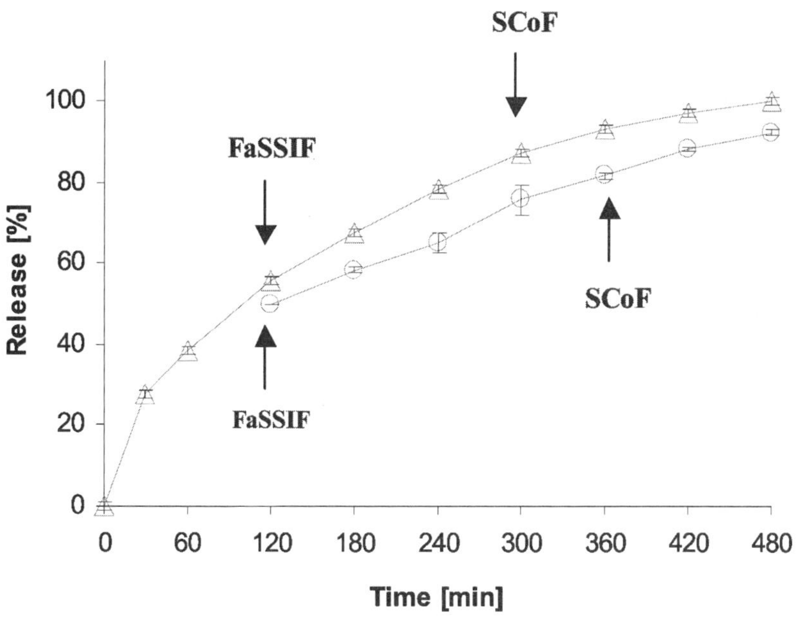

Figure 12 Comparison of release of salbutamol from an MR product under simulated fasted and fed conditions using the BioDis® (USP Type 3 tester) apparatus. Triangle corresponds to simulated fasted state and circles to simulated fed state.

corresponds to results in pharmacokinetic studies. Slower release in the fed state can be due to slower hydration of a film coating or a matrix, or inhibition of erosion, to name just a couple of possibilities. Of perhaps even greater concern would be very fast release of drug from an MR dosage form when given with food: so-called "dose-dumping." Limited results with the media set-up outlined in Table 8 suggest that these effects, too, can be predicted qualitatively in vitro with the BioDis (USP Type 3) dissolution tester using biorelevant media. As with the IR formulations of poorly soluble, weak bases, a lot of time and money can be saved in the development of an MR product if poor formulations can be weeded out prior to taking them into the clinic.

FUTURE DIRECTIONS OF BIORELEVANT DISSOLUTION TEST DESIGN

In the last 10 years, the use of biorelevant testing conditions has become standard in the characterization of new compounds and the development of formulations. With some care, they can also be used as the basis for developing appropriate quality control tests, under consideration of appropriate pH and buffer capacity, by substituting appropriate synthetic

Table 8 Biorelevant Media for Studying Food Effects on Release from MR Dosage Forms

Segment	Pre-prandial medium	Post-prandial medium
Stomach	FaSSGF	Ensure plus®
Duodenum	FaSSIF (pH 6)[a]	FeSSIF (pH 5)
Upper jejunum	FaSSIF (pH 6.5)	FeSSIF (pH 5)
Lower jejunum	FaSSIF (pH 6.8)[a]	FeSSIF (pH 6)[a]
Upper ileum	FaSSIF (7.2)[a] halved (bile components)	FaSSIF (7.2)[a]
Lower ileum	$FaSSIF_{blank}$ (7.5)[a]	$FaSSIF_{blank}$ (7.5)[a]
Proximal colon	SCoF	SCoF

[a]pH adjusted by adding sodium hydroxide or hydrochloric acid solution, as appropriate.

surfactants for the natural ones. Still, there are areas where the biorelevant media can be improved. For example, in the fed state lipid digestion products may also contribute to the solubilization of lipophilic compounds, so inclusion of lipid digestion products in the media would no doubt be of interest for prediction of fed vs. fasted state dissolution in vivo. Another continuing area of focus will be the refinement of efforts to predict food effects for MR formulations and to validate the media for various types of MR formulations (hydrogels, osmotic pumps, coated pellets etc.). In addition, the use of hydrodynamics (through changes in the dip rate in the apparatus) can be used to identify robustness of the formulation at the pylorus and ileocecal junction. All in all, we can be confident that the use of biorelevant media in formulation development will continue to expand and find new applications.

REFERENCES

1. FDA. Guidance for Industry: Waiver of In vivo Bioavailability and Bioequivalence Studies for Immediate-Release Solid Oral Dosage Forms Based on a Biopharmaceutics Classification System. Rockville MD, USA: U.S. Department of Health and Human Services, Food and Drug Administration, Center for Drug Evaluation and Research (CDER), 2000.
2. Glomme A, März J, Dressman JB. Comparison of a miniaturized shake-flask solubility method with automated potentiometric acid/base titrations and calculated solubilities. J Pharm Sci 2005; 94(1):1–16.
3. Stippler E. Development of BCS-conform Dissolution Testing Methods. Dissertation thesis, University of Frankfurt, 2004.
4. Lindenberg M, Dressman J, Kopp S. Classification of orally administered drugs on the WHO “Essential Medicines” list according to the BCS. Eur J Pharm Biopharm 2004; 58: 265–278.
5. Avdeef A, Berger CM, Brownell C. pH-metric solubility. 2: correlation between the acid–base titration and the saturation shake-flask solubility–pH methods. Pharm Res 2000; 17:85–89.

6. Oberle R, Chen T-S, Lloyd C, Barnett J, Owyang C, Meyer J, Amidon G. The influence of the interdigestive migrating motility complex on the gastric emptying of liquids. Gastroenterology 1990; 99:1275–1282.

7. FDA. Guidance for Industry: SUPAC-IR Immediate-Release Solid Oral Dosage Forms: Scale-Up and Post-Approval Changes: Chemistry, Manufacturing and Controls, In Vitro Dissolution Testing, and In Vivo Bioequivalence. Rockville MD, USA: U.S. Department of Health and Human Services, Food and Drug Administration, Center for Drug Evaluation and Research (CDER), 1995.

8. FDA. Guidance for Industry: Bioavailability and Bioequivalence Studies for Orally Administered Drug Products—General Considerations (Revised) (I). Rockville MD, USA: U.S. Department of Health and Human Services, Food and Drug Administration, Center for Drug Evaluation and Research (CDER), 2003.

9. World Health Organization. www.who.int.

10. Kalantzi L, Fürst T, Abrahamsson B, Goumas K, Kalioras V, Dressman J, Reppas C. Characterization of the human upper gastrointestinal contents under conditions simulating bioavailability studies in the fasting and fed states. Proceedings of the AAPS Annual Meeting, Salt Lake City, UT, 2003.

11. Vertzoni M, Dressman J, Reppas C. Dissolution testing in media simulating the gastric composition in the fasted state. Proceedings of the AAPS Annual Meeting, Toronto, Canada, 2002.

12. Marques M. Dissolution Media Simulating Fasted and Fed States. Dissolution Technol 2004; 11:16.

13. Nicolaides E, Symillides M, Dressman JB, Reppas C. Biorelevant dissolution testing to predict the plasma profile of highly lipophilic drugs after oral administration. Pharm Res 2001; 18(3):380–388.

14. Galia E, Nicolaides E, Hörter D, Löbenberg R, Reppas C, Dressman JB. Evaluation of various dissolution media for predicting in vivo performance of Class I and II drugs. Pharm Res 1998; 15:698–705.

15. Wunderlich M, Kostewicz E, Becker R, Brauns U, Dressman JB. Transfer model for the precipitation of weak bases in the gastrointestinal tract. J Pharm Pharmacol 2004; 56:43–51.

16. Charman W, Rogge M, Boddy A, Barr W, Berger B. Absorption of danazol after administration to different sites of the gastrointestinal tract and the relationship to single- and double-peak phenomena in the plasma profiles. J Clin Pharmacol 1994; 33:1207–1212.

17. United States Pharmacopeia. (USP 27). Rockville, MD: United States Pharmacopoeia Convention, Inc., 2004.

18. Klein S, Rudolph M, Dressman JB. Drug release characteristics of different mesalazine products using USP apparatus 3 to simulate passage through the GI tract. Dissolution Technol 2002; 9:6–12.

8

Orally Administered Drug Products: Dissolution Data Analysis with a View to In Vitro–In Vivo Correlation

MARIA VERTZONI,
ELEFTHERIA NICOLAIDES,
MIRA SYMILLIDES, and
CHRISTOS REPPAS

Laboratory of Biopharmaceutics & Pharmacokinetics, National & Kapodistrian University of Athens, Greece

ATHANASSIOS ILIADIS

Department of Pharmacokinetics, Mediterranean University of Marseille, France

DISSOLUTION AND IN VITRO–IN VIVO CORRELATION

In vitro–in vivo correlation (IVIVC) is a general term that refers to a relationship between a biological property produced by a dosage form and a physicochemical characteristic of the same dosage form (1). Establishment of an IVIVC could facilitate drug development by reducing the number of in vivo studies required for confirming either the safety and the efficacy of a drug product or the bioequivalence of products containing the same drug.

For drug products intended for systemic activity, the biological property produced by the dosage form is usually assumed to be related to the presence of the drug in the systemic circulation, i.e., the pharmacokinetic profile. As the elimination process is generally not affected by the dosage form, the arrival process of the drug into the general circulation is likely to govern the degree to which the biological property is produced by the dosage form. On the in vitro side, dissolution [or release, in case of products with extended-release (ER) characteristics] or some characteristic(s) of this process are the most frequently in vitro variables used to generate an IVIVC.

When Is It Possible to Forecast the In Vivo Behavior of an Orally Administered Product from In Vitro Dissolution Data?

Dissolution (or release) is the main process that limits the supply of the gastrointestinal (GI) fluids with the drug but only one of the processes that lead to the appearance of the drug into the systemic circulation (2). Therefore, in principle, there are three possibilities (3). The first is that dissolution has no practical influence on the arrival of the drug into the general circulation. For example, substances with low dose-to-solubility (D:S) ratio will exhibit fast and complete dissolution within a few minutes after administration of an immediate-release (IR) dosage form. A second possibility is that the arrival of the drug in the general circulation is limited by more than one process, including dissolution. This applies, for example, to substances with low-solubility and low-permeability properties. The third possibility is that dissolution is the only process that limits the arrival of the drug in the systemic circulation. Examples include drugs with little or no stability problems in the GI lumen (3) or first-pass metabolism, which are either of low solubility or housed in ER dosage forms.

Development of a robust IVIVC is possible when absorption is limited by lumenal dissolution, provided lumenal dissolution (or release) is adequately simulated in vitro.

If not in an ER product, a drug is likely to exhibit dissolution-limited absorption if it is poorly soluble in the GI lumen. Usually, identification of a compound with dissolution-limited GI absorption is based on D:S ratio (4); when D:S is about < 250 mL over the pH range of 1–7.5, the compound is usually considered to have less than ideal lumenal dissolution characteristics (3,5), with 250 mL being a conservative estimate of the total volume of fluids that will be in contact with the dose in the upper GI tract under fasting conditions. However, this approach has several weaknesses:

i. early in drug development, the dose is often unknown;
ii. a 250 mL cutoff may be too conservative, especially for fed-state conditions (6);
iii. consideration of only pH and volume effects only may lead to incorrect classification of some lipophilic substances as poorly soluble compounds that, in presence of naturally occurring solubilizing agents, would be classified as highly soluble substances; and
iv. compounds with low doses may be incorrectly classified as highly soluble; for example, digoxin has D:S ~ 21 mL (3), but this drug is known to exhibit a particle size-dependent absorption (7).

Therefore, early in drug development, the definition of a compound that is poorly soluble in the GI lumen might be better based on its solubility characteristics in biorelevant media. This is similar to the procedure that Pharmacopeias worldwide suggest for assessing the ability of a compound to dissolve in a given solvent (1). In cases where the dose is known, a poorly soluble drug can be more reliably identified by considering D:S under biorelevant conditions. Assessment of solubility characteristics with biorelevant media and evaluation of permeability and lumenal stability characteristics [again under biorelevant conditions (8)] will provide the basis for deciding whether or not an IVIVC (with dissolution data used as the in vitro data) is possible.

The design of a biorelevant in vitro dissolution test requires consideration of two key factors affecting the concentration along the gut wall, i.e., composition of the gut contents and hydrodynamics. Composition of the lumenal contents may affect the kinetics by affecting the dissolution rate constant or coefficient or by affecting the solubility. Hydrodynamics refers to both the type and the intensity of agitation and affects the kinetics directly. Issues relevant to the intralumenal composition and hydrodynamics are covered in detail in other chapters of this book. It should be noted, however, that as the mechanism of release from ER products is often less dependent on the local physiology (e.g., highly soluble drugs housed in osmotic pumps) than the dissolution of poorly soluble drugs from IR dosage forms, precise simulation of the lumenal environment may be of less importance when such dosage forms are considered. This, in conjunction with the fact that release occurs at slow rates, constitutes the main reason for the more facile establishment of IVIVCs for ER products than for IR dosage forms. Only recently has it been possible to obtain IVIVCs a priori for various lipophilic drugs housed in IR dosage forms, by combining dissolution data collected in various biorelevant media (9).

Approaches for Correlating In Vitro Dissolution Data with Plasma Data

IVIVCs can be divided into non-quantitative and quantitative. In non-quantitative correlations the two variables are not related to each other via a mathematical relationship. A characteristic example is the rank-order correlation that was popular in the 1970s (10–15). In quantitative correlations the in vitro variable correlates with the in vivo variable via a linear or a non-linear equation. A quantitative IVIVC can be established, with or without the framework of a model, by using estimated values of characteristic parameters of the in vitro dissolution process and estimated values of the characteristic parameters of the in vivo arrival-in-bloodstream process. Some of the parameters used in single-point correlations are presented in Table 1.

However, single-point correlations are of limited value for two reasons. The first relates to the choice of the specific parameters to be correlated. Although there are some procedures in the literature that could be used for selecting the most appropriate parameter [e.g., the quadrant analysis (16,17)], these are not easy to apply in practice and the choice is usually based on a best-result basis. Another reason is that two processes having the same value of the chosen characteristic parameter can be different in terms of their overall shape. Consequently, a quantitative IVIVC is much more informative if established using all available in vitro and in vivo raw data: these are termed multiple-point or point-to-point correlations.

Point-to-point IVIVCs can be established by using two approaches. The first approach is to establish a relationship between the actual time course of the in vitro dissolution and the time course of the lumenal dissolution or arrival into the general circulation (Fig. 1), as estimated by deconvolution of the observed concentration in the bloodstream vs. time profile. The second approach is to establish a relationship

Table 1 Parameters Used for Correlating In Vitro Dissolution with Plasma Data

In vitro parameters	In vivo parameters
Time for specific amount dissolved (e.g., 50% of the dose dissolved)	Area under the concentration- in-bloodstream vs. time curve Maximum concentration in bloodstream
Amount dissolved at a specific time point	Fraction absorbed, absorption rate constant
Mean dissolution time	Mean residence time, mean dissolution time, mean absorption time
Parameter estimated after modeling the dissolution process	Concentration at time *t* Amount absorbed at time *t*

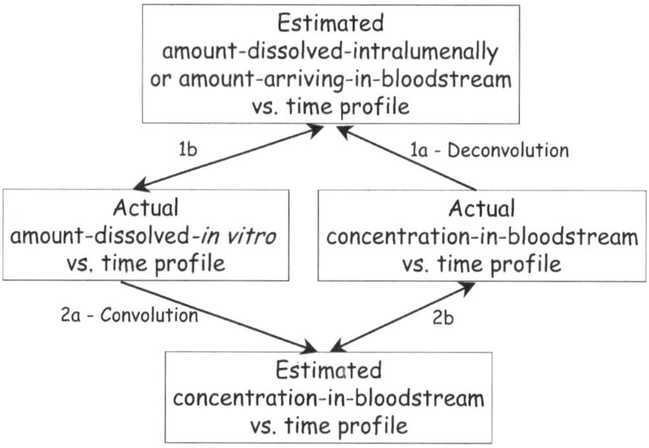

Figure 1 Schematic of the two approaches usually followed for developing a point-to-point IVIVC. Procedure 1 has two steps (a and b) and involves deconvolution of a concentration-in-bloodstream vs. time profile. Procedure 2 has also two steps (a and b) but involves convolution of a concentration-in-bloodstream vs. time profile.

between the observed time course of plasma drug concentration and the time course of plasma levels (Fig. 1) estimated by convolution of the in vitro dissolution data. To be applicable, both approaches require the availability of intravenous or oral solution data or, in case of an ER product of a highly soluble drug, oral data from a solid IR dosage form. Exceptions to this requirement are limited to cases where the entire dose reaches the general circulation, and drug absorption and disposition can be described by an open one-compartment pharmacokinetic model (18).

Regardless of the approach, a point-to-point IVIVC should be evaluated to demonstrate that predictability of in vivo performance of a drug product from its in vitro dissolution characteristics is maintained over a range of dosage forms with similar physicochemical characteristics [when IR dosage forms are considered (9)] or over a range of in vitro release rates [usually three (19)] of related formulations (when ER products of a specific drug are being considered).

At both the evaluation and the application level of a point-to-point IVIVC, in vitro dissolution data sets need to be treated and/or compared with each other. Appropriate methods vary with the data collection procedure and whether or not a model is to be fitted to the data.

ANALYSIS OF DISSOLUTION DATA SETS

In vitro dissolution data can be collected in closed systems (e.g., in compendial dissolution vessels) or by using open (flow-through) systems (1).

Closed systems are currently the more frequently used, perhaps for practical reasons, as they are not expensive and can be easily operated. A disadvantage, however, is that, apart from a few specific setups (e.g., the reciprocating disk apparatus), media changes within a single run cannot be easily performed. Open systems are less frequently used, possibly because the maintenance of specific flow rates requires the use of expensive pumps even if simple dissolution media are used. Compared with closed systems, however, flow-through systems are more useful when media changes and/or maintenance of sink conditions are required. In addition, the principle of their operation is more physiologically relevant than that of the closed systems. A major issue relevant to the analysis of the collected data is that with closed systems it is the cumulative dissolved drug that is measured, whereas with open systems the amount dissolved within specific time intervals (differential amount dissolved) is measured (20).

Analysis of Cumulative Data Sets

A review of methods frequently used in the analysis of cumulative dissolution profiles has been recently published (21).

In this chapter, the emphasis is on producing physiologically relevant dissolution data sets. Compared to dissolution profiles obtained according to relevant compendia requirements for quality control purposes, biorelevant dissolution data sets collected in closed systems often do not reach 100% dissolved and frequently are associated with higher variability (22).

Characterization of the Dissolution Process

Complete characterization of the cumulative profile can be considered only with modeling (21). Nicolaides et al. (9) compared the first-order model, the Weibull function, and a model based on the Noyes–Whitney theory for dissolution using individual data sets for the dissolution of various lipophilic compounds in physiologically relevant media. On the basis of the correlation matrix of estimates [that is obtained from the inverse of the Fischer-information matrix (23)], the Weibull model was over-parameterized in some cases where data were highly variable and/or data points prior to the plateau level were limited. Therefore, in contrast to previously reported results for cumulative dissolution profiles obtained in simpler media and with more data points prior to the plateau level (24), the Weibull model may not be always applicable in biorelevant cumulative dissolution testing. However, using the model selection criterion (MSC) [a criterion that takes into account the goodness of fit and the number of model parameters (18)], in cases where fitting was successful with all three tested functions, MSC values favored the Weibull function (9).

Comparison of Two Cumulative Dissolution Data Sets

Model-dependent methods

Various model-dependent methods for the comparison of two cumulative dissolution data sets have been proposed (21). Usually, these methods involve prior characterization of both profiles by one to three parameters per profile. In some models, these parameters can be interpreted in terms of the kinetics, the shape, and/or the plateau, but in other instances, they have no physical meaning. One issue that requires some attention is that, in cases where more than one parameter is estimated, a multi-variate procedure for the comparison of the parameters must be applied (9,21).

Model-independent methods

In recent years, the comparison of two profiles with an index has become very popular mainly because it does not require the use of a model. Models used in the analysis of drug dissolution/release data are usually empirical and multi-parametric. Therefore, even when they are successfully fitted to the data, the subsequent profile comparison frequently requires a complicated multi-variate procedure (21).

Vertzoni et al. (30) recently clarified the applicability of the similarity factor, the difference factor, and the Rescigno index in the comparison of cumulative data sets. Although all these indices should be used with caution (because inclusion of too many data points in the plateau region will lead to the outcome that the profiles are more similar and because the cutoff time per percentage dissolved is empirically chosen and not based on theory), all can be useful for comparing two cumulative data sets. When the measurement error is low, i.e., the data have low variability, mean profiles can be used and any one of these indices could be used. Selection depends on the nature of the "difference" one wishes to estimate and the existence of a reference data set. When data are more variable, index evaluation must be done on a confidence interval basis and selection of the appropriate index, depends on the number of the replications per data set in addition to the type of "difference" one wishes to estimate. When a large number of replications per data set are available (e.g., 12), construction of nonparametric or bootstrap confidence intervals of the similarity factor appears to be the most reliable of the three methods, provided that the plateau level is 100. With a restricted number of replications per data set (e.g., three), any of the three indices can be used, provided either non-parametric or bootstrap confidence intervals are determined (30).

Analysis of Non-Cumulative Dissolution Data Sets

The analysis of non-cumulative dissolution data sets has not been considered in detail in the literature, presumably due to the limited use of in vitro setups that lead to collection of this type of data.

Characterization of the Dissolution Process

To date, whenever the open flow-through apparatus is used, the differential release data obtained are usually converted to their cumulative form, and characterization of the dissolution process is then performed on the cumulative data using various models (25–28). A problem that arises with this procedure is that the least squares criterion for drawing the best-fitted curve through a set of errant data is only valid if errors are independent (23,29). By converting the data from the differential to the cumulative form, any error associated with a specific observation is added to all subsequent observations and, therefore, the fundamental assumption of independence of errors is violated. Characterization of the kinetics must, therefore, be made using the raw data without transformation. A procedure for characterizing the kinetics from non-cumulative data sets is illustrated in what follows with simulated data obtained using the Weibull function:

$$W(t) = W_0 \cdot \left\{ 1 - \exp\left[-\left(\frac{t}{b} \right)^c \right] \right\} \tag{1}$$

where b and c are the scale and shape parameters, respectively. As, in this case, one does not measure cumulative amount dissolved at a specific time point, $W(t)$, but rather the amount dissolved between two consecutive sampling times, $W(t_{j-1},t_j)$, the Weibull function had to be appropriately adjusted:

$$\begin{aligned} W(t_{j-1}, t_j) &= W(t_j) - W(t_{j-1}) \\ &= W_0 \left\{ \left\{ 1 - \exp\left[-\left(\frac{t_j}{b} \right)^c \right] \right\} \right. \\ &\quad \left. - \left\{ 1 - \exp\left[-\left(\frac{t_{j-1}}{b} \right)^c \right] \right\} \right\} \end{aligned} \tag{2}$$

with $j = 1, \ldots, n$, where n is the number of time points. To investigate the applicability of the Weibull function on the characterization of the dissolution process when differential dissolution data are available, simulations were performed according to a recently published procedure (30) using SigmaPlot® (version 4.0 for Windows® 95, SPSS Inc., Illinois, USA) and assuming a dose of $W_0 = 100$. Three shape parameters were considered, $c = 0.5$, 1, and 3. Each c was matched with three scale parameters, i.e., $b = 0.5$, 1, and 1.5.

Simulations were performed as follows: for at least 90% of the process to be complete, observation periods of 8, 4, and 2 hr were used for $c = 0.5$, 1, and 3, respectively. The simulated sampling schedule had nine sampling points that varied according to the c value:

$$\begin{array}{ll} \text{for } c = 0.5: & 0.25, 0.5, 1, 1.5, 2, 3, 4, 6, 8 \\ \text{for } c = 1: & 0.167, 0.333, 0.5, 0.75, 1, 1.5, 2, 3, 4 \\ \text{for } c = 3: & 0.083, 0.167, 0.25, 0.5, 0.75, 1, 1.25, 1.5, 2 \end{array}$$

A total of nine simulated data sets were generated by assuming the earlier sampling schedules, exploring all combinations of b and c values and applying Eq. (2). Because in most real dissolution profiles, the coefficient of variation (CV) decreases with time, the simulated data sets were perturbed by an additive homoscedastic measurement error, resulting in simulated data that would be closer to usual experimental observations. The added error had a net mean of 0 for each data set and a standard deviation (SD) of either 2 or 4. At each SD level and for every [c, b] pair, six replicated profiles were generated. Equation (2) was fitted to the data sets with built-in error. All fitting procedures were performed and evaluated using Mathematica® (Wolfram Research Europe Ltd., Oxfordshire, U.K.). Equation (2) was identified as being over-parameterized in only two out of 54 cases with a built-in SD = 2 and in only six out of 54 cases with a built-in SD = 4. It may be argued, therefore, that the Weibull function appears to be a useful model to characterize the kinetics of dissolution/release from non-cumulative data. An example of the graphical presentation of a data set and its corresponding successfully fitted line using Eq. (2) is shown in Figure 2.

It should be emphasized that models other than the Weibull function represented in Eq. (2) could also be proposed and tested. For these models, the possibility of over-parameterization should first be checked using the correlation matrix of the estimates. Of those tested, the best model can be

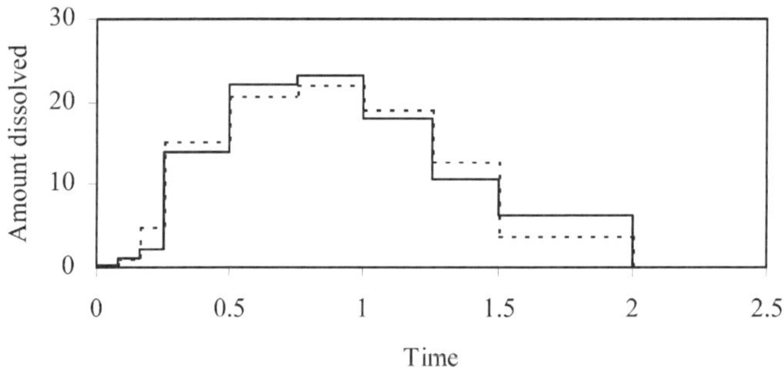

Figure 2 Example of graphical presentation of a % dissolved vs. time simulated data set obtained by using Eq. (2) ($W_0 = 100$, $b = 1$, $c = 3$), assuming a specific sampling scheme (indicated in the text) and perturbing the data with homoscedastic error with a mean of 0 and SD = 4 (dotted line) and the corresponding fitted line obtained by fitting Eq. (2) to the specific data set (continuous line).

selected by means of various criteria suggested in the literature [e.g., the Akaike's criterion or MSC (18)].

In the work described earlier, the applicability of the Weibull model was further tested by assessing the precision of estimation [expressed by the CV defined as the standard error of estimates divided by the estimated value] and the relative accuracy of estimation of the model parameters (based on the difference of the estimates from the actual value, divided by the actual value). Regarding the precision of estimates, for data with SD = 2 the maximum CV value for W_0, b, and c was 13%, 52%, and 16%, respectively, whereas the corresponding numbers for data with SD = 4 were 33%, 151%, and 34%, respectively. As expected, the precision of the estimates decreases as the SD of the data increases, with the poorest precision for the b estimates and the best for the W_0 estimates. Additionally, the maximum CV values were associated with low c values ($c = 0.5$).

The relative accuracy of estimation is illustrated in Figure 3 by the box plots obtained from the individual data sets. On the basis of Figure 3, the accuracy of the estimates decreases with the data variability.

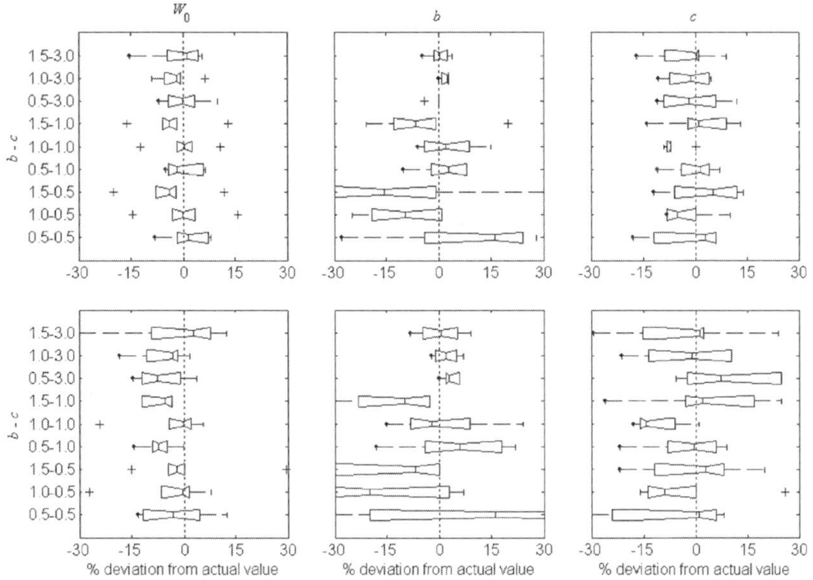

Figure 3 Box plots of W_0, b, and c values estimated after fitting Eq. (2) to individual (simulated, 6-fold replicated) errant data sets and their deviation from the actual parameter values. Upper graphs refer to data with SD = 2 and lower graphs refer to data with SD = 4. The actual W_0 value was always 100.

In general, W_0 estimates are the most accurate, whereas the b estimates are the least accurate. For the c parameter, the poorest accuracy was observed at low c values.

Comparison of Two Non-Cumulative Data Sets

Model-dependent methods

Using the data with built-in error generated in the previous section (six replications per data set), for every c value, two test data sets ($b = 0.5$ and $b = 1.5$) were separately compared with a reference data set ($b = 1$). The estimated total amount dissolved (W_0) of the test and the reference data sets were compared by constructing confidence intervals at the 0.05 level for their mean differences. Estimated shape parameter, c, and scale parameter, b, of the test and the reference data set were compared using a multi-variate model-dependent technique (9,24). Estimated total amount dissolved, W_0, as found to be *not* different in 12 out of 12 cases (six for each SD level). The estimated shape parameter, c, was found to be *not* different in 10 out of 12 cases. In both cases, where shape parameters were found to be different, the shape parameter was $c=1$ (one at each SD level). In contrast, the estimated scale parameter, b, was found to be *different* in nine out of 12 cases; not different was found only in three case where the profiles had $c=0.5$ (one at SD = 2 and two at SD = 4 level). These data suggest that the applied multi-variate comparison procedure using the Weibull function may lead to wrong conclusions in some cases where dissolution follows first-order or faster than first-order kinetics. However, as confirmed in Figure 3, such problems usually occur due to imprecise estimates of the corresponding parameters.

Model-independent methods

As with cumulative data sets, indices such as the difference factor and the Rescigno index can be used to compare two non-cumulative dissolution data sets. However, the application of these indices to non-cumulative data sets is different in two key ways.

The first difference is that non-cumulative data refer to amount of drug dissolved within a certain time period and not at a specific time point, i.e., in this case the observed variable is the amount dissolved, $W(t_1,t_2)$, between the time points t_1 and t_2 ($t_2 > t_1$). Consequently, in contrast to their application to cumulative data (30) where the difference factor and the Rescigno index refer to area differences, for non-cumulative data these indices refer to the difference between the dissolved amount of the test and the reference product in a given time interval.

Mathematically, if the successive time points are designated $t_1,t_2,\ldots,t_n$ (with $t_1=0$ and $t_n\rightarrow\infty$) the time course of the experiment can be partitioned according to the time at which samples were taken, $[t_{j-1},t_j, j=2,n$, with associated measurement of the dissolved amount $W(t_{j-1},t_j)$. The following equations are, therefore, appropriate for the evaluation of the indices:

$$f_1^* = \frac{\sum_{j=2}^{n} |W_T(t_{j-1}, t_j) - W_R(t_{j-1}, t_j)|}{\sum_{j=2}^{n} W_R(t_{j-1}, t_j)} \tag{3}$$

$$\xi_i^* = \left[\frac{\sum_{j=2}^{n} |W_T(t_{j-1}, t_j) - W_R(t_{j-1}, t_j)|^i}{\sum_{j=2}^{n} |W_T(t_{j-1}, t_j) + W_R(t_{j-1}, t_j)|^i}\right]^{1/i} \tag{4}$$

The asterisk denotes that the difference factor, f_1 (31), and of the Rescigno index, ξ_i (32), have been adjusted to apply to non-cumulative data; T and R denote the test and the reference data set, respectively; and i is usually set equal to 1 or 2 (30,32).

The second difference relates to the definition of a cutoff time point for the evaluation of the difference factor and the Rescigno index. When cumulative data are available, evaluation of the difference factor or the Rescigno index usually requires a reference data set in order to define the cutoff time point for index evaluation (30). For the evaluation of f_1^* and the ξ_i^*, i.e., when the difference factor and the Rescigno index are evaluated from non-cumulative data, this difficulty does not exist, provided that the release process has been monitored up to the end (i.e., until dissolution of the drug is complete). At this point, it is worth mentioning that a similar conclusion cannot be drawn for the similarity factor (31) because application of this index to non-cumulative data is set apart by the careful scaling procedure required, in addition to the existence of a reference data set. The reason is that this index can continue to change even after dissolution of both products is complete.

Using the non-cumulative data sets generated in the previous section and a methodology recently used for addressing the problem of the comparison of two highly variable cumulative data sets (30), we additionally assessed the potential for using f_1^*, ξ_1^*, and ξ_2^* in the comparison of two data sets collected with the flow-through apparatus. Indices were evaluated using Eqs. 3 and 4. Bootstrap confidence intervals were constructed (30), assuming 3, 6, and 12 replications per data set. The results are summarized in Table 2. Comparison of the 50th percentiles of the bootstrap samples with the value of the index corresponding to the errorless data sets, it can be concluded that in most cases bootstrapping leads to overestimation of the index. In two specific scenarios, this overestimation becomes so substantial that the confidence intervals do not include the "observed" value. In the first, where $b_{test} = 1.5$ and $c = 0.5$, for all indices and in all but one case the confidence intervals did not include the "observed" value. In the second scenario where $b_{test} = 1.5$ and $c = 1$, in most cases for ξ_2^* and in one case for f_1^* and ξ_1^* the confidence intervals did not include the "observed" index value. Table 2 further indicates that indices values increase with the SD level. Finally, as expected, for a given index, as the number of replications increases the confidence range becomes narrower.

CONCLUSIONS

As simulation of intralumenal conditions in in vitro dissolution testing becomes closer to actual conditions in the GI tract, the resulting dissolution data will most likely show increased variability. At high inter-"individual" variability (expected both in vivo and in vitro) the development of an IVIVC will most likely have to be based on model-independent approaches. This will also apply to the application of the resulting IVIVC to the comparison of in vitro dissolution profiles. Depending on the type of data, various indices to assess the difference between two profiles will be appropriate. When the data are highly variable, it is necessary to estimate the index on a confidence interval basis. In this case, the index can only be as good as the procedure used to construct the confidence interval. When cumulative data sets are available, none of the proposed indices is ideal for general use because they all change continuously with time. However, if an acceptable cutoff time is used, the similarity factor estimated from the mean data sets (when data show low variability) or from bootstrap confidence intervals (when data show high

Table 2 Index Values from the Simulated Non-cumulative Data Sets with No Built-in Error, and 50th (5th–95th) Percentiles of Each of the 1000-Sized Bootstrap Index Sample Constructed from 3-fold, 6-fold, and 12-fold Replicated Data Sets with Built-in Error

Bootstrap 1000			b_{Test}/c					
Ind	Rpl	SD	0.5/0.5	1.5/0.5	0.5/1	1.5/1	0.5/3	1.5/3
f_1^*	3	0	0.223	0.115	0.490	0.244	1.243	0.745
		2	0.269 (0.140–0.471)	0.258 (0.176–0.326)	0.449 (0.356–0.565)	0.264 (0.222–0.306)	1.273 (1.102–1.389)	0.734 (0.582–0.807)
		4	0.280 (0.172–0.530)	0.399 (0.300–0.605)	0.380 (0.313–0.791)	0.422 (0.302–0.577)	1.314 (1.035–1.496)	0.787 (0.605–0.882)
	6	2	0.240 (0.174–0.316)	0.202 (0.156–0.273)	0.453 (0.400–0.526)	0.239 (0.202–0.291)	1.236 (1.144–1.324)	0.687 (0.611–0.758)
		4	0.320 (0.194–0.449)	0.313 (0.227–0.445)	0.421 (0.324–0.564)	0.310 (0.237–0.419)	1.213 (1.047–1.352)	0.667 (0.521–0.807)
	12	2	0.235 (0.191–0.286)	0.184 (0.151–0.227)	0.492 (0.428–0.546)	0.248 (0.200–0.296)	1.213 (1.140–1.297)	0.704 (0.651–0.754)
		4	0.275 (0.200–0.378)	0.289 (0.219–0.397)	0.440 (0.356–0.563)	0.316 (0.236–0.408)	1.224 (1.109–1.342)	0.716 (0.597–0.832)
ξ_1^*	3	0	0.109	0.059	0.242	0.126	0.622	0.390
		2	0.131 (0.068–0.206)	0.132 (0.090–0.177)	0.216 (0.176–0.274)	0.139 (0.114–0.165)	0.609 (0.560–0.634)	0.383 (0.312–0.420)
		4	0.133 (0.086–0.233)	0.215 (0.157–0.260)	0.187 (0.155–0.352)	0.215 (0.156–0.277)	0.636 (0.520–0.675)	0.436 (0.276–0.465)
	6	2	0.116 (0.084–0.154)	0.103 (0.080–0.139)	0.222 (0.197–0.256)	0.124 (0.105–0.148)	0.605 (0.576–0.633)	0.361 (0.316–0.393)
		4	0.155 (0.094–0.217)	0.159 (0.116–0.221)	0.209 (0.160–0.274)	0.159 (0.125–0.212)	0.599 (0.535–0.650)	0.347 (0.267–0.418)
	12	2	0.114 (0.092–0.137)	0.094 (0.076–0.117)	0.241 (0.212–0.265)	0.127 (0.101–0.153)	0.600 (0.573–0.628)	0.367 (0.341–0.392)
		4	0.130 (0.095–0.175)	0.141 (0.108–0.187)	0.211 (0.174–0.260)	0.155 (0.116–0.197)	0.602 (0.558–0.645)	0.361 (0.304–0.414)
ξ_2^*	3	0	0.120	0.073	0.255	0.139	0.602	0.422
		2	0.130 (0.072–0.163)	0.124 (0.090–0.150)	0.243 (0.193–0.267)	0.172 (0.144–0.201)	0.618 (0.581–0.642)	0.413 (0.355–0.448)
		4	0.101 (0.058–0.171)	0.177 (0.133–0.222)	0.195 (0.147–0.335)	0.253 (0.204–0.318)	0.631 (0.508–0.657)	0.433 (0.316–0.498)
	6	2	0.121 (0.091–0.145)	0.099 (0.077–0.125)	0.245 (0.219–0.273)	0.147 (0.126–0.171)	0.611 (0.586–0.628)	0.401 (0.363–0.431)
		4	0.122 (0.072–0.172)	0.137 (0.100–0.181)	0.234 (0.181–0.294)	0.193 (0.151–0.246)	0.605 (0.529–0.644)	0.392 (0.320–0.454)
	12	2	0.127 (0.108–0.141)	0.089 (0.073–0.108)	0.255 (0.230–0.277)	0.144 (0.118–0.170)	0.603 (0.585–0.618)	0.413 (0.388–0.435)
		4	0.122 (0.086–0.157)	0.115 (0.089–0.147)	0.236 (0.193–0.289)	0.176 (0.134–0.225)	0.605 (0.565–0.640)	0.407 (0.351–0.455)

Note: The shape parameter was the same for the test and the reference data sets. In all cases $b_{\text{reference}} = 1$.

variability) can be used. At high variability levels and when the number of replications per data set is small (e.g., when $n = 3$), other indices such as the difference factor or the Rescigno index are equally useful (30). In contrast, as shown in this chapter, when non-cumulative data are available, the difference factor or the Rescigno index is more convenient than the similarity factor because their estimation does not require a specific cutoff time rule.

REFERENCES

1. United States Pharmacopeial Convention, Inc. The United States Pharmacopeia (USP 26). Rockville, Maryland, USA, 2003.
2. Macheras P, Reppas C, Dressman JB. Biopharmaceutics of orally administered drugs. In: Series in Pharmaceutical Technology (ISBN 0-13-108093-8). England: Ellis Horwood, 1995.
3. Amidon GL, Lennernäs H, Shah VP, Crison JRA. Theoretical basis for a biopharmaceutics drug classification: the correlation of in vitro drug product dissolution and in vivo bioavailability. Pharm Res 1995; 12:413–420.
4. Dressman JB, Amidon GL, Fleisher D. Absorption potential: estimating the fraction absorbed for orally administered compounds. J Pharm Sci 1985; 74:588–589.
5. Guidance for Industry, waiver of in vivo bioavailability and bioequivalence studies for immediate release solid oral dosage forms based on a biopharmaceutics classification scheme. Food and Drug Administration, Center for Drug Evaluation and Research, Rockville, Maryland, USA, August 2000.
6. Dressman JB, Amidon GL, Reppas C, Shah VP. Dissolution testing as a prognostic tool for oral drug absorption: immediate release forms. Pharm Res 1998; 15:11–22.
7. Dressman JB, Fleisher D. Mixing—tank model for predicting dissolution rate control of oral absorption. J Pharm Sci 1986; 45:109–116.
8. Ingels F, Deferme S, Destexhe E, Oth M, Van den Mooter G, Augustijns P. Simulated intestinal fluid as transport medium

in the Caco-2 cell culture model. Int J Pharm 2002; 232: 183–192.

9. Nicolaides E, Symillides M, Dressman JB, Reppas C. Biorelevant dissolution testing to predict the plasma profile of lipophilic drugs after oral administration. Pharm Res 2001; 18: 380–388.
10. Nelson E. Comparative dissolution rates of weak acids and their sodium salts. J Amer Pharm Assoc Sci Ed 1958; 47:297–300.
11. Sawardeker JS, McShefferty J. Effect of formulation of anagestone acetate on progestational proliferation of rabbit uterus after oral administration. J Pharm Sci 1971; 60:1591–1592.
12. Schirmer RE, Kleber JW, Black HR. Correlation of dissolution, disintegration, and bioavailability of aminosalicylic acid tablets. J Pharm Sci 1973; 62:1270–1274.
13. Needham TE, Shah K, Kotzan J, Zia H. Correlation of aspirin excretion with parameters from different dissolution methods. J Pharm Sci 1978; 67: 1070–1073.
14. Mendes RW, Masih SZ, Kanumuri RR. Effect of formulation and process variables on bioequivalency of nitrofurantoin II: in vivo–in vitro correlation. J Pharm Sci 1978; 67:1616–1619.
15. Sjovall J, Sjoqvist R, Huitfeldt B, Nyqvist H. Correlation between the bioavailability of microencapsulated bacampicillin hydrochloride in suspension and in vitro microcapsule dissolution. J Pharm Sci 1984; 73:141–145.
16. Fairweather WR. Investigating relationships between in vivo and in vitro pharmacological variables for the purpose of prediction. J Pharmacokinet Biopharm 1977; 5:405–418.
17. Concheiro A, Vila-Jato JL, Martinez-Pacheco R, Seijo B, Ramos T. In Vitro–In Vivo correlations of eight nitrofurantoin tablet formulations: effect of various technological factors. Drug Dev Ind Pharm 1987; 13:501–516.
18. Wagner JG. Pharmacokinetics for the Pharmaceutical Scientist. New York: Technomic Publishing, 1993.
19. Guidance for Industry, extended release oral dosage forms: development, evaluation, and application of in vitro/in vivo

correlations. Food and Drug Administration, Center of Drug Evaluation and Research, Maryland, USA, September 1997.

20. Langenbucher F, Rettig H. Dissolution rate testing with the column method: methodology and results. Drug Dev Ind Pharm 1977; 3:241–263.
21. Reppas C, Nicolaides E. Analysis of drug dissolution data. In: Dressman JB, Lennernaes H, eds. Oral Drug Absorption. New York: Marcel Dekker (ISBN: 0-8247-0272-7), 2000.
22. Nicolaides E, Galia E, Efthymiopoulos C, Dressman JB, Reppas C. Forecasting the in vivo performance of four low solubility drugs from their in vitro dissolution data. Pharm Res 1999; 16:1876–1882.
23. Fukunaga K. Introduction to statistical pattern recognition. In: Booker HG, DeClaris N, eds. Monographs and Texts in Electrical Science. New York: Academic Press, 1972:369.
24. Sathe PM, Tsong Y, Shah VP. In vitro dissolution profile comparison: statistics and analysis, model dependent approach. Pharm Res 1996; 13:1799–1803.
25. Joergensen K, Jacobsen L. Factorial design used for ruggedness testing of flow through cell dissolution method by means of Weibull transformed drug release profiles. Intl J Pharm 1992; 88:23–29.
26. Menon A, Ritschel A, Sakr A. Development and evaluation of a monolithic floating dosage form for furosemide. J Pharm Sci 1994; 83:239–245.
27. Ishii K, Saito Y, Takayama K, Nagai T. In vitro dissolution test corresponding to in vivo dissolution of sofalcone formulations. STP Pharm Sci 1997; 7:270–276.
28. Costa P, Sousa Lobo JM. Influence of dissolution medium agitation on release profiles of sustained-release tablets. Drug Dev Ind Pharm 2001; 27:811–817.
29. Draper NR, Smith H. Applied Regression Analysis, Wiley Series in Probability and Mathematical Statistics. In: Bradley RA, Hunter JS, Kendall DG, Watson GS, eds. New York: John Wiley and Sons, Inc., 1966: 407.

30. Vertzoni M, Symillides M, Iliadis A, Nicolaides E, Reppas C. Comparison of simulated cumulative drug vs. time data sets with indices. Eur J Pharm Biopharmac 2003; 56:421–428.

31. Moore JW, Flanner HH. Mathematical comparison jof dissolution profiles. Pharm Tech 1996; 20:64–74.

32. Rescigno A. Bioequivalence. Pharm Res 1992; 9:925–928.

9

Interpretation of In Vitro–In Vivo Time Profiles in Terms of Extent, Rate, and Shape

FRIEDER LANGENBUCHER
BioVista LLC, Riehen, Switzerland

INTRODUCTION

The quantitative analysis of dissolution profiles and the comparison of such profiles have found increasing interest in the recent literature. A comprehensive survey was given in a previous textbook of this series (1). The purpose of this chapter is to discuss the same topic from a more systematic point of view, with a critical judgment as to which analytical methods are most adequate in certain specific situations and which methods are less adequate for general application.

Dissolution/release profiles in vitro, as well as body response profiles in vivo (e.g., plasma concentrations or urinary excretion), belong to a common category of mathematical functions, namely, *distribution* functions (2). Various distributions, based on the exponential distribution as the most simplest approach, are applicable; but the Weibull distribution is the most versatile extension to cover various profiles in vitro and in vivo.

Many methods are available to characterize single profiles or to compare two profiles, whether these are given numerically as observed data or in the advanced format of fitted functions. Semi-invariants ("moments") are the most adequate metrics for this purpose, as they provide a systematic procedure in terms of the following descriptors:

- *Extent* characterizes the profile vertically in terms of its final plateau.
- *Rate* characterizes the process as fast or slow, i.e., along the horizontal time axis, in terms of its mean time.
- *Shape* provides additional information about the profile, in terms of the variance or another equivalent metric.

CHARACTERIZATION OF TIME PROFILES

Distribution Functions

Time profiles in vitro and in vivo represent distribution functions in a mathematical and statistical sense. For example, a release profile $F_D(t)$ in vitro expresses the distribution of drug released at time t; the corresponding probability distribution function (PDF) profile $f_D(t)$ characterizes the rate of release. Similarly, a plasma concentration profile $f_P(t)$ represents the distribution of drug in the plasma at any time t, i.e., absorbed but not yet eliminated; its cumulative distribution function (CDF) equivalent $F_P(t)$ represents the drug absorbed and already eliminated.

Figure 1 illustrates the general behavior of distribution functions, where the time abscissa is constricted to positive values $t \geq 0$. Two typical formats must be distinguished. In

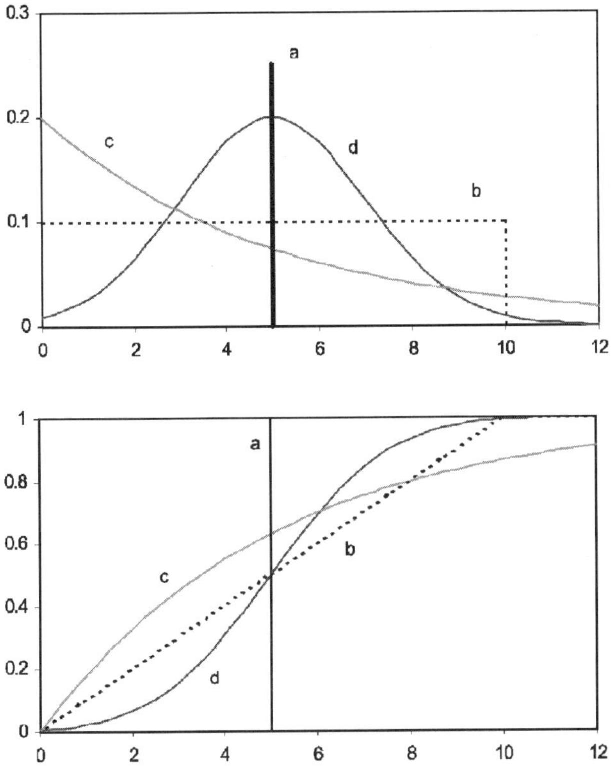

Figure 1 Four elementary distribution functions, displayed as PDF (*top*) or CDF (*bottom*). All functions are relative to $F_\infty = 1$ and time scaled to a mean $\mu = 5$. (a) Unit pulse ($\sigma = 0$); (b) rectangular ($\sigma = 2.89$); (c) exponential ($\sigma = 5$); (d) normal ($\sigma = 2$).

absolute terms, the CDF represents an amount or concentration profile $F(t)$, which reaches a final plateau F_∞; the corresponding PDF represents the rate $f(t)$ of the process, and the area AUC under this profile is identical with F_∞. In *relative* terms, as is well known from statistical applications, the ordinates of CDF and PDF are divided by F_∞. Hence, both represent dimensionless fractions with range $0 \leq F(t) \leq 1$. In other words, the absolute format includes the *extent* in the function itself, whereas in the relative format this aspect is separated out.

The Weibull distribution, illustrated in Figure 2, is most attractive, as it permits characterization of all typical cases of a PDF and CDF with only three parameters (2–4):

$$f(t) = F_\infty\left[\left(\frac{\alpha}{\beta}\right)\left(\frac{t}{\beta}\right)^{\alpha-1} \mathrm{e}^{-(t/\beta)^\alpha}\right] = F_\infty\left[\left(\frac{\alpha}{\beta^\alpha}\right)t^{\alpha-1}\mathrm{e}^{-(t/\beta)^\alpha}\right] \tag{1a}$$

$$F(t) = F_\infty[1 - \mathrm{e}^{-(t/\beta)^\alpha}] \tag{1b}$$

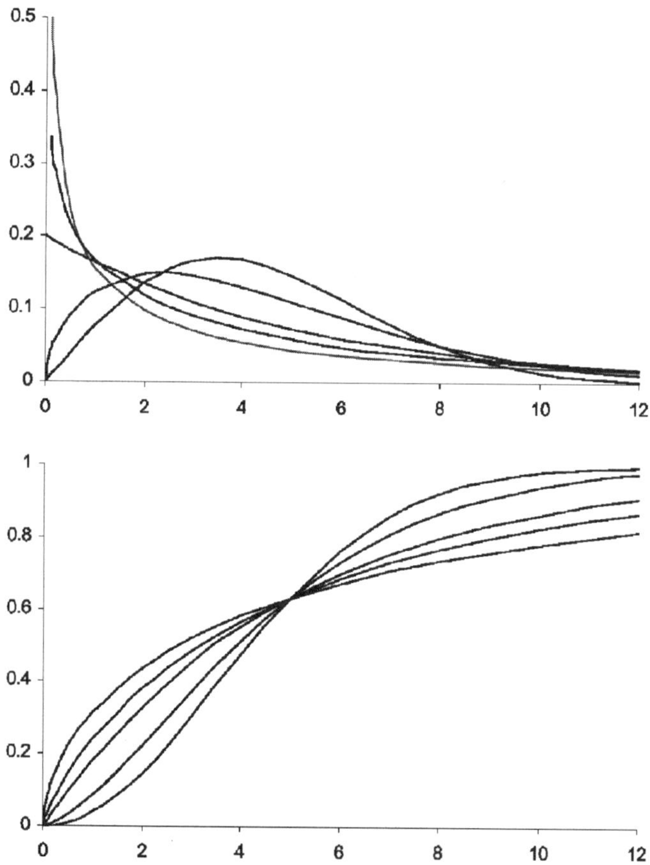

Figure 2 Relative Weibull time profiles shown as PDF (*top*) and CDF (*bottom*). Parameters: $\beta = 5$ and $\alpha = 0.6, 0.8, 1, 1.5, 2$.

The *scale* parameter β characterizes the overall rate; the dimensionless *shape* parameter α raises the time scale to a power other than 1. While $\alpha = 1$ represents a mono-exponential, $\alpha > 1$ describes a "sigmoid" profile retarded in the beginning, $\alpha < 1$ represents a profile faster in the beginning but retarded in the tail. Figure 2 illustrates the performance for $\beta = 5$ and five differing values of α. All CDF profiles intersect at a point ($t = 5$, $F = 0.632$), which closely reflects the mean of the distribution.

Characterization by Semi-invariants ("Moments")

Data known to belong to a distribution function are best summarized in terms of *semi-invariants* κ_0, κ_1, κ_2 (5,6), the first five of which are compiled in Table 1. In the pharmaceutical literature, the first three have been introduced in Ref. 7; since then, the first two, area and mean, are discussed in many papers (8–14). In this context, they are usually referred to as "moments," which is not strictly speaking correct but should not lead to serious confusion.

All semi-invariants are defined in terms of integrals of the profiles between $t = 0$ and $t = \infty$. For given mathematical functions such as the Weibull or the polyexponential distribution, they are computed from the parameters of the function (2,4). Alternatively, they can be computed numerically from the experimental data pairs, e.g., by means of the trapezoidal

Table 1 Compilation of the First Five Semi-invariants

κ_0	AUC, F_∞	Area	Extent	AUC of PDF, F_∞ of CDF
κ_1	MRT, μ	Mean	Rate	Gravity center of PDF and CDF
κ_2	VRT	Variance	Shape	Width about the mean
κ_3		Skewness	Shape	Symmetry around the mean
κ_4		Kurtosis	Shape	Proportion of tails in relation to center

rule. In the latter case, care must be taken to not truncate the profile but to extrapolate its time course until the true plateau is essentially reached; truncated curves will necessarily yield misleading results.

Area(κ_0, AUC, F_∞)

The most important statistic represents the final plateau of the CDF and the area AUC_∞ of the corresponding PDF between $t=0$ and $t=\infty$. It clearly quantifies the *extent* of the relevant process, which is in proportion to the applied dose D, or a constant fraction or multiple $f\,D$ of this, in case of overdose, chemical degradation, etc. Proportionality with dose is violated only if the process contains nonlinear or time-dependent steps such as early loss by defecation, absorption windows, chemical degradation, or non-linear presystemic (first-pass) elimination.

For a CDF, κ_0 is the final plateau value F_∞, extrapolated if necessary. For a PDF, it is defined as

$$\kappa_0 = \text{AUC} = \int_0^\infty f(t)\,\mathrm{d}t \tag{2}$$

An outstanding feature of the Weibull distribution is that it provides a clear separation of this parameter from the exponential part reflecting rate and shape of the profile.

Numerically, the area of PDF data is computed by means of the trapezoidal formula

$$\kappa_0 = \text{AUC} = \sum_{n=1}^{N} \frac{f_n + f_{n-1}}{2}(t_n - t_{n-1}) \tag{3}$$

where the summation starts with the first interval from $t=0$ to $t=t_1$ and continues over all following intervals; usually, an exponential extrapolation term is added to account for the partial area after the last observation. If desired, other convenient algorithms such as Simpson's rule or integration by splines may be used in place of the trapezoidal formula (see mathematical textbooks).

Mean(κ_1, MRT, μ)

The mean represents the overall *rate* of the relevant process and corresponds to the abscissa of the center of gravity of the PDF and the mean value of the CDF. It is exactly reflected by the rate parameter of the Weibull distribution; $t_{63.2\%}$ is exact for mono-exponential and may be used as a shorthand estimate for any CDF of similar shape.

The mean of a PDF is defined as

$$\kappa_1 = \mu = \frac{\text{AUMC}}{AUC} = \frac{\int_0^\infty tf(t)\mathrm{d}t}{\int_0^\infty f(t)\mathrm{d}t} \tag{4}$$

The numerator of Eq. (4) is the integral of the derived function $t \times f(t)$, usually called the "area under the moment curve (AUMC)" (7,8). The denominator is the AUC according to Eq. (2). As visualized by the top plot of Figure 2, the mean as center of gravity represents the time value where the profile (when cut from cardboard) would be in perfect balance.

For a CDF, the mean is defined as

$$\kappa_1 = \mu = \frac{\text{ABC}}{F_\infty} \tag{5}$$

where the denominator is the (extrapolated) final plateau F_∞. The numerator is the so-called "area between the curves (ABC)," i.e., between $F(t)$ and the plateau F_∞ (14)

$$\text{ABC} = \int_0^\infty [F_\infty - F(t)]\mathrm{d}t \tag{6}$$

If $F(t)$ is reported in relative units with range 0–1, F_∞ equals "1" by definition and Eq. (6) directly computes the mean. An interesting alternative definition is obtained by reversing abscissa and ordinate. If the cumulative fraction F is taken as abscissa and $t(F)$ as ordinate, integration of $t(F)$ from $F=0$ to $F=1$ gives

$$\text{ABC} = \int_0^\infty t(F)\mathrm{d}F \tag{7}$$

The bottom plot of Figure 2 illustrates this interpretation of the mean as the average of the time values associated with all cumulative fractions.

Higher-Order Semi-invariants

According to Table 1, semi-invariants of higher order characterize the shape of the profile in terms of variance, skewness, and kurtosis. The outstanding merit of the Weibull distribution is that its shape parameter α provides a summarizing measure for this property. For other distributions, the characterization of the shape is less obvious.

Variance κ_2 characterizes the sharpness of the profile, i.e., whether it changes abruptly or smoothly from "0" to "1" at the mean time. The smaller is its value, the more are the residence times centered about the mean, and the sharper is the profile. It is usual practice to report its square root, the standard deviation (STD), as this gives a measure on the same scale.

Skewness κ_3 characterizes the symmetry of the distribution. A value of 0 characterizes the distribution as symmetric; for asymmetric (skewed) distributions, it will be positive or negative, depending on whether the larger deviations from the mean are in the positive or negative direction (5).

Kurtosis κ_4 characterizes the proportion of the tails in relation to the center. When compared with the normal distribution, platykurtic distributions have more values in the tails and leptokurtic distribution have less.

Descriptive Metrics

Many other metrics are used for the characterization of distribution functions. Most of these can be easily computed or immediately read from the raw data. However, they are based on a single observation and/or they cannot distinguish properly between extent and rate of the process.

For PDF profiles in vivo, the peak co-ordinates are frequently used, because they are immediately read from the tabulated observations or from a corresponding plot. In statistical terms, t_{max} is the "modus" of the distribution, i.e., the most frequent value of the PDF; its value is close to the mean and may be used as a shorthand estimate for this. C_{max} is the corresponding maximum value, which may be used as a crude estimate of the extent. For plasma concentration profiles, C_{max} is useful to characterize whether a therapeutic or toxic level is reached or not. However, the dependence on only a single observation is the inherent weakness of this characterization (15).

For (differential) plasma concentration profiles, the initial slope f'_0 is frequently used as metric reflecting the rate of absorption. Again, it must be realized that this metric is affected by extent as well as by rate. Only when extent is proven as complete, may the initial slope be used as measure of rate of the input.

For cumulative dissolution profiles, the following set of metrics is frequently used

$$F(t_i) = F(t_1),\ F(t_2),\ F(t_3),\ \ldots \tag{8a}$$

$$t(F_i) = t(F_1),\ t(F_2),\ t(F_3),\ \ldots \tag{8b}$$

Equation (8a) lists the cumulative fractions observed at given time points, e.g., at 10, 20, 60 min; these are directly given by the raw data. Equation (8b) records the times to reach specified fractions, e.g., 20%, 60%, 80%; these must be computed by interpolation or curve fitting.

COMPARISON OF TIME PROFILES

The comparison of two time profiles, e.g., a *test* $T(t)$ vs. a *reference* $R(t)$, can be handled by many techniques. (1,16,17). Before looking at them in more detail, it seems useful to briefly discuss a few general aspects.

General Aspects

Model Dependent/Independent Comparison

In IVIVC, it has become common practice to define methods as model dependent, if they take into account that data points represent a time profile according to a distribution function; model-independent methods do not rely on this assumption.

Model-independent techniques compare data pairs observed at corresponding time values, where time is only a class effect, as in a paired *t*-test or in an ANOVA. A "data-poor" set of only two or three observations, originating from routine quality control of an immediate-release dosage form, cannot be treated other than model independent.

Model-dependent techniques are superior in that they assume the observed data pairs to belong to a general distribution function; as a consequence, the time dimension is taken into account. In order to substantiate the model, a "data-rich" set of observations is required, i.e., a larger number, well placed over the entire time range including the final plateau. At the lowest level (a), no attention is paid to the specific function, but general properties of a distribution are regarded; examples are numerical techniques (e.g., the numerical form of the Rescigno index). At a higher level (b), a specific distribution function, e.g., a Weibull distribution, is fitted to the data points, and the further comparison is made in terms of the fitted parameters or derived statistics. At the highest level (c), which is beyond the scope of this chapter, the distribution function is interpreted in terms of a mechanistic model (compartment models in vivo; cube-root law or Higuchi formula in vitro).

Horizontal/Vertical Comparison

Data belonging to distribution profiles may be compared either vertically along the release/response ordinate or horizontally along the time abscissa. The semi-invariants (moments) provide a complete set of metrics, representing both aspects in logical sequence: AUC accounts (vertically) for the difference of the extent, the mean compares (horizontally) the rates, and higher-order moments and higher-order statistics (variance, etc.) characterize the shape aspect from coarse to finer.

Vertical comparison answers the question "what value is obtained at a given time," i.e., the *extent* characteristic of the process. This approach is natural because observations are usually reported for given time values, which in most cases are identical for the profiles to be compared. For "data-poor" experiments, which do not permit a reliable estimation of the full time profile, this is indeed the only possible analysis. However, it stresses only the *extent* aspect of the profile, as expressed by AUC or C_{max}.

Horizontal comparison answers the question "what time is required to reach a certain ordinate value." This approach stresses the *rate* aspect of the process, i.e., its property of being faster or slower. Typical parameters are t_{max} or time parameters t_{f} for a given fraction (percentile).

This distinction becomes clear from a comparison of two cumulative profiles shown in Figure 3. The left panel displays both profiles in the original $F(t)$ plot, with common scales. With t as independent variable, it is easy to compare F values for any given time t. With the same ease, one can compare time values at which a certain F value is reached.

The right panel illustrates a "correlation" (sometimes termed "Levy") plot of the same data, which is widely used in IVIVC. Here, fractions $F_{\mathrm{T}}(t)$ and $F_{\mathrm{R}}(t)$, dissolved at the same time, are plotted against each other, which ease vertical comparison. An equally justifiable alternative would be to stress the horizontal aspect by plotting time values $t_{\mathrm{T}}(F)$ and $t_{\mathrm{R}}(F)$ for the same F value against each other. In both

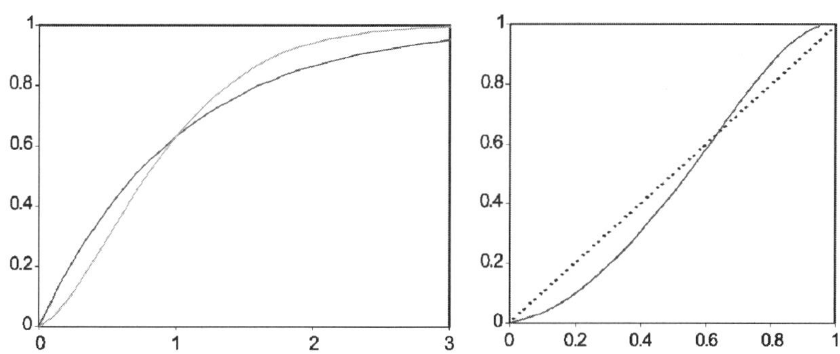

Figure 3 Graphical comparison of time profiles, in the original $F(t)$ presentation (*left*) and a "*correlation*" plot (*right*).

cases, the main diagonal, shown by a dashed line, represents complete identity between both profiles, and differences between time profiles show up as deviations from the diagonal. Such plots must be interpreted with care, bacause time as an essential variable is lost completely. In addition, the axes do not represent independent and dependent variables; hence usual regression techniques cannot be applied.

Comparison by Semi-invariants

Model-dependent comparison of two time profiles is best achieved in terms of the semi-invariants discussed earlier in the section on Characterization of Semi-invariants ("Moments"). This treatment is in accordance with the "Level B" definition of IVIVC, as proposed in several official guidelines. It makes full use of the underlying model that the data are presented by a distribution function, but no specific function is required. Although derived function parameters (e.g., Weibull, polyexponential, etc.) may be used, the computation may also be performed numerically on the observations as such.

Fundamental characteristics are illustrated in Figure 4. Obviously, the difference between both profiles is best estimated from the area enclosed by the two profiles, as it would be obtained directly by graphical planimetry. When summing over various parts of the profiles, it is important to distinguish between actual differences (keeping the sign) and absolute differences (disregarding the sign). If the profiles intersect at some time, areas before and behind the intersection point are added to estimate the overall dissimilarity or subtracted to give a more specific characterization.

Three cases have been constructed by Weibull functions according to Eqs. (1a) and (1b), as these best reflect systematic differences in the sequence. In all cases, a reference profile is defined by extent $F_{\infty} = 1.0$, scale parameter $\beta = 2.0$, and shape parameter $\alpha = 1.5$. In each case, one parameter is altered to illustrate its influence.

1. $F_{\infty} = 0.8$ illustrates the change of 20% in extent while rate and shape are the same. In such a situation, it is almost impossible to assess details of either

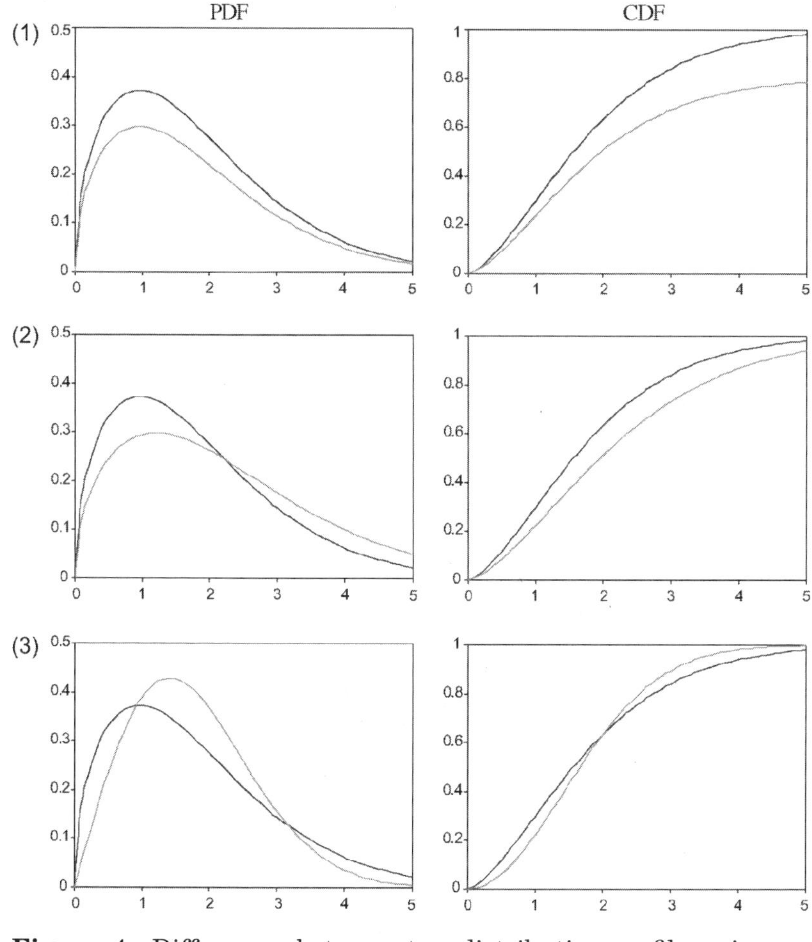

Figure 4 Differences between two distribution profiles, given as PDF (*left*) or CDF (*right*), and differing by extent (1), rate (2), shape (3).

rate or shape: both profiles must be adjusted beforehand by vertical multiplication to identical values of F_∞ or AUC.

2. $\beta = 2.5$ illustrates the situation where only the rate differs between the two profiles. Because the AUCs are the same for both profiles, the difference of rates is indicated by the difference of the two wedge areas

of the PDFs before and after the intersection. For the CDFs, the single wedge is a direct measure of the difference of rates, i.e., the means of the time profiles.

3. $\alpha = 1.5$ illustrates the situation where extent and rate are the same, but the shape differs between the two profiles as indicated by the wedges. The PDFs intersect twice, resulting in three wedges. The CDFs intersect once, resulting in two wedges.

Rescigno Indices ξ_i and ξ_i^*

For the comparison of two differential plasma concentration profiles $R(t)$ and $T(t)$, Rescigno (18) proposed a dimensionless "index of bioequivalence"

$$\xi_i = \left[\frac{\int_0^\infty |R(t) - T(t)|^i \mathrm{d}t}{\int_0^\infty |R(t) + T(t)|^i \mathrm{d}t}\right]^{1/i} \tag{9}$$

In Eq. (9), the numerator sums differences without respect to their sign. The exponent i specifies the weighting of the deviations, e.g., mean absolute error (ME) ($i = 1$) or mean squared error (MSE) ($i = 2$). The denominator represents the mean $\sum(R + T)$ of both profiles. The result is a "coefficient of variation" that quantifies the dissimilarity between both profiles, according to cases I(b) and II(b). $\xi = 0$ characterizes complete identity; $\xi = 1$ characterizes complete dissimilarity where one profile is "1" while the other is "0."

Equation (9) is clearly model dependent, because the difference of both profiles is integrated between $t = 0$ and $t = \infty$, in a manner very similar to the definition of moments. This analogy is most obvious for the case $i = 1$: if the numerator terms were entered as signed differences $R(t)–T(t)$ rather than the absolute differences $|R(t)–T(t)|$, the recognition of the sign would compute the difference of the areas between the two profiles. The denominator calculates the "relating factor"

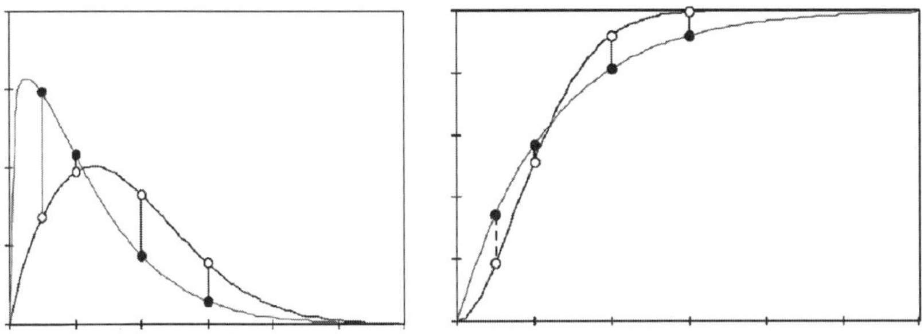

Figure 5 Numerical computation of the difference between two profiles (*left*: PDF, *right*: CDF), from four actual data points, observed at corresponding times.

as the sum of the two areas; this scales the results so that possible values are in the range 0–1.

The application of Eq. (9) to differential profiles is illustrated in the left-hand plot of Figure 5. With $i = 1$, it represents the wedge area between the two curves. If the curves do not cross each other, the nominator directly represents the difference of the two AUCs. If they intersect as shown in the example, the choice of absolute differences computes a general dissimilarity index; the area difference would be obtained by using signed differences instead of absolute differences.

Cumulative Profiles

The index according to Eq. (9) may likewise be applied to *cumulative*-release profiles, as can be seen from the right-hand plot in Figure 5. Once both profiles have been converted to the same final plateau F_{∞}, the "wedge" area between both can be computed directly from the profiles. Note that in contrast to the computation of single profiles, it is not necessary to use the indirect procedure of calculating ABC and $1-F(t)$ according to Eq. (6).

If the two profiles do not cross, the nominator of Eq. (9) directly represents the difference of the two mean times. If they intersect at some time, signed differences compute the difference of the means and absolute differences provide the more general dissimilarity index.

Numerical Definition

On a numerical level, the integrals in Eq. (9) are substituted by numerical integration, e.g., by means of the trapezoidal rule:

$$\xi^* = \left[\frac{\sum (|R - T|)^i \Delta t}{\sum (R + T)^i \Delta t} \right]^{1/i} \tag{10}$$

This simply means that multiple straight-line sections replace the smooth profiles in Figure 5.

The corresponding definition on p. 926 of Ref. (18) is somewhat confusing; in that, it prescribes a weighting coefficient w_j in the place of Δt in Eq. (10). This coefficient is characterized as "an appropriate coefficient representing the weight that the sampling time t_j has in the determination of the whole function," from which it is clear that w_j has the same significance as Δt in the trapezoidal formula. Hence, Eq. (10) estimates the difference between the two profiles numerically as the sum of all wedges between the profiles, irrespective of their signs. However, this careless notation has led to the misunderstanding of the Rescigno index as profile-independent comparison.

Model-Independent Indices

It may happen that experimental data are recorded with an insufficient number of observations or at inappropriate time points. In such cases, it is not possible to obtain insight into the curve profile and to compute metrics such as semi-invariants or even a Rescigno index. Amazingly, this situation is discussed mainly with respect to in vitro-release data, although modern equipment easily permits automatic recording of complete time profiles; obviously, the problem does not exist for in vivo data despite the more pretentious experimentation.

If the data are recorded at corresponding time values, an alternative is to treat them in a way similar to "paired differences" as in a paired t-test or in an ANOVA, where time is not considered as continuous independent variable but only as a class effect. The result is a "model-independent" index, which

Table 2 Classification of "Model-Independent" Indices According to the Power Used in Summation and the Relating Factor

	Relating factor	Absolute differences (I)	Squared differences (II)	Meaning
(a)	N	$[\sum\|R-T\|]/N$	$[\sum(R-T)^2]/N$	ME, VAR, SD
(b)	$\sum(R+T)$	$[\sum\|R-T\|]/[\sum(R+T)/2]$	$[\sum(R-T)^2]/[\sum(R+T)/2]$	CV
(c)	$\sum R$	$[\sum\|R-T\|]/[\sum R]$	$[\sum(R-T)^2]/[\sum R]$	CV

compares the observations in terms such as mean error (ME), MSE, SD, coefficient of variation (CV), etc.

These indices may be described in various terms: "difference" is neutral in that both values are considered on the same level; "deviation" and "error" imply that one value is a reference and the other a deviation from this. However, all quantify *dissimilarity*: "0" denotes identity and, for properly scaled distribution functions, a value of "1" expresses complete dissimilarity (18). Various possibilities to define such indices are shown in Table 2.

Summation of *absolute* differences (I) results in an ME in which all differences have the same statistical weight. Summation of *squared* differences (II) is the more common practice and gives an MSE in which large deviations have higher weight than small ones. In order to make the metric independent of the number N of observations, the error sum must be related to N or an equivalent sum of the observations:

- Case (a) divides by the number N of observations, which represents an ME or an SD.
- Case (b) divides by the (halved) sum of both profiles, i.e., the mean of both profiles, to give a CV.
- Case (c) also computes a CV, now related to the reference profile.

Cases (a) and (b) are symmetric with respect to exchanging of R and T. Case (c) is asymmetric with respect to the two profiles and justified only if R represents a "reference" in a strict sense; different results are obtained depending on whether R is the larger or smaller of the two. Another kind of symmetry applies to an exchange of $T = R + \delta$ and $T = R - \delta$, i.e., a positive or negative deviation of same size, from the reference. With respect to this, cases (a) and (c) are symmetric while (b) is asymmetric.

Moore–Flanner Index f_1

In 1996, Moore and Flanner (19) proposed an index

$$f_1 = \frac{\sum |R - T|}{\sum R} \tag{11a}$$

$$f_1 = 100 \left\{ \frac{\sum |\hat{R} - \hat{T}|}{\sum \hat{R}} \right\} \tag{11b}$$

This index clearly corresponds with case I(c) of Table 2, i.e., an ME computed as the sum of absolute deviations and related to the sum of the reference data. In the original definition according to Eq. (11b), R and T are supplied as percentages and a factor of 100 is included so that the results can be expressed on a percentage scale.

A formal point of objection is the improper use of percentage notation, which is open to cumbersome handling as well as to error of interpretation. In "good mathematical practices," the percentage symbol is the abbreviation of a dimensionless factor ($\% = 1/100 = 0.01 = 10^{-2}$). The abbreviation should never be used in the definitions of formulas and calculations; these must be carried through in terms of fractions. Only in the final presentation may a percentage (99.5%) be used in place of the actual fraction (0.995).

Moore-Flanner Index f_2

Another index, also proposed by Moore and Flanner (19), is defined as

$$f_2 = 0.5 \log \left\{ 1 / \sqrt{0.0001 + \sum (R - T)^2 / N} \right\} \tag{12a}$$

$$f_2 = 50 \log\left\{\frac{100}{\sqrt{1 + \sum(\hat{R} - \hat{T})^2/N}}\right\} \tag{12b}$$

Both definitions are identical, but Eq. (12a) expresses all relative quantities (R, T, f_2) as fractions, whereas the original definition according to Eq. (12b) expresses them as percentages. This index has found much attention in the subsequent literature (20,21), but some objections have been raised against the use of percentages and the *similarity* scale in the definition of f_2, which is in opposition to the "dissimilarity" scale used generally in statistics (22).

Basically, f_2 is defined as an SD $[(\sum(R-T)^2)/N]^{1/2}$ according to case II(a) of Table 2. A special feature is the "similarity" transformation, which reverses the scale and makes it curvilinear to pass through three pivotal points:

	Identity	Borderline	Complete dissimilarity
RMSE, SD	0	0.1 (=10%)	1 (=100%)
$f2$	1 (=100%)	0.5 (=50%)	0

Table 3 and Figure 6 illustrate the transformation in a plot similar to Figure 1 of Ref. 19, by plotting transformed indices against f_1 values, assuming identical deviations for all observations.

A first question is whether a "similarity" scale is beneficial at all. On the one hand, many users are familiar with statistical reasoning and have to translate an f_2 value back to the underlying RMSE scale for better understanding. On the other hand, if the transformation were proven to be scientifically sound and useful, it should not be restricted to f_2 but generalized to f_1 and all other indices of similar structure.

A second question is whether the f_2 transformation is the best way to establish a similarity scale. Despite the clumsy definition, Eq. (12) gives only approximated values; although

Table 3 Alternative Definitions of a Similarity Index, Computed for an Equivalent Value of f_1

F_1	f_1^2	f_2	f_2'	f_2''
0	*0*	*1.0000*	∞	*1.0000*
0.01	0.0001	0.9247	1.0000	0.7500
0.02	0.0004	0.8253	0.8495	0.6920
0.05	0.0025	0.6463	0.6505	0.5942
0.1	*0.0100*	*0.4989*	*0.5000*	*0.5000*
0.2	0.0400	0.3492	0.3495	0.3840
0.5	0.2500	0.1505	0.1505	0.1883
1	*1.0000*	*−1.1E−5*	*0.0000*	*0.0000*

this may not affect practical applications, it is considered to be a mathematical deficiency. Hence, consideration should be given to replacement with a more flexible and mathematically correct transformation. A first alternative is to drop the additive constant in the square root of Eq. (12a), which defines a simplified index as

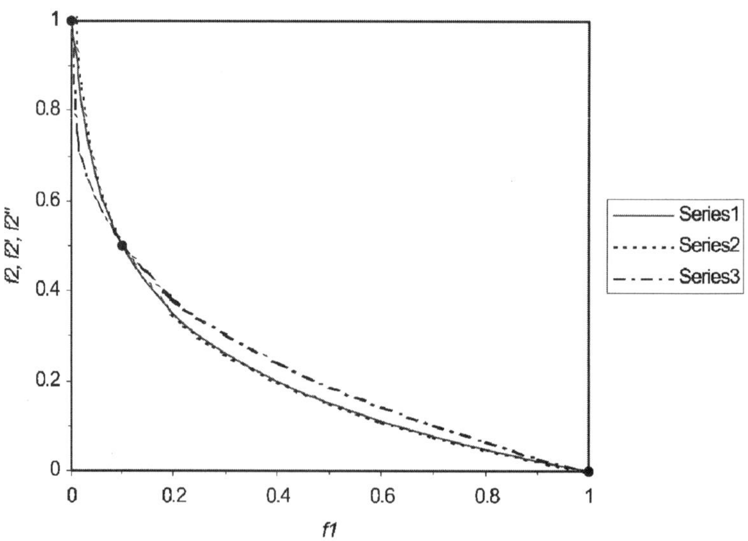

Figure 6 Alternative similarity transformations according to Table 3: f_2 (series 1), f_2' (series 2), f_2'' (series 3).

$$f_2' = 0.5 \log\left\{\frac{1}{\sum(R-T)^2}/N\right\} \tag{13}$$

Equation (13) gives exact values of 0.5 for $f_1 = 0.1$ and 0 for $f_1 = 1$ and is almost indistinguishable over the entire transformation region: only at extremely small values it does deviate considerably, whereas at $f_1 = 0$ it gives ∞ rather than 1. Another simple and flexible alternative is a logarithmic transformation such as

$$f_2'' = 1 - f_1^{\log 2} = 1 - f_1^{0.30103} \tag{14}$$

For the three pivotal points, this transformation has the same effect as Eq. (12) but with exact values and simpler handling; deviations between the pivots are remarkable but without interest for the intended goal. An interesting property of Eq. (14) is that it may be adapted to any other decision point f_1 by simply altering the value of the exponent c. While the two extreme pivots remain unchanged, the exponent of the break-even point $f_2'' = 0.5$ is found as $c = \log(0.5)/\log(f_1)$:

Decision point f_1	0.05	0.1	0.2	0.5
Exponent c	0.23138	0.30103	0.43068	1.00000

Alternative Metrics

In a series of papers (23–26), Polli and colleagues proposed alternative "direct curve comparison" metrics on this level. In their papers, attention was focused on two aspects: (i) are means or medians more suitable for comparison? and (ii) how can symmetric confidence intervals be constructed that are invariant when exchanging reference and test? In addition, this work was devoted to bioavailability and bioequivalence, i.e., time profiles in vivo, but the conclusions apply likewise to in vitro-release profiles.

Marston and Polli (24) compared the performance of the Rescigno index ξ_i and the Moore–Flanner index f_1 with a metric originally proposed by Chinchilli and Elswick (27). The latter is defined in terms of "lower and upper boundaries of the test region": $T_L = \min[T,(R/T)R]$ and $T_U = \max[T,(R/T)R]$. Polli and McLean (26) defined and compared four additional metrics, in which the denominator is $\sum(R+T)$, and which are symmetric about R and T:

- $\delta_a = 2 \times \sum[|R-T|]$ is obviously equivalent to the numerical Rescigno index for absolute differences ($i = 1$), apart from a constant factor of "2", and disregarding the time dimension.
- $\delta_s = 4 \times \sum[(R-T)^2/(R+T)]$ appears to be equivalent to the Rescigno case of squared differences ($i = 2$), but expressed in an unfortunate way.
- $\rho = \sum[(R+T) \times max\{T/R;R/T\}]$ represents a differing approach that "considers the ratio of the profiles at the same time points." It is claimed that the goal is achieved by a weighting factor, which "is the larger of T/R and R/T."
- $\rho_m = \Sigma[(R+T) \times (max\{T/R;R/T\} - 1)]$ is similar to ρ, but the ratio is diminished by 1.

The definition of these metrics appears somewhat arbitrary and is hard to understand in the framework of statistical reasoning. In particular, the meaning of maximum and minimum terms in the definition of the "Chinchilli" and the "rho" metrics cannot be easily verified. The fact that an arbitrarily defined index performs better for an arbitrarily selected set of experimental data cannot be accepted as a general proof of validation.

Statistical Considerations

The comparison of time profiles involves many statistical aspects, some of which were touched upon in the previous discussion, where appropriate. In particular, it was stressed that, with the Moore–Flanner index f_2 as the sole exception, statistical comparisons are generally made in terms of dissimilarity rather than similarity. If computed statistics or indices exceed a pre-defined decision limit, both specimens are considered as different; if this limit is not reached, they are considered as "similar" (a better term would be "indistinguishable"). In this section, some additional aspects, which have found attention in the recent literature, are briefly summarized.

Decision Intervals and Limits

The statistical significance of a computed difference is best quantified in terms of confidence intervals for the means (CLM). If the mean of a profile "T" falls into the CLM of profile "R", both may be regarded as equivalent. For in vivo data, an acceptance limit of $\pm 20\%$ seems to be generally accepted; for in vitro data, this would be unnecessarily wide and $\pm 5\%$ appears more reasonable.

A frequently discussed question is whether equivalence or acceptance limits are better defined on a *linear* or a *logarithmic* scale. Although discussed in many papers, it is felt that this question does not have much practical importance. It is recommended to decide pragmatically on the environment in which the comparison is made. For in-vivo data, logarithmic modeling seems to be a generally accepted practice, and logarithmic limits such as "0.8 ... 1.25" appear reasonable. On the other hand, no model demands such a transformation for in-vitro data, hence no objection can be risen against treating them on a linear scale with limits such as "0.8 ... 1.2."

Several special decision intervals and limits have been proposed in the recent literature. Two of them should be mentioned for completeness, although their general usefulness appears rather doubtful:

- Chow and Ki (16) proposed "equivalence limits" $\delta_L = (Q-\delta)/(Q+\delta)$ and $\delta_U = (Q+\delta)/(Q-\delta)$ for a single time point, on the basis of an official specification Q (e.g., $0.75 = 75\%$) and a reasonable tolerance δ (e.g., $0.05 = 5\%$). The limits are centered about the specification Q, not symmetric but in reverse proportion such that $\delta_U = 1/\delta_L$. They are not helpful for comparing two experimental-release profiles.

- The Chinchilli metric (24,27), defined as "the ratio of the test region area over the reference region area," uses a "reference region area" specified by $R_L = 0.8R$ and $R_U = 1.2R$ as upper/lower acceptance (bioequivalence) limits for the reference. This is compared with the "test region area" mentioned earlier. The procedure as such appears rather complicated.

Multi-variate Aspects

Unless two profiles are compared with a single observation or a summarizing index, the comparison involves a set of metrics; these may be specific observation points such as F_{10}, F_{20}, and F_{30}, fitted function parameters such as α and β of a Weibull distribution, or estimated semi-invariants AUC, MDT, and VDT. In this situation, each metric can be compared separately, resulting in a manifold of independent "local" comparisons; alternatively, all relevant metrics may be summarized in a common "global" model by means of multi-variate techniques (16).

Tsong et al. (28) illustrated the principle by an example, where two batches are compared by means of eight time points and six tablets for each. These data constitute two vectors X_T and X_R of size eight for the sample means, which summarize the two profiles; X_T–X_R is a measure of the overall difference. "Variance-covariance" matrices S_T and S_R, each with eight rows and columns according to the time points, describe the variability of the data: variances are shown on the main diagonal, and the off-diagonal elements show the covariances as measure of the mutual dependence. The final comparison may be summarized by single-value index, e.g., the "Mahalanobis" distance D defined by this matrix equation

$$D^2 = [X_T - X_R][(S_T + S_R)/2]^{-1}[X_T - X_R] \qquad (15)$$

The approach was extended to the function parameters α and β of a Weibull distribution (29,30). This, however, appears less appropriate, as these metrics have a distinct meaning and are better compared individually.

Dependence of Observations

For cumulative data, a frequently heard, but not well documented, argument is that these are not independent of each other because any observation $i+1$ depends on the previous observation i. It cannot be seen how this could invalidate the usual statistical analysis.

- When observed directly, as in a dissolution test in a closed vessel, all observations are in fact independent, without any propagation of previous observations or errors.
- When computed from a corresponding PDF, the PDF clearly represents independent observations; any analysis of these is also valid for the corresponding CDF.

An "*autoregressive time series*" model (16) seems to be less suitable for cumulative distribution data. This technique is primarily designed for finding trends and/or cycles for data recorded in a time sequence, under the null-hypothesis that the sequence has no effect.

Bootstrap Techniques

Bootstrap and similar statistical techniques have been applied to IVIVC and related problems. These techniques, as summarized in Ref. (31), are intended to validate statistics estimated from a small data sample (e.g., mean, SD, correlation coefficient) with respect to their bias and/or confidence intervals. *Cross-validation* splits observations into two groups and validates "internally" one group against the other. Other techniques substitute additional experimental data by pseudo samples simulated randomly from the original data, typically with 100–1000 repetitions: *bootstrap* samples are generated by randomly choosing samples from the raw data; *Jackknife* samples by repeating an original sample and omitting a value by chance from the original data set. From this large data set, reliable estimates may be obtained again from simple "plug-in" formulas.

Within the scope of biopharmaceutics and IVIVC, bootstrap techniques have been applied to several specific problems related to the estimation of confidence intervals of, e.g., the similarity factor f_2 (21), the "Chinchilli" metric (27), parameters of an open two-compartment system (32), and the SD in general (33). From these few applications, it cannot be judged how much is actually gained from these new techniques.

Notations

ABC	Area between the curves, used to integrate CDFs
AUC	Area AUC under a PDF, final value F_∞ of a CDF
AUMC	Area under the first moment curve
CDF	Cumulative distribution function, $F(t)$
CLI, CLM	Confidence limit for a single observation or a mean
CV	Coefficient of variance, SD/mean
ME	Mean error
MEAN	Mean time of distribution function
MSE	Mean squared error
MRT, MDT	Mean time of response, dissolution
PDF	Probability density function, $f(t)$
RMSE	Root mean squared error
SD	Standard deviation
VAR	Variance

REFERENCES

1. Reppas C, Nicolaides E. Analysis of drug dissolution data. In: Dressman JB, Lennernnäs H, eds. Oral Drug Absorption. New York: Marcel Dekker, 2000.

2. Langenbucher F. Handling of computational in vitro/in vivo correlation problems by Microsoft Excel, part II. Distribution functions and moments. Eur J Pharm Biopharm 2002; 53: 1–7.

3. Langenbucher F. Linearization of dissolution rate curves by the Weibull distribution. J Pharm Pharmacol 1972; 24:979–981.

4. Langenbucher F. Parametric representation of dissolution rate curves by the RRSBW distribution. Pharm Ind 1976; 38:472–477.
5. Bennett CA, Franklin NL. Statistical Analysis in Chemistry and the Chemical Industry. New York: John Wiley & Sons, 1963.
6. Johnson NL, Leone FC. Statistics and Experimental Design in Engineering and the Physical Science. New York: John Wiley & Sons, 1977.
7. Yamaoka K, Nakagawa T, Uno T. Statistical moments in pharmacokinetics. J Pharmacokinet Biopharm 1978; 6:547–558.
8. Riegelman S, Collier P. The application of statistical moment theory to the evaluation of in vivo dissolution time and absorption time. J Pharmacokinet Biopharm 1980; 8:509–534.
9. Brockmeier D. Die Rekonstruktion der Freisetzungsprofile Mikoverkapselter Arzneiformen durch den Mittelwert und die Varianz der Freisetzungsprofile. Arzneim Forsch 1981; 31: 1746–1751.
10. Tanigawara Y, Yamaoka K, Nakagawa T, Uno T. Moment analysis for the separation of mean in vivo disintegration, dissolution, absorption, and disposition time of ampicillin products. J Pharm Sci 1982; 71:1129–1133.
11. von Hattingberg HM, Brockmeier D, Vögele D. Momentenanalyse und in vitro-/in vivo-Korrelation. Acta Pharm Technol 1984; 30:93–100.
12. Chanter DO. The determination of mean residence time using statistical moments: is it correct? J Pharmacokinet Biopharm 1985; 13:93–100.
13. Veng-Pedersen P, Gillespie W. The mean residence time of drugs in the systemic circulation. J Pharm Sci 1985; 74: 791–792.
14. Brockmeier D. In vitro/in vivo correlation of dissolution using moments of dissolution and transit times. Acta Pharm Technol 1986; 32:164–174.
15. Khoo KC, Gibaldi M, Brazzell RK. Comparison of statistical moment parameters to c_{max} and t_{max} for detecting differences in in vivo dissolution rates. J Pharm Sci 1985; 74:1340–1342.

16. Chow SC, Ki FYC. Statistical comparison between dissolution profiles of drug products. J Biopharm Stat 1997; 7:241–258.
17. Freitag G. Guidelines on dissolution profile comparison. Drug Inf J 2001; 35:865–874.
18. Rescigno A. Bioequivalence. Pharm Res 1992; 9:925–928.
19. Moore JW, Flanner HH. Mathematical comparison of dissolution profiles. Pharm Tech 1996; 20:64–74.
20. Liu JP, Ma MC, Chow SC. Statistical evaluation of the similarity factor f_2 as a criterion for assessment of similarity between dissolution profiles. Drug Inf J 1997; 31:1255–1271.
21. Shah VP, Tsong Y, Sathe P, Liu JP. In vitro dissolution profile comparison—statistics and analysis of the similarity factor, f_2. Pharm Res 1998; 15:889–896.
22. Langenbucher F. IVIVC indices for comparing release and response profiles. Drug Dev Ind Pharm 1999; 25:1223–1225.
23. Polli JE, Rekhi GS, Shah VP. Methods to compare dissolution profiles. Drug Inf J 1996; 30:1113–1120.
24. Marston SA, Polli JE. Evaluation of direct curve comparison metrics applied to pharmacokinetic profiles and relative bioavailability and bioequivalence. Pharm Res 1997; 14:1363–1369.
25. Polli JE, Rekhi GS, Augsburger LL, Shah VP. Methods to compare dissolution profiles and a rationale for wide dissolution specifications for metoprolol tartrate tablets. J Pharm Sci 1997; 86:690–700.
26. Polli JE, McLean AM. Novel direct curve comparison metrics for bioequivalence. Pharm Res 2001; 18:734–741.
27. Chinchilli VM, Elswick RK. The multivariate assessment of bioequivalence. J Biopharm Stat 1997; 7:113–123.
28. Tsong Y, Hammerstrom T, Sathe P, Shah VP. Statistical assessment of mean differences between two dissolution data sets. Drug Inf J 1996; 30:1105–1112.
29. Sathe PM, Tsong Y, Shah VP. In vitro dissolution profile comparison: statistics and analysis, model dependent approach. Pharm Res 1996; 13:1799–1803.

30. Sathe P, Tsong Y, Shah VP. In vitro dissolution profile comparison and IVIVR, carbamazepine case. In: Young D, Devane JG, Butler J, eds. In Vitro-in Vivo Correlations. New York: Plenum Press, 1997:31–42.
31. Efron B, Tibshirani RJ. An Introduction to the Bootstrap. NewYork: Chapman & Hall, 1994.
32. Hunt CA, Givens GH, Guzy S. Bootstrapping for pharmacokinetic models: visualization of predictive and parameter uncertainty. Pharm Res 1998; 15:690–697.
33. Broberg P. Estimation of relative SD. Drug Dev Ind Pharm 1999; 25:37–43.

10

Study Design Considerations for IVIVC Studies

THERESA SHEPARD, COLM FARRELL, and MYRIAM ROCHDI

GloboMax, A Division of ICON plc, Marlow, Buckinghamshire, U.K.

INTRODUCTION

The usefulness of an in vitro/in vivo correlation (IVIVC) during product development depends on how accurately it can predict resultant plasma concentrations from any given set of in vitro data. This, in turn, is heavily dependent on the design of the in vitro and in vivo studies used to develop and validate the IVIVC. The design of in vitro studies is covered in another chapter, but the temporal aspect of the in vitro study as it relates to the IVIVC will be covered here. The major emphasis of this chapter, however, will be the design of the in vivo study.

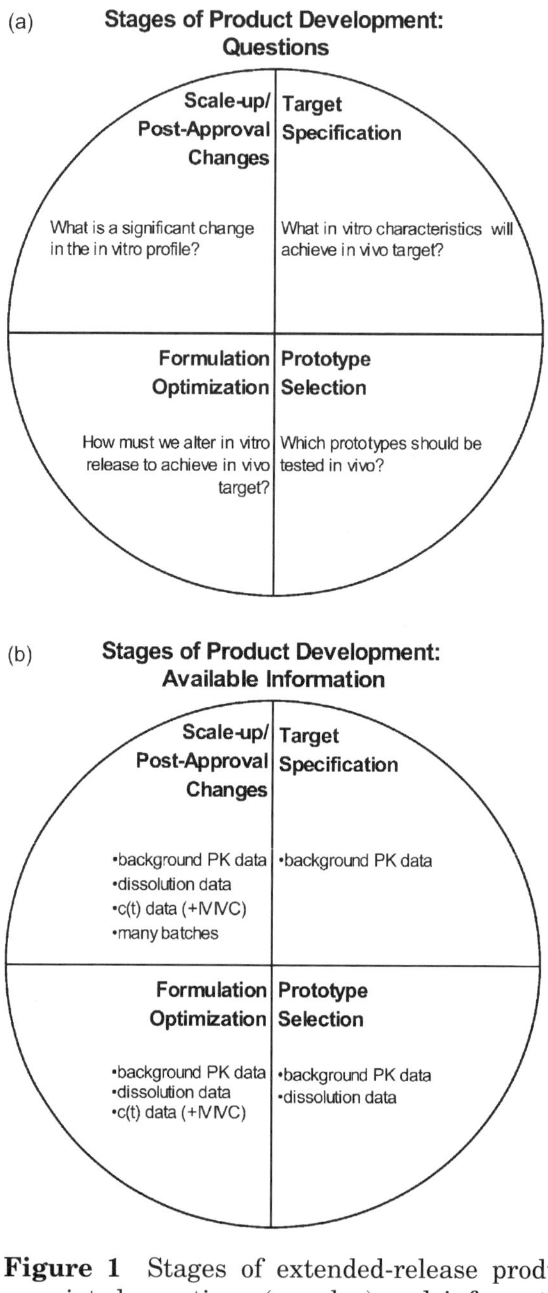

Figure 1 Stages of extended-release product development and associated questions (panel a) and information available at each stage (panel b).

For perspective, it is useful to start with the role of IVIVC in the development of extended-release (ER) formulations. Modeling and simulation, including IVIVC, can be used throughout formulation development to improve the quality of decision-making. The questions of interest during each stage of product development are shown in Figure 1 (panel a). During *target specification*, the development team decides on the type of formulations to develop and specifically what in vitro release profile is likely to achieve the therapeutic objective for the product. This is a critical stage and the thoroughness of the approach here can have a large impact on the success of later stages of development. Once the target is agreed, the responsible formulation team develops numerous formulations, hopefully covering the entire range of dissolution behaviors possible, given the drug and the formulation technology. After this is done, the next stage, *prototype selection*, involves selecting a few formulations (ideally at least three for any one release mechanism) to be tested in a pilot pharmacokinetic (PK) study. After the first PK study, *formulation optimization* may be necessary if the desired target profile has not been achieved. Once the ideal formulation has been identified through one or more PK studies, the formulation is scaled up and may go through other pre- and/or post-approval changes. A reliable IVIVC is especially useful during this stage (*scale-up and post-approval changes*, SUPAC) to predict the impact of any resultant changes in the in vitro profile on plasma concentration (and possibly, on the therapeutic effect).

At each of these stages, not only do the questions of interest change, but so also does the quality of the information available to answer these questions (Fig. 1; panel b). During *target specification*, all available pharmacokinetic characteristics are used to build a suitable model (e.g., disposition of the drug after administration of an immediate-release (IR) tablet, oral solution, or intravenous dose; dose-proportionality; time-dependence; metabolism and pharmacological activity of metabolites; efficiency of absorption from various sites; etc.). However, since no formulations have yet been developed, the in vitro release behavior is unknown, as is the relationship between in vitro release and in vivo release/absorption (IVIVC). Thus, the shape of the in vitro release profile (e.g., constant rate, first order, Weibull, etc., as described in Chapter 9) must be assumed as well as the IVIVC. Often at this stage, it is assumed (either explicitly or without notice) that in vitro release will exactly mimic in vivo release, that the IVIVC follows a 1:1 relationship. At *prototype selection*, in vitro data are now available and can be used as an input into the model. However, the IVIVC is still unknown. The quantum leap in the reliability of the simulation procedure comes after the first PK study. It is only at this point that the relationship between in vitro release and in vivo release can finally be defined and from this point forward, the derived IVIVC is an integral part of the simulation model. Once at the stage of *SUPAC*, many more batches have been manufactured, critical manufacturing variables and the normal range of dissolution characteristics for the formulation are known and also, additional data may have been added to the initial data set used to develop the IVIVC, giving even more confidence in the model.

The modeling discussed here depends on being able to describe the entire concentration–time curve. This can only be done using a Level A IVIVC (i.e., a point-to-point relationship between in vitro release and in vivo release/absorption). In fact, the U.S. Food and Drug Administration (FDA) defines a Level A IVIVC as a predictive mathematical model for the relationship between the entire in vitro dissolution-release time course and the entire in vivo response time course.

REGULATORY GUIDANCE DOCUMENTS

There are a number of FDA regulatory guidances that are associated with IVIVC development and validation, as well as the application of IVIVC to SUPAC. The specific IVIVC guidance for oral modified-release formulations was first published in September 1997 (1). There are several guidances on SUPAC, including those for both modified release (2) and immediate-release solid oral dosage forms (3). The recent guidance on bioavailability (BA) and bioequivalence (BE) studies for oral products (4) also provides information on the application of IVIVC models.

The Committee for Proprietary Medicinal Products (CPMP) within the European Agency for the Evaluation of Medicinal Products (EMEA) has also issued a Note for Guidance on the pharmacokinetic and clinical evaluation of modified-release oral products, which provides some information on the development and evaluation of an IVIVC (5).

This chapter focuses primarily on the development and evaluation of IVIVC for ER oral products in accordance with the 1997 FDA Guidance. However, as the CPMP guidance provides almost identical information on these topics, the dis-

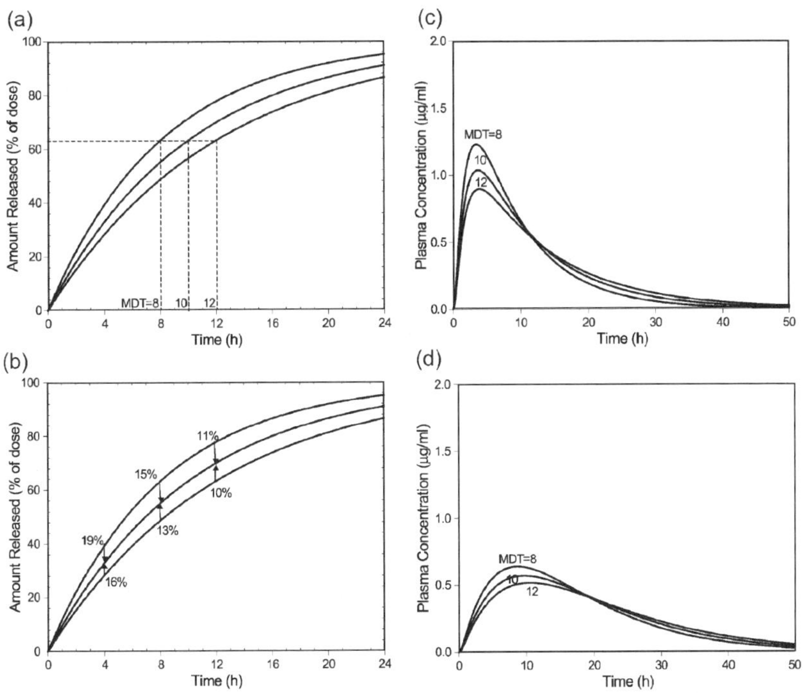

Figure 2 Simulated in vitro drug-release profiles (panels a and b) and resultant plasma concentration–time profiles for a drug with a 1–hr half-life (panel c) and a 6–hr half-life (panel d).

cussions should also serve those working outside of the U.S. regulatory environment.

STUDY DESIGN ELEMENTS

Prototype Selection

In the FDA guidance for development and validation of IVIVCs, it is stated that ideally three formulations of "different" release rates should be used to develop the IVIVC. "Different" is defined as at least 10% difference in the in vitro release profiles between the slow and medium formulations (refers to an absolute difference; e.g., 40–60% if the target is 50%) and between the medium and fast formulations, *and* at least 10% difference in the resultant plasma concentration–time profiles ($C_{\max}$ and/or AUC). This is an important concept. The aim of an IVIVC study is not to show bioequivalence. Formulations should be as different from one another as practically possible, while maintaining the same mechanism of release. The range of dissolution behavior selected is an important determinant of the usefulness of the IVIVC for later stages of development (including setting dissolution specifications and biowaivers for post-approval changes), because the IVIVC can legitimately only be used to make predictions over the range of dissolution data that were used in its development and validation.

Prototype selection is never wisely made based solely on in vitro dissolution data. This is because the resultant plasma concentration–time profiles are dependent not only on this input rate, but also on the pharmacokinetics of the particular drug. This is illustrated in Figure 2.

Here (simulated) in vitro release profiles that differ by at least 10% are shown (panels a and b), as well as the (simulated) resulting plasma concentration–time profiles for a drug with a 1–hr half-life (panel c) and 6–hr half-life (panel d). The simulated-release profiles are described by the following Weibull equation:

$$x_{\text{vitro}}(t) = F_{\text{inf}}\left[1 - e^{-(t/\text{MDT})^{b}}\right]$$

Table 1 Comparison of Predicted Pharmacokinetic Parameters for Two Different Drugs with Identical In Vitro Drug Release Profiles, But Different Drug Disposition Characteristics ($t_{1/2} = 1$ or 6 hr)

$t_{1/2}$(hr)	MDT (hr)	C_{max} (μg/mL)	AUC (μg.hr/mL)	Percentage difference[a]
1	8	1.23	14.4	
	10	1.04	14.4	18.7
	12	0.896	14.4	15.5
6	8	0.637	14.4	
	10	0.569	14.4	11.9
	12	0.516	14.4	10.3

[a]Percentage difference in C_{max} values between the 8 and 10 hr formulations and the 10 and 12 hr formulations.

where $x_{vitro}(t)$ is the amount of drug released from the formulation at time, t (percentage of dose), F_{inf} is the fraction of drug released at time infinity (percentage of dose), MDT is the mean dissolution time (corresponds to time for 63.2% dissolution) and b is the slope factor, which describes the sigmoidicity of the release profile.

Only the mean dissolution time differs among the profiles (MDT = 8, 10, and 12 hr; panel a). The release profiles fulfill the FDA criteria of showing at least a 10% difference in release between the slow and medium and medium and fast formulations (16% and 19% at 4 hr, 13% and 15% at 8 hr, and 10% and 11% at 12 hr respectively; panel b). The resultant plasma concentrations for two different drugs with exactly the same dissolution profiles are shown in panel c for a rapidly eliminated drug ($t_{1/2}$ = 1 hr) and in panel d for a drug that is more slowly eliminated ($t_{1/2}$ = 6 hr). The associated derived pharmacokinetic parameters are listed in Table 1. For the rapidly eliminated drug, the in vivo differences in the formulations are predicted to be adequate (15.5% and 18.7% difference in C_{max}), but borderline for the more slowly eliminated drug (10.3% and 11.9%). As will be shown in a later example, observed differences are often less than predicted, and so erring on the high side when choosing formulations is prudent. These simulations assumed a 1:1

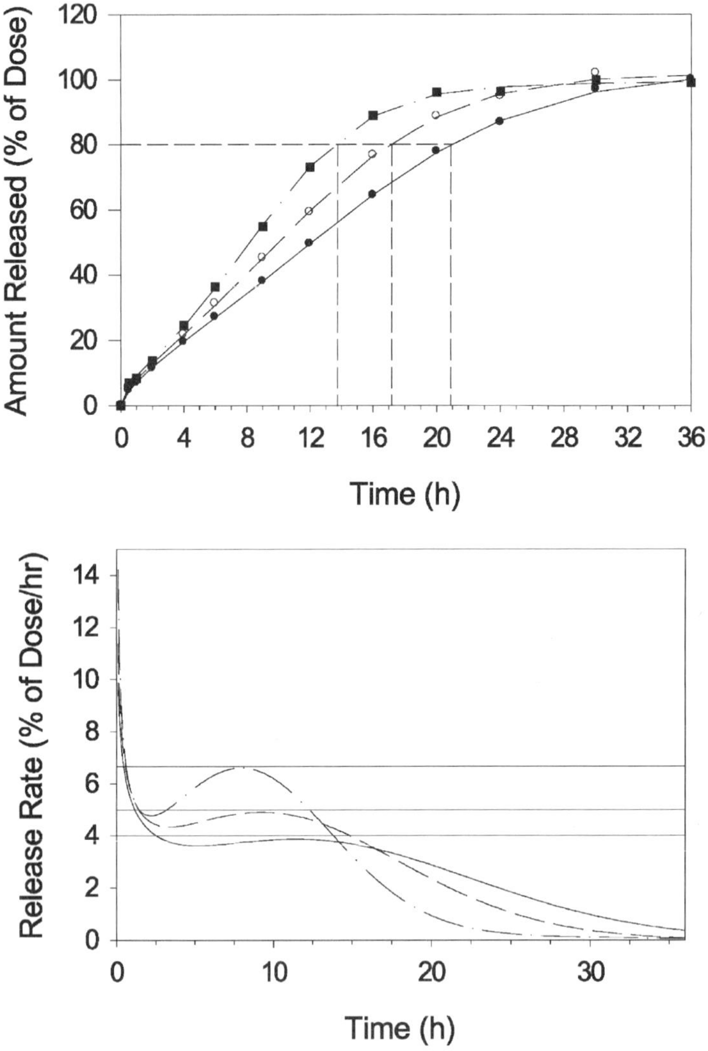

Figure 3 Observed in vitro dissolution data for three ER formulations (panel a): fast (■ target $t_{80\%}$=12 hr), medium (○; target $t_{80\%}$=16 hr), and slow (•; target $t_{80\%}$ = 20 hr). Also shown are the predicted lines corresponding to fitting the data to the double Weibull equation (fitted parameter values are listed in Table 2). The associated rate plot for the three formulations is shown in panel b (fast, ————; medium, — — —; slow, ——).

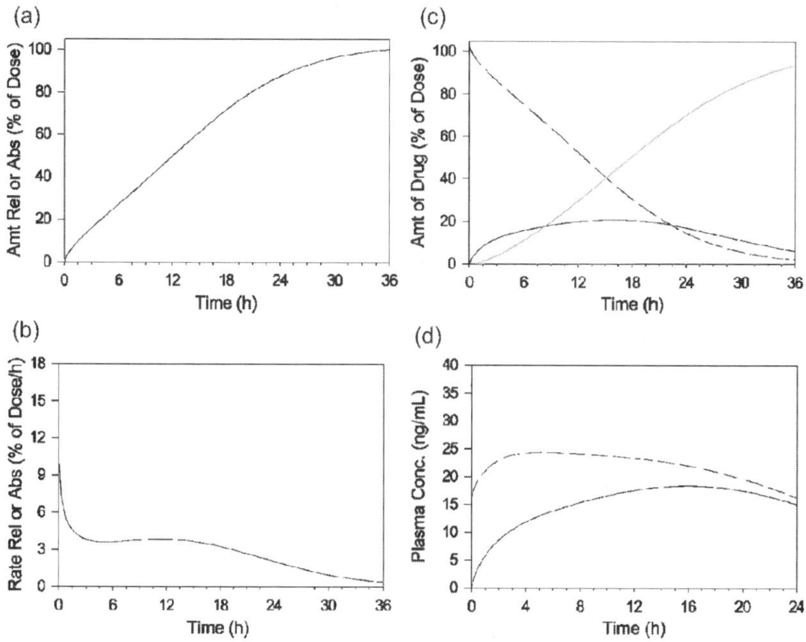

Figure 4 Simulation output for the slow formulation whose dissolution behavior is shown in Figure 3. Pharmacokinetic parameters: $F=1$, $k_a=1000\,\text{hr}^{-1}$, $k_{10}=0.17\,\text{hr}^{-1}$, $V_1=114\,\text{L}$, $f_{col}=1$, $t_{col}=9\,\text{hr}$, $t_{abs}=96\,\text{hr}$. Dosing parameters: dose $=10\,\text{mg}$, $\tau=24\,\text{hr}$. IVIVC equation: $x_{vivo}=x_{vitro}$ (1:1 IVIVC). Double Weibull (drug release) parameters: $F_{inf}=102\%$, $f_1=0.349$, $\text{MDT}_1=6.85\,\text{hr}$, $b_1=0.783$, $\text{MDT}_2=18.7\,\text{hr}$, and $b_2=2.11$ (Table 2). Panel a shows two lines, one for in vitro release (——) and the other for in vivo absorption (— — —). Panel b shows two lines, one for the in vitro release rate (——) and the other for the in vivo absorption rate (— — —). Panel c shows the amount of drug in the drug delivery system (— — —), GI tract (follows x-axis), central compartment (——), and the total in all compartments (for mass balance, ——; cumalative line). Panel d shows the simulated plasma concentration after single dose (——) and at steady state (— — —).

IVIVC. However, if some *a priori* information suggests a different relationship (perhaps technology-specific) or a range of relationships, then it would make sense to use these to aid the formulation–selection decisions.

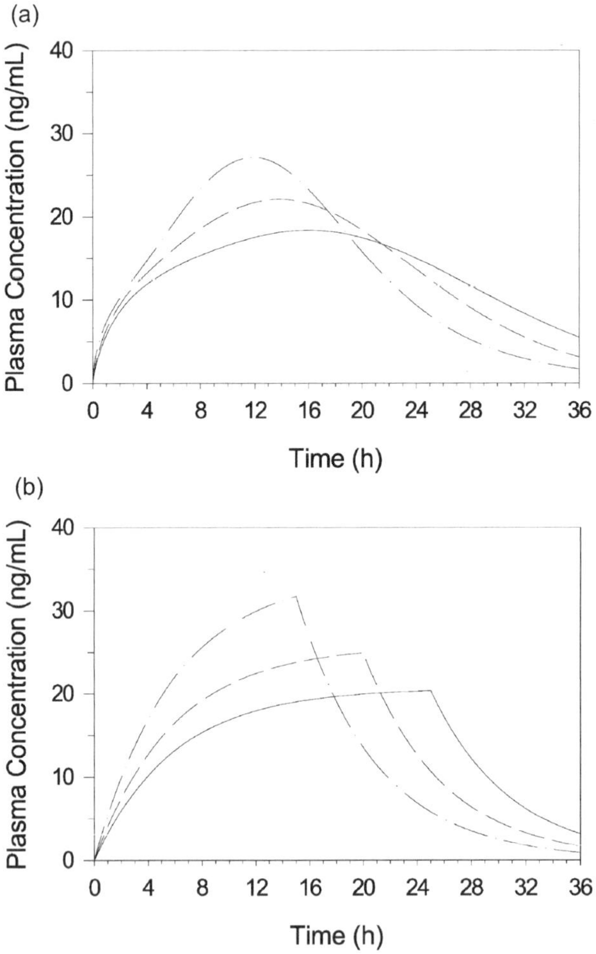

Figure 5 Predicted concentration–time profiles for the three extended-release formulations (fast, —·—·— ; medium, — — —; slow, ——), using the pharmacokinetic model shown in Appendix A, the fitted Weibull parameters listed in Table 2 and the remaining model parameters as listed in Figure 4 (panel a) or the assumed zero order release rates of 4%, 5%, and 6.7% per hour (panel b).

A specific example showing the application of these principles within a development program for an ER dosage form is shown in Figures 3–5. A generalized pharmacokinetic model that can be used to support prototype selection is shown in

Table 2 Fitted Weibull Parameters for the Three In Vitro Drug-Release Profiles Shown in Figure 3

Formulation	Fast	Medium	Slow
f_1	0.317	0.273	0.349
$F_{\text{inf}}(\%)$	100	102	102
MDT_1 (hr)	5.60	4.39	6.85
b_1	0.646	0.759	0.783
MDT_2 (hr)	11.1	14.9	18.7
b_2	2.24	2.03	2.11

Appendix A. The model is the simplest body model, assuming instantaneous drug distribution (one compartment body model), thus needing no peripheral distribution compartment, and first order drug elimination. Modifications according to what is known about a particular drug and its absorption, distribution, and elimination characteristics would be necessary to make it appropriate for a particular drug entity. This model has been used to simulate the resulting concentration–time profiles for the dissolution profiles shown in Figure 3 and the output from the model (simulated mass balance and concentration–time profiles) is shown in Figure 4 for the slowest formulation.

The target-release durations for prototype development were 12, 16, and 20 hr for 80% drug release. The observed release profiles for the three formulations that most closely met these targets are shown in Figure 3, along with reference lines for actual time for 80% drug release (panel a). For all three formulations, the $t_{80\%}$ values were somewhat longer than the target values (14, 17, and 21 hr vs. 12, 16, and 20 hr, respectively). The cumulative profiles show a close to zero order release profile until between 70% and 80% release, after which the rates of release decline. The release profiles were well described by the double Weibull function

$$x_{\text{vitro}}(t) = f_1 \cdot F_{\text{inf}} \cdot \left[1 - e^{-(t/\text{MDT1})^{b1}}\right] + (1 - f_1) \cdot F_{\text{inf}} \cdot \left[1 - e^{-(t/\text{MDT2})^{b2}}\right]$$

Table 3 Comparison of Predicted Pharmacokinetic Parameters for the Three In Vitro Drug-Release Profiles Shown in Figure 3

Formulation	C_{max} (ng/mL)	AUC (ng.hr/mL)	Percentage difference $(C_{max})^a$	Percentage difference (AUC)[a]
Fast	28.69	493.14	18.45	0.74
Medium	24.22	496.83		
Slow	20.65	498.43	14.77	0.32

[a]Percentage difference in C_{max} and AUC values between the fast and medium formulations and the medium and slow formulations.

where f_1 is the fraction of drug release described by the first Weibull component. The double Weibull equation splits the total drug release between two different fractions with different mean dissolution times and slope factors. The fitted lines are shown in the cumulative release plot in Figure 3 (panel a) and the parameter values are listed in Table 2.

The rates of drug release as a function of time are shown in panel b along with the target-release rates for the three formulations (4%, 5%, and 6.7% per hour for the slow, medium, and fast formulations, respectively). The "observed" rate profiles correspond to the first derivative of the cumulative release and are constructed using the fitted parameter values. The order of drug release is best judged from these rate plots. The release pattern for all three formulations deviates obviously from zero order (constant rate) release. All have an initial "burst" in the release with the initial rate about twice the target rates. The slowest formulation comes closest to maintaining a constant release rate with little fluctuation in the release rate up to 15 hr.

The predicted concentration–time profiles for all three formulations are shown in Figure 5 (panel a). These simulations use the fitted in vitro profiles for input to the model. For comparison, the simulations assuming zero order release are shown in panel b. Although the zero order simulations may be useful for initial specification of target profiles, they offer little of value for selecting specific formulations for the in vivo study or for study design (e.g., selection of sampling times), since the predicted peak concentrations tend to be higher and the decline in concentrations at later times more precipitous.

The expected $C_{\max}$ and AUC for each of the profiles are listed in Table 3. The profiles are predicted to show an acceptable range of $C_{\max}$ values with around 20% difference between the fast and medium formulations and between the medium and slow formulations. The predicted differences in AUC are only related to the slightly different content of the three formulations, reflected in the F_{inf} values (100% for the fast formulation and 102% for the other formulations). Normally, AUC is not expected to be rate-dependent unless there is some non-linear process involved in the disposition of the drug or drug release or absorption is very slow compared to gastrointestinal transit time. Given the predicted $C_{\max}$ differences, these three formulations are appropriate choices for an IVIVC study as they show acceptable in vitro and predicted in vivo differences.

Sampling Times

As mentioned above, sampling time decisions are best made based on simulations using the actual (or modeled) in vitro release data for the clinical batches manufactured for the IVIVC study. Assumed zero order release profiles are likely to be misleading in terms of the shape and duration of plasma profiles (compare panels a and b in Figure 5). If the in vitro dissolution is pH or rotation-speed dependent, it is useful to do simulations using the range of in vitro dissolution profiles in order to design a sampling regimen to cover the range of potential in vivo behaviors. Also, if there is some a priori understanding of the likely IVIVC relationship, this is best built into the initial simulation. For example, for injectable ER formulations, in vitro release testing is often designed to be complete within 24–48 hr, while the in vivo delivery is designed to continue for 1–2 months. Thus, a time-scaling factor (or range of factors) can be anticipated a priori and built into the model to provide a more realistic picture of the expected in vivo behavior and better guide the choice of appropriate sampling times for the test formulations.

In vitro sampling times are also critical to the quality and predictability of the developed IVIVC. Best practice is to characterize the entire in vitro release profile until a definite plateau has been reached (judged by three consecutive points within 5% of each other). On-line detection systems are particularly useful for this purpose, but may not always be possible. If not, in vitro tests for early formulations covering a wide range of in vitro behavior should be oversampled and then modeling techniques can be used to identify critical sampling time points. These time points can then be used with confidence for clinical batches (assuming these are within the range of dissolution behaviors initially tested). The plateau is particularly important to characterize because it determines the ultimate amount of drug delivered by the system. That is, if sampling is carried out only up to 90% release, this leaves 10% of the dose unaccounted for, with a predicted AUC 10% lower than it should be given the tablet content.

Role and Choice of Reference Formulation

The reference formulation is used to correct for differences in drug clearance between study populations when data from more than one study are combined. The reference formulation is chosen so that when it is used in deconvolution with the ER formulation, the in vivo drug release or absorption from the ER formulation is obtained. Appropriate reference formula-

Table 4 Formulations and Studies for ISMN GEOMATRIX

	Study number			
	194.573	196.581	372.05/ 196.638	372.02
Number of subjects	12	8	8	25
Batch number	R4K21F R4K22F R4K23F	S6H32E	R6M12E2 R6M12E3	N970039
IMDUR batch no.	3-DJC-6	3-DJC-16	3-DJC-16	3-DJC-16

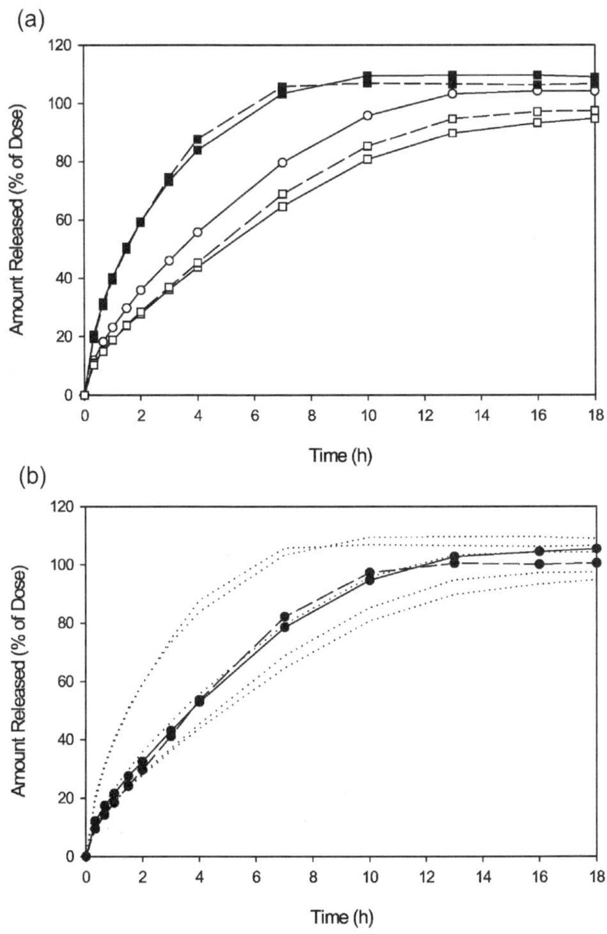

Figure 6 In vitro release profiles for ISMN GEOMATRIX formulations. The small-scale batches used for IVIVC development and validation are shown in panel a, and the large-scale batches used for external validation are shown in panel b, with dotted line tracings for the small-scale batches. IVIVC development included two fast (■), one medium (○), and two slow batches (□), while external validation included two medium batches (●).

tions include IV solutions, immediate-release formulations, or oral solutions.

Including a reference formulation is always recommended, even if a specifically designed IVIVC study is

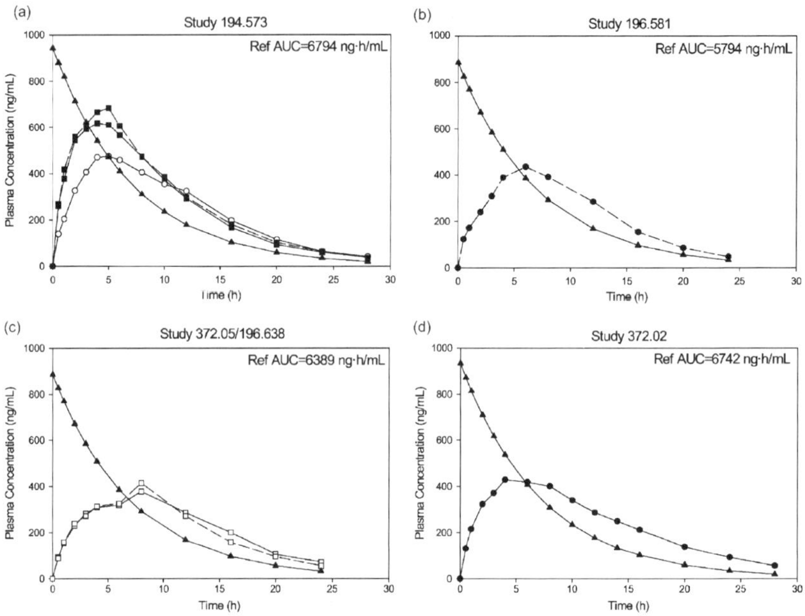

Figure 7 Observed concentration–time data for ISMN from the test extended-release formulations included in the four PK studies. The profile for the reference formulation (▲) is represented as an intravenous injection with the same AUC as the reference extended-release formulation (IMDUR) and the literature elimination half-life of 3.77 hr. IVIVC development included the two fast (■) and one medium (○) batch from Study 194.573 and two slow batches (□) from Study 372.05/196.638 and external validation included the two medium batches (•) in Studies 196.581 and 372.02.

included in the development program, as it is useful to leave open the possibility of adding other formulations to the IVIVC at a later date (either for inclusion in the IVIVC itself or for external validation of the IVIVC). The impact of the reference formulation on the validation statistics for an IVIVC is illustrated with the example of an ISMN GEOMATRIX^{TM} formulation developed using a patented hydrophilic matrix technology (SkyePharma AG, Muttenz, Switzerland). A total of seven batches were studied in vivo. The batches differed in the number of barrier layers used, the quality of HPMC used and the blend and supplier of active material. The details of the pharmacokinetic studies are listed in Table 4. There were a total of five small-scale batches and two large-scale batches. There was no one study that contained three formulations that differed sufficiently in their release rates, so it was necessary to combine data from at least two studies for IVIVC development. The small-scale batches (R4K21F, R4K22F, R4K23F, R6M12E2, and R6M12E3) were used for IVIVC development and internal validation and the large-scale batches (S6H32E and N970039) for external validation. A common reference formulation was included in all studies, an ER reference formulation, IMDUR™. The in vitro dissolution data for five batches used in IVIVC development and internal validation are shown in Figure 6 (panel a), while those for two large-scale batches used for external validation are shown in panel b. The small-scale batches differ sufficiently in vitro (i.e., $>10\%$) for IVIVC development and validation according to FDA guidelines. The observed mean concentration–time data are shown in Figure 7 for all ER formulations and for an IV reference concentration–time curve (constructed using the data from the reference ER formulation and a literature elimination rate constant of $0.1836\,\text{hr}^{-1}$ for ISMN). This choice of reference is an atypical one and is not absolutely ideal because of the need for construction of an impulse response function from it. More appropriate reference formulations include IV, oral solution, or oral immediate-release formulations. However, reference ER formulations fit naturally into the development program for generic ER products and do give an indication of clearance differences across studies. Their usefulness depends very much on the variability of the product in question relative to an immediate-release formulation and in this case was very low (intrasubject CV% approximately 4%).

The AUC associated with the mean profile for the reference, IMDUR, differs by a maximum of 17% across the studies. The reference IV profiles, constructed on an individual subject basis, were used to deconvolve the GEOMATRIX formulation data to derive the percentage absorbed for each formulation relative to the reference, from which the mean absorption profiles were calculated. The derived mean absorption vs. time profiles are shown in Figure 8. Mean absorption vs. the percentage released in vitro at the same time was plotted for each of the formulations and a time-scaled IVIVC equation applied. These plots are shown in Figure 9. The left-hand side panel is for the analysis where the reference formulation in each study is used for deconvolution and the right-hand side panel where only the reference data for Study 194.573 are used. Although there is more variability when the study-specific reference data are not used, the derived IVIVC equation is very similar.

However, the real test of an IVIVC is whether it can accurately predict plasma concentration. This involves convolution of the predicted absorption data with those of the unit impulse response function derived from the reference product data. And this is where the reference data are crucial. The prediction errors for the small- and large-scale batches used for internal and external validation, respectively, are listed in Table 5. The FDA acceptance ranges for prediction errors for $C_{\max}$ and AUC are $\leq 15\%$ for internal validation of individual batches, and $\leq 10\%$ on average and $\leq 10\%$ for external validation. The left-hand side columns for study-specific reference and right-hand side columns list the results of convolution using the Study 194.573 reference data across all studies. This disregard for cross-study differences in study populations has turned an acceptable IVIVC, with all its inherent advantages, into an unacceptable one. Thus, prospective use of a reference formulation in studies to be included in IVIVC analysis greatly improves the probability of being able to successfully validate and reliably use the IVIVC.

Often the first few pilot PK studies in formulation development are not aimed for the specific purpose of IVIVC. However, it is normally in these first studies that the greatest difference in release rates is seen, before settling on a target profile, making them very valuable for IVIVC development. Prospective inclusion of an appropriate reference formulation can allow these valuable data to be used retrospectively for the purpose of IVIVC.

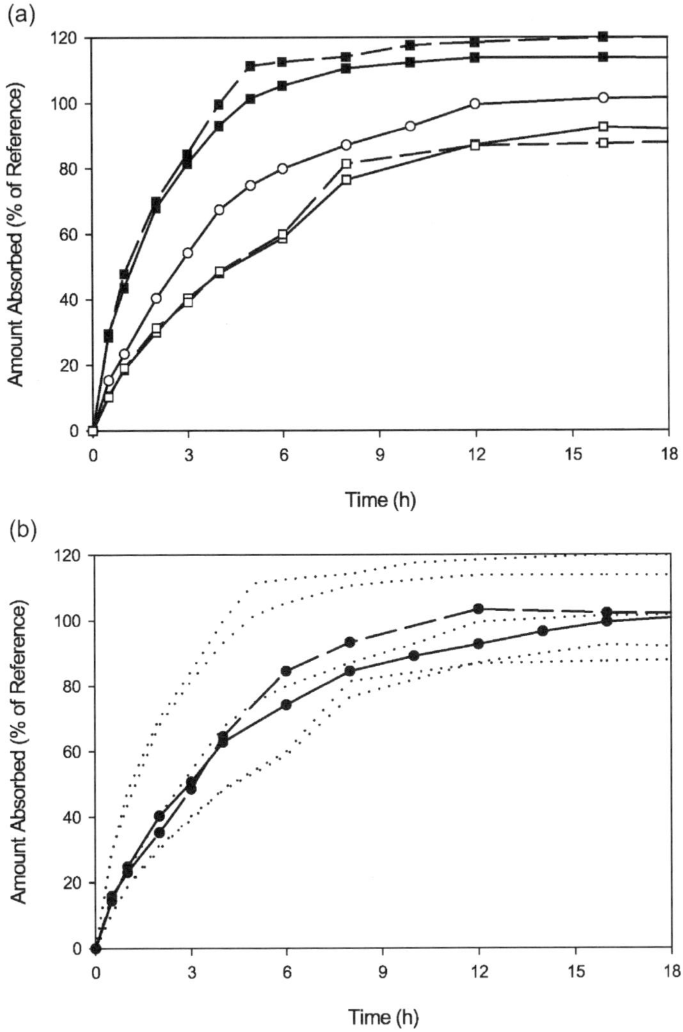

Figure 8 In vivo absorption profiles for ISMN GEOMATRIX formulations. The small-scale batches used for IVIVC development and validation are shown in panel a and the large-scale batches used for external validation are shown in panel b, with dotted line tracings for the small-scale batches. IVIVC development included two fast (■), one medium (○), and two slow batches (□), while external validation included two medium batches (•).

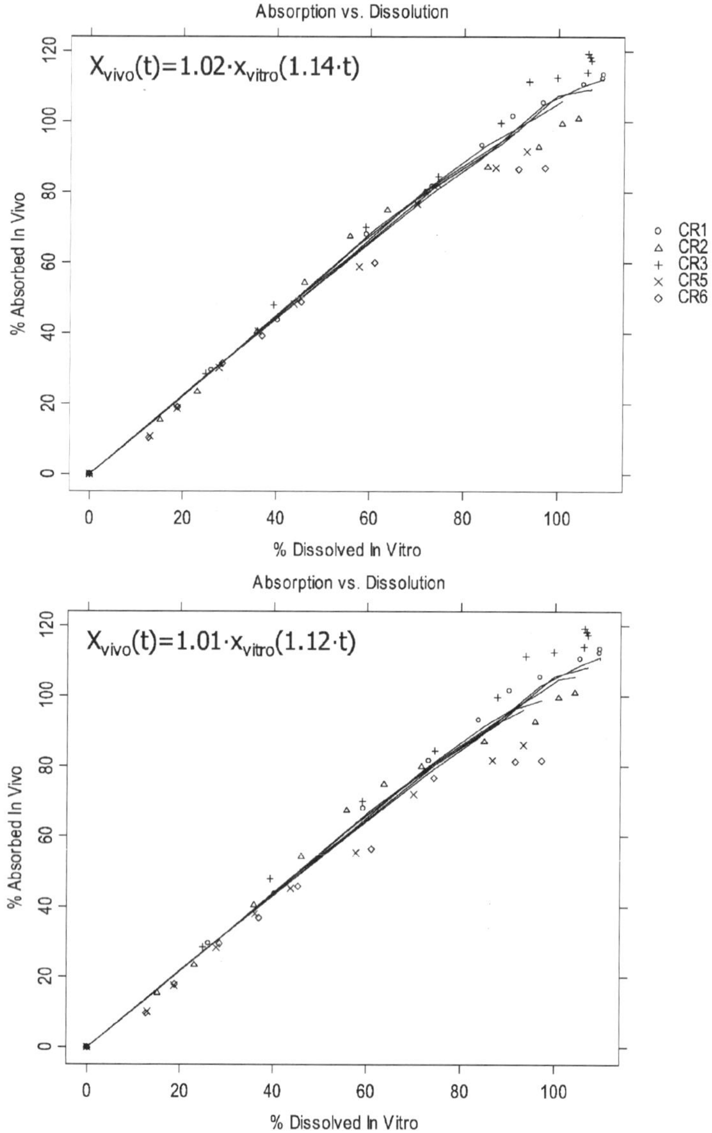

Figure 9 Observed data (amount absorbed in vivo vs. amount released in vitro) for the five ISMN test formulations included in IVIVC development and internal validation. The fitted IVIVC equations are shown as well as the corresponding predicted lines. Panel a shows the analysis where the study-specific reference was used for deconvolution and panel b where the reference for Study 194.573 was used for the deconvolution analysis of all study data.

Table 5 Prediction Errors Associated with ISMN GEOMATRIX IVIVC Developed Using the Study-Specific Reference Data for Deconvolution or the Reference Data from Study 194.573 for Deconvolution of All Study Data. Prediction Errors Outside of the FDA Acceptance Criteria Are Indicated in Bold

	Reference in every study		Reference in Study 194.573 only	
Batch	C_{max} PE(%)	AUC PE(%)	C_{max} PE(%)	AUC PE(%)
Internal validation				
R4K22F	4.63	2.91	6.09	3.86
R4K23F	9.42	9.61	10.67	10.5
R4K21F	0.569	4.32	2.93	3.27
R6M12E2	1.27	4.78	2.92	14.3
R6M12E3	3.91	12	0.0229	**22.0**
Average	3.96	6.73	4.53	**10.8**
External validation				
S6H32E	1.03	0.131	**13**	16.2
N970039	9.1	4.92	7.46	4.65
Average	5.07	2.53	10.2	10.4

PE, absolute value of the prediction error.

Crossover Study Design

The FDA guidance on IVIVC development and validation states that crossover studies are preferred; however, parallel studies or cross-study analyses may be acceptable. The advantage of a crossover study is that it avoids bias to any one particular treatment as a result of a period effect. A crossover study also provides the highest probability of successfully validating the IVIVC, since it avoids the variability introduced by cross-study comparisons.

IVIVC studies normally involve two to four ER formulations and a reference formulation (e.g., IV solution, immediate release, or oral solution). Data analysis involves deconvolution of each ER formulation, using the reference data for each subject. Thus, if a subject drops out of the study prior to the IR arm, none of that subject's data can be used for IVIVC development. To address this, the reference formulation can be dosed to all subjects during the first study period and the remainder of ER treatments randomized across the remaining study periods. The advantage of this approach is that it maximizes the number of subjects that can be included in the deconvolution analysis for the ER formulations. The disadvantage is that the same subjects are not contributing to the mean absorption data for all treatments. The choicc of design must be judged based on number of subjects in the study, the anticipated drop-out rate and the variability of the drug in both the reference and ER formulations.

For a product where it is desired or necessary to show external predictability (e.g., to bridge to the commercial product for a low therapeutic index product), the external validation batch can be included in the same study as the IVIVC batches, normally in a separate study arm (i.e., not randomized). This reduces the probability of failing to fulfill the strict external validation criteria (prediction errors for $C_{\max}$ and AUC of $\leq 10\%$), as the data are collected in the same study population as those used to develop and validate the IVIVC.

Parallel group studies are not particularly useful for IVIVC development, as by definition, subjects receive only one treatment and so there would be no reference for each subject for individual deconvolution. This becomes less problematic as the variability of the drug declines. Thus, it may be acceptable for a low variability drug to use a mean reference profile for deconvolution of the mean profile for each ER treatment.

Cross-study comparisons are common at some stage during IVIVC development and indeed are to be encouraged during the duration of formulation development, through scale-up and production of commercial batches. As an illustration, early formulations may be included in a crossover study for IVIVC model development and validation. Later changes to the formulation may prompt another PK study, which can then also be incorporated into the IVIVC or at least used for external validation, depending on the impact on dissolution (i.e., if extending the dissolution range, then it is useful to include in the IVIVC, otherwise may be used for external validation).

Retrospective IVIVC development, using studies not designed for this purpose, reduces the probability of successful IVIVC development and validation. Normally such studies are compromised by not including a reference formulation and do not have a large enough range of release rates, thereby requiring cross-study comparisons where subjects have different clearance characteristics that could have been accounted for had a reference formulation been incorporated.

Systematic inclusion of an IVIVC study in the development plan for ER formulations is a wise strategy for such products, given the usefulness of this relationship throughout the development process.

Number of Subjects

The current guidelines for IVIVC development and validation state that studies for IVIVC development should be performed with enough subjects to adequately characterize the performance of the drug product under study. Acceptable data sets have ranged from 6 to 36 subjects.

Unless a product has particularly low variability, a minimum of 12 subjects is advised. A higher number will be necessary if the drug/drug product is highly variable.

Fasting vs. Fed IVIVC Study

IVIVC studies are normally conducted in the fasted state. Where a product is not tolerated in the fasted state, studies may be conducted in the fed state (1). Some drugs are labeled to be administered with food, either to take advantage of greater bioavailability or lessen the incidence of adverse events. For such formulations, it could be argued that the IVIVC model should be developed using in vivo data obtained under fed conditions, so that the model predicts the in vivo performance under the intended condition of administration. We have had recent experience in successfully correlating the in vivo performance of an ER product administered with food, as intended, and the corresponding in vitro dissolution profile, obtained using modified simulated gastric fluid.

USEFULNESS OF AN IVIVC

Product Development

The value that an IVIVC can add to the accuracy of translating in vitro data to expected in vivo behavior is illustrated in Figures 10–13.

Figure 10 shows, in panel a, the observed mean concentrations for the fast medium and slow formulations, whose in vitro dissolution data are shown in Figure 3 and predicted concentration–time curves in Figure 4. The mean absorption profiles obtained by deconvolution are shown in Figure 10 (panel b). A rank order correlation is seen between in vitro and in vivo, whereby the fast, medium, and slow ordering is the same in both. The relationship between in vitro release and in vivo absorption is shown in Figure 11. A 1:1 relationship is shown by the dotted line. For this product, absorption is faster than in vitro release. The IVIVC relationship is described as a 4th order polynomial, but other functions (i.e., Hill equation, time-scaling model) could also be used. The impact of the IVIVC on the simulations of cumulative absorption, absorption rate, mass balance, and plasma concentration is shown for the slow formulation in Figure 12. In contrast to the simulations assuming a 1:1 IVIVC (Fig. 4), there is now a differentiation of release and absorption, in that the absorption is faster but plateaus at less than 100% (panel a) and has a shorter period of nearly constant rate input (panel b). Mass balance shows less than 100% absorption (panel c). Steady-state concentrations are expected to show a larger peak to trough difference than would be predicted given the in vitro profile and no knowledge of its IVIVC (compare panel d in Fig. 12 and in Fig. 4).

The predicted concentration–time profiles with and without an IVIVC are shown in Figure 13. Here, it can be seen

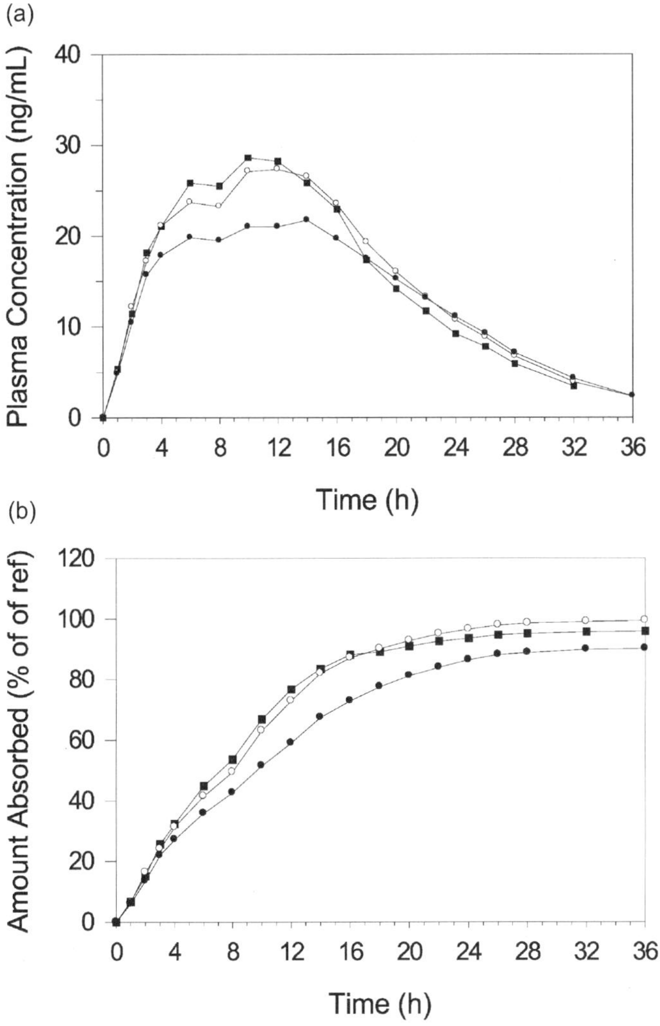

Figure 10 Mean observed concentration–time profiles for the three extended-release formulations, fast (■), medium (○), and slow (•), whose in vitro dissolution data are shown in Figure 3 (panel a) and the derived mean absorption–time profiles (panel b).

that without the IVIVC (particularly for the medium and slow formulations), the shape of the concentration–time profiles is badly predicted. The impact on the BE parameters can be

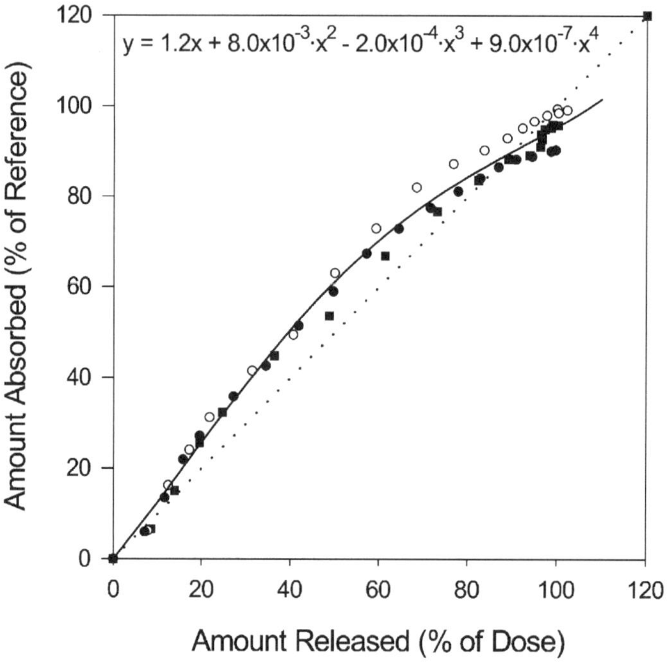

Figure 11 Observed data (amount absorbed in vivo vs. amount released in vitro) for the three ER formulations whose dissolution data are shown in Figure 3 and absorption–time profiles in Figure 10. The fitted IVIVC equation is shown as well as the corresponding equation and predicted line. The dotted line represents a 1:1 relationship.

particularly important for specification setting and assessing the risk associated with running a BE study between two formulations, etc. The impact of the IVIVC on these prediction errors is listed in Table 6. In this example, the IVIVC reduces the prediction errors by a factor of 2 on average.

Once a product is developed that meets a company's needs in terms of efficacy and safety, no one wants to change it. This is particularly true once in phase 3 trials, where there is a risk of compromising the safety and efficacy database. However, for many reasons, changes are inevitable. The key is to manage any changes so that they do not impact negatively on efficacy and safety. In the absence of an IVIVC,

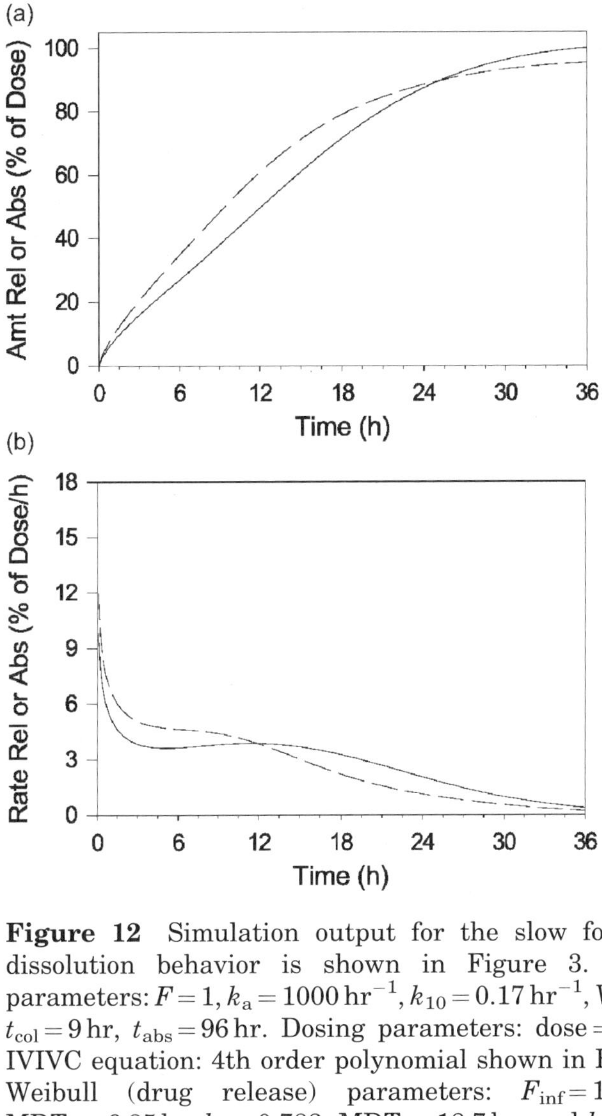

Figure 12 Simulation output for the slow formulation whose dissolution behavior is shown in Figure 3. Pharmacokinetic parameters: $F = 1$, $k_a = 1000\,\mathrm{hr}^{-1}$, $k_{10} = 0.17\,\mathrm{hr}^{-1}$, $V_1 = 114\,\mathrm{L}$, $f_{col} = 1$, $t_{col} = 9\,\mathrm{hr}$, $t_{abs} = 96\,\mathrm{hr}$. Dosing parameters: dose = 10 mg, $\tau = 24\,\mathrm{hr}$. IVIVC equation: 4th order polynomial shown in Figure 11. Double Weibull (drug release) parameters: $F_{inf} = 102\%$, $f_1 = 0.349$, $\mathrm{MDT}_1 = 6.85\,\mathrm{hr}$, $b_1 = 0.783$, $\mathrm{MDT}_2 = 18.7\,\mathrm{hr}$, and $b_2 = 2.11$ (Table 2). Panel a shows two lines, one for in vitro release (——) and the other for in vivo absorption (— — —). Panel b shows two lines, one for the in vitro release rate (——) and the other for the in vivo absorption rate (— — —). Panel c (see p. 308) shows the amount of drug in the drug delivery system (— — —), GI tract (follows x-axis), Central Compartment (——) and the total in all compartments (for mass balance, ——; cumulative line). Panel d shows the simulated plasma concentration after single dose (——) and at steady state (— — —). (*Continued.*)

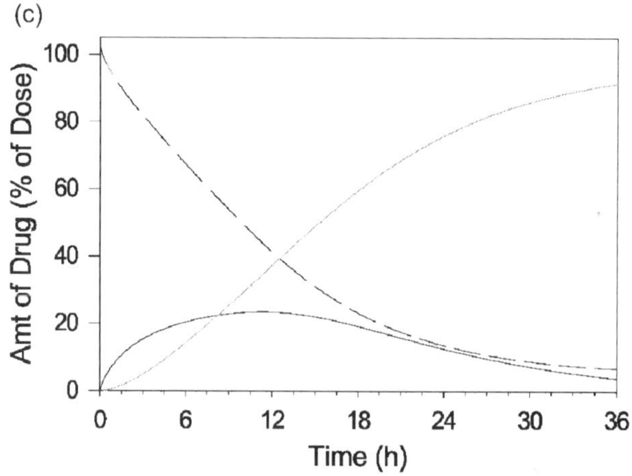

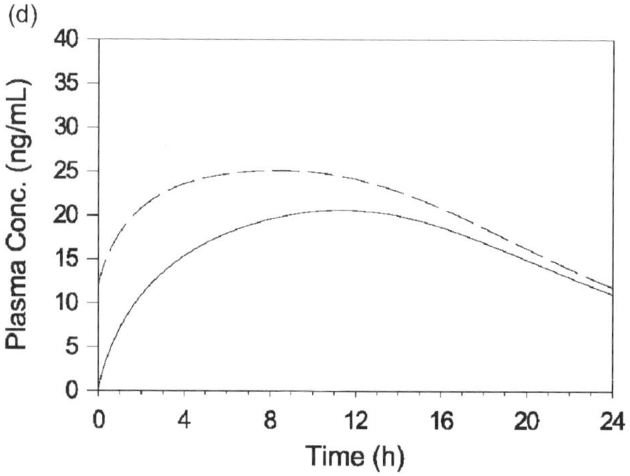

Figure 12 *(Continued)*

this is typically done according to the procedure shown on the left-hand side of the diagram in Figure 14. The new process (or formulation) is used to produce GMP material, which is subjected to dissolution testing. If the in vitro data are acceptable, then a semiquantitative/qualitative decision is made as to whether to progress to a BE study between batches produced with the new process vs. the old. If the two products are shown to be bioequivalent, then the new process is substituted for the old in the development program. If not,

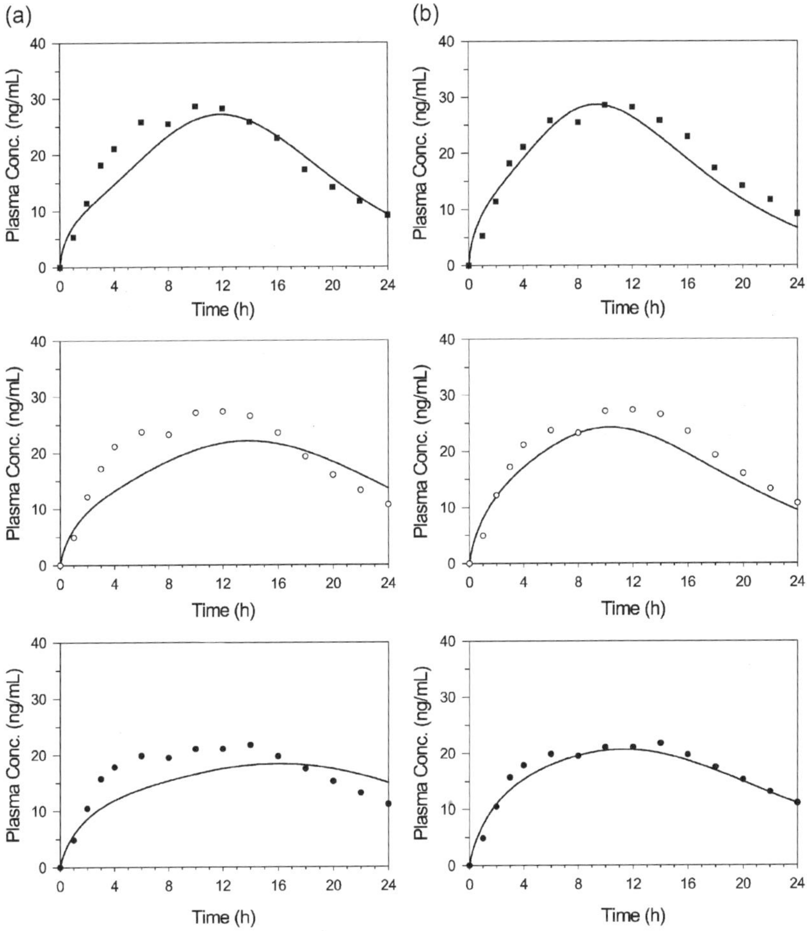

Figure 13 Comparison of the mean observed and predicted concentration–time profiles for the three ER formulations, fast (■), medium (○), and slow (●), whose dissolution behavior is shown in Figure 3. Pharmacokinetic parameters: $F = 1$, $k_a = 1000\,\text{hr}^{-1}$, $k_{10} = 0.17\,\text{hr}^{-1}$, $V_1 = 114\,\text{L}$, $f_{\text{col}} = 1$, $t_{\text{col}} = 9\,\text{hr}$, $t_{\text{abs}} = 96\,\text{hr}$. Dosing parameters: dose $= 10\,\text{mg}$, $\tau = 24\,\text{hr}$. IVIVC equation: $x_{\text{vivo}} = x_{\text{vitro}}$ (1:1 IVIVC; panel a) or 4th order polynomial shown in Figure 11 (panel b). Double Weibull (drug release) parameters for each of the three formulations are listed in Table 2.

Table 6 Prediction Errors Associated with an Assumed 1:1 IVIVC and the Derived 4th Order Polynomial IVIVC Shown in Figure 11

Batch	With IVIVC		Without IVIVC	
	C_{max}(%)	AUC PE (%)	C_{max}(%)	AUC PE (%)
Fast	0.556	9.41	5.19	12.4
Medium	11.4	11.4	19.1	13.3
Slow	4.69	1.50	15.3	10.6
Average	5.55	7.44	13.2	12.1

PE, absolute value of the prediction error.

the cycle starts over again. With an IVIVC (right-hand side), the process is similar, but now the bioequivalence decision is taken on the basis of the in vitro test and validated IVIVC (by predicting concentration–time profiles for new and old and calculating BE differences). The major difference between the two approaches is not the money saved on the BE study, but the time saved. This is particularly important in modern drug development as it avoids decisions taken at risk pending the results of a BE study a few months down the line. Thus, the value of the systematic inclusion of an IVIVC in the pro-

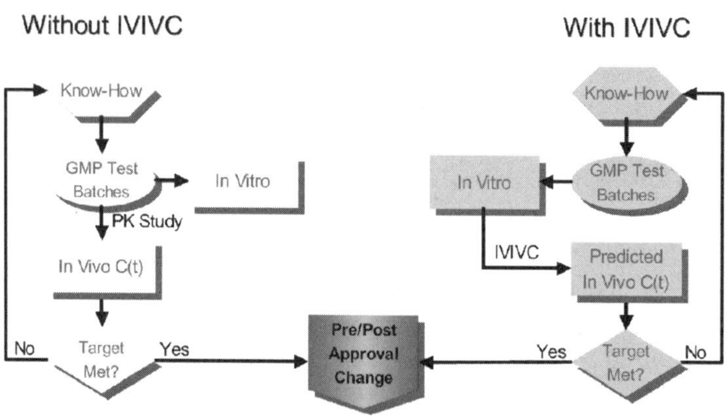

Figure 14 Schematic showing the decision-making process for pre- and post-approval changes with and without an IVIVC.

gram for product development is more timely and reliable decisions.

Regulatory Applications

The FDA guidance on IVIVC development and validation defines a number of circumstances where an IVIVC can be used to justify a biowaiver request: in support of (1) level 3 process changes, (2) complete removal or replacement of non-release-controlling excipients, (3) level 3 changes in release-controlling excipients, (4) approval of lower strengths, and (5) approval of new strengths. Additionally, use of the IVIVC to justify "biorelevant" dissolution specifications is cited as the optimal approach.

CONCLUSION

IVIVC is a valuable tool to be used along with other modeling techniques to improve the efficiency and quality of development decisions for ER dosage forms, to support SUPAC, and to provide a basis for "biorelevant" dissolution specifications. The probability that IVIVC development will be successful can be greatly enhanced by prospective design of the IVIVC strategy at the start of a development program and periodic re-evaluation throughout the development. Informed study design decisions should be an integral part of this strategy.

APPENDIX A

Pharmacokinetic Model for Simulation of Concentration–Time Profiles for Orally Administered Extended-Release Dosage Forms

A generalized pharmacokinetic model that can be used to support prototype selection is shown below.

This model consists of a total of five compartments, the drug delivery system (DDS), the gastrointestinal tract (GIT), the central compartment (Central), and two elimination compartments denoted with a dashed box outline, one for pre-systemic elimination (Unavailable) and one for

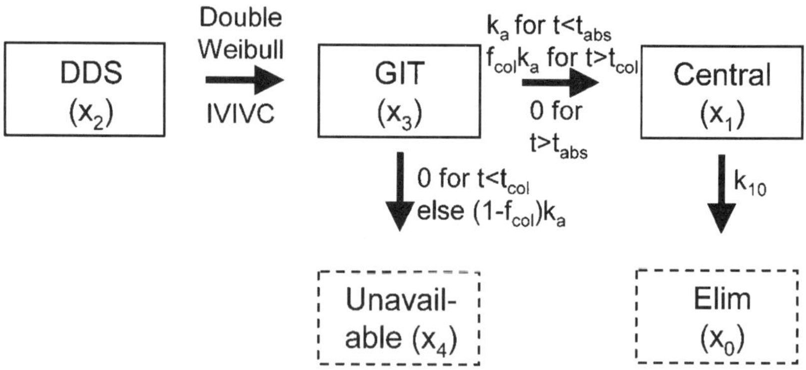

systemic elimination (Elim). Strictly speaking, these elimination compartments are not absolutely necessary, but they are useful as a mass balance check for the system, particularly with complicated IVIVC models. Input from the DDS to the GIT first involves drug release according to the in vitro dissolution time course, followed by a transformation involving the IVIVC, which translates the input into in vivo dissolution. In this particular model, a double Weibull function is used to describe in vitro dissolution; however, any suitable function found to describe the in vitro data can be used. The most common functions include Weibull, sigmoid, Hill, and double Weibull functions. Polynomials are not particularly useful for this purpose, because they do not reach plateaus. Thus, even though they can be used to describe the observed in vitro data, they can give anomalous simulation results. The IVIVC can be any function, but is typically expressed as a direct proportionality, a linear relationship, a polynomial or may be more sophisticated, incorporating time-shifting and/or time-scaling (e.g., PDx-IVIVC®, GloboMax, A Division of ICON plc, Hanover, Maryland, U.S.A.). The model shown above incorporates the possibility of reduced colonic absorption of drug and finite GI transit of the formulation (i.e., fecal excretion). For this, two time switches are included in the model, one for the arrival of the formulation in the colon (t_{col}) and one for the total absorption duration (t_{abs}; i.e., the time of fecal excretion of the formulation). Colonic absorption is reduced through the term, f_{col}, which is the efficiency of absorption from the colon, relative to the upper part of the GIT.

REFERENCES

1. Food and Drug Administration Guidance for Industry. Extended Release Oral Dosage Forms: Development, Evaluation, and Application of In Vitro/In Vivo Correlations, September 1997.
2. Food and Drug Administration Guidance for Industry. SUPAC-MR: Modified Release Solid Oral Dosage Forms Scale-Up and Postapproval Changes: Chemistry, Manufacturing, and Controls, In Vitro Dissolution Testing and In Vivo Bioequivalence Documentation, October 1997.
3. Food and Drug Administration Guidance for Industry. SUPAC-IR: Immediate-Release Solid Oral Dosage Forms: Scale-Up and Post-Approval Changes: Chemistry, Manufacturing and Controls, In Vitro Dissolution Testing, and In Vivo Bioequivalence Documentation, November 1995.
4. Food and Drug Administration Guidance for Industry. Bioavailability and Bioequivalence Studies for Orally Administered Drug Products—General Considerations, March 2003.
5. Committee for Proprietary Medicinal Products (CPMP). Note For Guidance on Quality of Modified Release Products: A. Oral Dosage Forms; and B. Transdermal Dosage Forms; Section I (Quality), July 1999.

11

Dissolution Method Development with a View to Quality Control

JOHANNES KRÄMER, RALF STEINMETZ, and ERIKA STIPPLER

Phast GmbH, Biomedizinisches Zentrum, Homburg/Saar, Germany

IMPLEMENTATION OF USP METHODS FOR A U.S.-LISTED FORMULATION OUTSIDE THE UNITED STATES

All FDA-approved drugs products must meet the quality requirements described in the U.S. Pharmacopeia (USP) (1,2). If a drug product is to be manufactured elsewhere in the world but marketed in the United States, compliance with existing USP–NF monographs is crucial. Non-compliance may result in the FDA blocking entry of the product into the U.S. market or removing the product from the market. For other markets compliance with USP standards is not binding. For example, the European Pharmacopoeia (Ph. Eur.) has jurisdiction in Europe, the Japanese Pharmacopeia in Japan, etc. Another compendium that serves as a worldwide reference is the International Pharmacopeia (IntPh), which is published by the World Health Organisation (WHO). But the degree of specificity of the various pharmacopeias with respect to setting specifications for drug products varies considerably. Unlike the USP, Ph. Eur., for example, does not include individual monographs of drug products, so applicants have to develop their own methods. As a result, the USP provides a valuable source of information for the European as well as the American pharmaceutical industry, with monographs for drug products that include dissolution methods with test result specifications. In practice, development of biopharmaceutical procedures regarding the choice of apparatus, dissolution media, agitation speed, and even acceptance criteria is often greatly influenced by the USP monograph, if one exists. With the addition of more and more USP monographs over the years, the USP has faced mounting criticism in Europe that the monographs do not follow a clear structure that is primarily based on the drug substance but also reflects the required biopharmaceutical properties of the drug product. In order to meet these goals, alternative attempts have been undertaken to implement Biopharmaceutical Classification Scheme (BCS) concepts in dissolution method development for the characterization of multi-source drug products (3). Although standard apparatus compliant with USP, JP, and Ph. Eur. are used, the media pH, volume, and stirring rate have been adjusted to address biopharmaceutical issues. However, these methods have only recently been accepted by the WHO (4), and to date have only been developed for a limited number of compounds. For these reasons and because of the legal status of the USP for the United States and the fact that USP is a recognized standard in many countries, following an available USP monograph, which describes dissolution test conditions for the intended drug product, continues to be the recommended procedure at the time of writing.

Sometimes certain aspects of the dissolution test suggested by the USP are not suitable for a particular drug product, and in these cases the sponsor may propose a different (changed) procedure, which, if accepted, would be incorporated into the relevant monograph as an alternative to the original procedure. Proposal of alternative procedures for apparatus, dissolution media, agitation, and analytical method for the drug in the dissolution samples can be submitted. But until the alternative method has been accepted for inclusion into the USP, the current compendial method will continue to be applied by the FDA to determine compliance or lack thereof with the requirements for the U.S. market.

Apart from the dissolution methodology itself, USP specifications also provide acceptance criteria, which are applied at three different testing stages as stated in the USP General Chapters ⟨711⟩ Dissolution for IR and ⟨724⟩ Drug release for MR formulations. In these acceptance tables, Q represents the amount of dissolved active ingredient at a given time point. Note that Q is always expressed as percentage of label claim. As an example, the USP acceptance table for IR solid oral dosage forms is given in Table 1.

This acceptance scheme describes a stepwise procedure. If each of the six dosage units initially tested shows a dissolution rate of not less than $Q + 5\%$, the test has passed at Stage

Table 1 USP Dissolution Acceptance Criteria for IR Formulations

Stage	Number of dosage units tested	Complies if
1	6	Each *single* dosage unit is not less than $Q + 5\%$
2	6	The *arithmetic mean* of the 12 dosage units (all units tested in Stages 1 and 2) is not less than Q and no single dosage unit is less than $Q - 15\%$
3	12	The *arithmetic mean* of the 24 dosage units (all units tested in Stages 1–3) is not less than Q and not more than two single dosage units are less than $Q - 15\%$ and no single dosage unit is less than $Q - 25\%$

1. Otherwise, six additional units must be tested. If the arithmetic mean of the 12 dosage units (all units tested in Stages 1 and 2) is not less than Q and no single dosage unit is less than $Q - 15\%$, the test is passed at Stage 2. If the product fails at both of the above-described stages, a further 12 units are to be tested. The product complies at Stage 3 if the arithmetic mean of the 24 dosage units (all dosage units tested in Stages 1–3) is not less than Q and not more than two of the 24 single dosage units are less than $Q - 15\%$ and no single dosage unit is less than $Q - 25\%$. The application of the three-stage dissolution testing and acceptance criteria as a method for how to proceed when the product is out of specification (OOS) in Stage 1 has been adopted by the European Pharmacopoeia for implementation (6). It is important that the standard operating procedure (SOP) for the dissolution test clearly states when replicate testing (i.e., Stage 2 and 3 testing) is to be used for products that are OOS in Stage 1. The SOP should provide the possibility to search for physical errors, which may have caused the failure to comply with specifications in Stage 1 testing (e.g., errors in media preparation). Identification of such failure would lead to discarding the first set of results and starting a new at Stage 1, rather than automatically proceeding to Stage 2 and 3 testing.

From a statistical point of view, it should be noted that the Stage 1 criteria consider the dissolution rates of individual units, whereas Stage 2 and 3 both the arithmetic mean and individual results are taken under consideration. Therefore, the discriminative power of Stage 1 testing is much greater than subsequent stages. As demonstrated by Hoffer and Gray (7), if (90% of the individual units show dissolution rates greater than or equal to $Q + 5\%$, the probability p of passing Stage 1 testing is 59% (0.96). And even if 96% of the individual results are estimated to be greater than or equal to $Q + 5\%$, p for passing the dissolution test at Stage 1 is only about 78%. Therefore, the choice of the Q value has an important impact on the frequency with which Stage 2 testing will be necessary. The authors indicated that to achieve an acceptable probability of Stage 2 testing (20% of batches), the true average release rate should be

$Q + 5 + 1.75\sigma$, where σ is the standard deviation of drug release at the given time point. From this analysis, it is clear that, in addition to the average drug release, the homogeneity of drug product is of great relevance.

For U.S. submissions, the dissolution specification must be based on these general acceptance criteria schemes. In cases of a generic drug product, where a USP monograph is already available, the applicable quantity, Q, and the respective sampling interval are stated in the USP monograph. For new chemical entities or in cases where no USP monograph is available, the sponsor must submit a proposal for Q and sampling time point, which will be reviewed by FDA's CMC staff at the Office of Pharmaceutical Sciences.

For generics of U.S.-listed drug products, sponsors should apply the acceptance criteria tables provided in the two USP general chapters during the initial phases of drug development and clinical trials, when in vivo verification of acceptance criteria is still outstanding. In other cases, Q is to be defined by the sponsor. Values for Q normally vary between 75% and 80% of label claim. As outlined by Hoffer and Gray (7), if a new drug application (NDA) is successful, the dissolution method submitted in regulatory filings will be subsequently transferred to an official method in USP–NF. This transfer is coupled to the availability of a verified reference standard material (8).

Sample Size

Independent of existing intra-lot variability, a sample size of six dosage units is generally recognized to suffice the needs of quality control (QC). In very early development less than six specimens may be used to create data, but as soon as possible tests should be run with at least $n = 6$. It is advisable to create statistically valid and sound data for manufacturing prototypes even at very early phases of development, in order to be able to identify formulations/batches with unwanted dissolution behavior. In the early phases of a drug product's development, formulations may not be of acceptable stability. This means that stability phenomena may mask the underlying biopharmaceutical properties. For this reason, it is important to analyze samples with a stability-indicating method as early as possible in the development process.

In later phases of the drug product's lifecycle, the generation of statistically valid dissolution data continues to be very important. In establishing an in vitro–in vivo correlation (IVIVC), where data generated in pharmacokinetic studies are compared and correlated to in vitro data, every effort should be made to produce data of at least the same quality on the in vitro side as in the generation of the in vivo data (see also Chapter 10). Latest at the point when clinical trials are started, the quality of the clinical trial material has to be proven according to GMP, which again will require a meaningful sample size (minimum $n = 6$). For pivotal and the so-called side batches, at least 12 dosage units per batch should be investigated in order to generate data, which can be compared using the f_2-algorithm. In the post-approval phase, statistically valid data on the influence of formulation changes is important to maintain product consistency.

Sampling

One point sampling is very common for immediate release (IR) products in the USP monographs. The choice of one time point to collect samples represents a substantial data reduction of the kinetic process of dissolution (time vs. amount released relationship). This reduction needs to be based on sound data generated in the formulation development phase, in which dissolution profiles should be generated. Formulation development should, of course, also include stability trials recommended by the International Committee on Harmonization (ICH). If the release mechanism from the product changes during storage, the data needed for a risk-based interpretation must be generated by taking several samples during the dissolution test and generating a percentage dissolved vs. time dissolution profile. A sampling grid consisting of sampling every 15 min in the case of IR dosage forms is often used, but deviation from this sampling schedule may be needed to fully characterize the biopharmaceutical properties of the formulation. For example, a film-coated tablet may require more precise observation at the early phase of the dissolution test to determine whether dissolution of the film is the rate-limiting step for subsequent release processes. Longer intervals between samples (e.g., every hour) are more typically used in early development for modified-release (MR) dosage forms. Here again, modification of the sampling procedure to examine the biopharmaceutical properties may be needed. An example would be in the development of a MR dosage form used for therapy of large bowel diseases, where it is important to characterize the time of onset of drug release (see also Chapter 5).

Aliquots taken from the dissolution test of each individual specimen are usually analyzed individually. Using simple statistics, the true value of the population mean is approximated as the arithmetic mean for the sample (often $n = 6$) assuming a normal distribution. In a limited number of cases, such as when the stability of the analyte is not adequate over the time span needed to analyze six individual samples, pooled sampling may be considered. Pooling the samples essentially creates a physical mean by mixing aliquots sampled for individuals prior to chemical analysis. The gains in terms of time saved and accuracy of the chemical analysis for % released must be weighed against the loss of information in terms of variability in the dissolution characteristics of the individual dosage units. It goes without saying that the standard USP acceptance table procedure for determining compliance to specifications is no longer applicable.

For further information on sampling and automation of sampling, including a discussion of apparatus suitability test acceptance criteria for IR or MR dosage forms, please refer to Chapters 2, 3, and 13.

HOW TO PROCEED IF NO USP METHOD IS AVAILABLE?

When the first dosage form of either of a new chemical entity or generic product is developed, a dissolution method will need to be developed. For some generic dosage form cases, the USP may offer a dissolution test as part of the relevant monograph. Some guidance can also be found in the British Pharmacopoeia (BP). Unlike the European Pharmacopoeia, the BP does contain some general guidelines about how to set up dissolution tests for various types of formulations. But for other generic products and all dosage forms of new chemical entities it will be necessary to design an appropriate dissolution test. A general discussion of the design of appropriate dissolution tests based on properties of the drug substance, GI physiology, and dosage form characteristics is given in Chapter 5. This chapter will focus more on the regulatory aspects of dissolution testing.

First Data for BCS Categorization

Dissolution testing is a technique, which is mainly dedicated to determining the influence of dosage form properties on the efficacy of the drug substance. Therefore, it is necessary prior to dissolution method development to determine whether drug substance-related characteristics and/or dosage form-related properties, i.e., factors that may affect release of the drug in vivo are likely to be rate-limiting to drug absorption and subsequently to efficacy. Therefore, BCS characterization should be the first step in developing the dissolution test.

One pre-requisite to achieving a dissolution rate, which does not restrict the rate or extent of drug absorption, is an adequate solubility of the drug in aqueous media representative of upper gastrointestinal (GI) conditions. The shake-flask method is widely recognized and of great precision (9). Shortly described, an excess mass of drug substance is added to a prescribed volume of the medium in which the solubility is to be tested. The suspension is shaken (preferably at 37°C) and the concentration of the drug substance in the supernatant is determined with a stability-indicating assay. Media with different pH values covering the physiological range should be used. To meet the requirements of the U.S.-FDA (which have been also been adopted conceptually by the EU and WHO) the media should be buffered at pH values in the range 1–6.8.

In Europe, the regulatory authority (EMEA) specifies the pH range from pH 1 to 8. If no stability/impurity indicating assay is available, the influence of impurities on the solubility can be detected by carrying out the experiments with various excesses of added substance. The resulting regression line can then be used to calculate the true solubility in such cases (10). To avoid misinterpretation caused by counterions or salting-out effects, NaOH/HCl mixtures may be used instead of or in parallel to the buffer systems described in pharmacopeiae such as USP, Ph. Eur., JP. However, if such mixtures are used, continual adjustment of the pH in the supernatant is necessary as NaOH/HCl typically have extremely small buffer capacities at the pH values of interest. The duration of the experiment should enable equilibrium to be reached. If the experiments are stopped too early, erroneous results may be reported—on the one hand, the medium may be supersaturated with the drug (if, e.g., a high-energy polymorph is present) leading to an overestimate of the true solubility, or, on the other hand, equilibrium may not have yet been reached, leading to an underestimate of the solubility. The use of the shake-flask method is limited to molecules that are reasonably stable in aqueous systems, and requires that the final concentration reached is above the (lower) limit of quantitation. An alternative method for ionizable substances is the pSol determination described by Avdeef (11), which is based on an acid/base titration.

According to BCS solubility, data are evaluated with regard to the highest dosage strength either already available on the market or envisaged for market introduction. The quotient of the highest (envisaged) dose to the solubility in a specific medium is called the dose–solubility ratio. According to the FDA criteria, this value must be 250 mL or lower across the entire pH range tested for the drug to be considered highly soluble. Note that this ratio does not take into account the influence of the dosage form and its transit through the upper GI tract, so a dose-solubility ratio of 250 mL or lower does not in and of itself guarantee that the amount dissolved and available for absorption at a certain time point in vivo will be adequate to ensure complete absorption (12).

A further pre-requisite for complete absorption of the drug is an adequate permeability. The permeability of the drug substance may be derived from data generated in clinical phase I studies. Absolute oral bioavailability (BA) requires data of an oral solution compared to an intravenous application. If a drug substance shows high absorption (according to FDA criteria a fraction absorbed $\geq 90\%$) it is considered to be highly permeable. Alternatively, data from human in vivo experiments performed in isolated gut segments can be used to directly generate the permeability, but this approach can be limited by a low solubility of the drug to be administered and practical limitations of the intubation technique itself. In vitro tissue models are widespread and provide a rough estimate of a drug substance's permeability on a relatively short turn-around basis. Well established is the CaCo-2 model (human colorectal carcinoma cell line model), which requires a lead time of 3 weeks to grow tissues into a monolayer, and which loses accuracy for molecules with a molecular mass greater than 400. Alternative models are available that do not show these disadvantages but still require proper validation with at least 15–20 marker substances (13). Once a drug has been categorized according to its permeability and solubility, one can determine what kinds of dissolution tests need to be run and how they can be used in product development to minimize the need to run pharmacokinetic studies. Table 2 summarizes the relationship between BCS classification and regulatory utility of dissolution testing.

According to Table 2, the likelihood of establishing an IVIVC for an IR dosage forms is greatest when the dissolution of the drug is slow enough to result in dissolution-limited drug absorption. A stepwise procedure is given in Table 3 (see also Chapter 5).

Variation of temperature is usually not an issue for solid oral dosage forms, since experiments are always conducted at body temperature (37°C). For dosage forms applied on the skin, this can be a further consideration: e.g., drug-release testing of transdermal products is typically performed at the average temperature of body surface 32°C (5).

Table 2 Rate-Limiting Step to Absorption and Requirements for Dissolution According to BCS Classification of the Drug Substance

BCS class	Solubility	Permeability	Major rate-limiting step	Requirement for dissolution
I	High	High	Gastric emptying	Fast over physiological range, 85% in 30 min in all media
II	Low	High	Dissolution	Specifications set on the basis of IVIVC
III	High	Low	Uptake across the intestinal mucosa	Very fast over physiological range, 85% in 15 min
I–V	Low	Low	Dissolution and uptake	Case by case evaluation; poor chance of IVIVC

WHAT ARE THE PRE-REQUISITES FOR A BIOWAIVER?

It is a general requirement for an optimal therapeutic effect that the active pharmaceutical ingredient (API) is delivered to the site of action in order to provide effective but not toxic concentration levels. Therefore, studies to measure BA are of great importance in order to support new drug product applications. Thus, data on the BA of orally administered drug products is a general requirement to the development

Table 3 Stepwise Approach to Developing a Dissolution Method

Step	Influencing factors	Experimental variation
1.	Well-defined physiological factors	pH value
		Concentration of salts
		Surfactants
		Enzymes
2.	Less well-defined physiological factors	Agitation
3.	Verification of method	Comparison to relevant in vivo data

pharmaceutics section of a common technical document (CTD) of a new drug application (NDA).

Additionally, proof of similar plasma concentration time courses, designated as bioequivalence (BE), will be necessary to ensure that BA is maintained between pivotal and early clinical trial formulations, among different formulations used in clinical trials and to demonstrate the comparability of therapeutic performance of a generic to the innovator product. Since for orally administered solid oral dosage forms BA and BE studies focus on determining the process by which a drug is released from the oral dosage form and moves to site of action, these studies will generally include in vitro dissolution studies as complementary data to prove the biopharmaceutical quality of the drug product, e.g., clinical trial formulation.

Typically, BA and BE are assessed by cumbersome and expensive studies in human volunteers. But, under certain circumstances, regulatory agencies may waive the requirement for the submission of evidence measuring the in vivo BA or establishing BE. This is referred to as a "biowaiver". The application of a biowaiver requires that supportive in vitro dissolution data are meaningful in terms of in vivo performance of the drug product.

Biowaivers Based on the Biopharmaceutics Classification System

In August 2000, FDA's Center for Drug Evaluation and Research (CDER) issued the Guidance for Industry "Waiver of In Vivo Bioavailability and Bioequivalence Studies for Immediate-Release Solid Oral Dosage Forms Based on a Biopharmaceutics Classification System"(14). This guidance provides recommendations for sponsors of investigational new drug applications (INDs), NDAs, and abbreviated new drug applications (ANDAs) who wish to request a waiver of the requirement of in vivo BE studies. Generally, these recommendations apply only to IR solid oral dosage forms and the possibility of a biowaiver is restricted to subsequent BE studies of IR oral drug products after initial establishment of BA during the IND period (in the case of a new chemical entity) or to further BE studies in the case of ANDAs and post-approval changes [e.g., SUPAC-IR Level 3 changes in components and composition (15)].

In July 2001, the European Agency for the Evaluation of Medicinal Products issued the "Note for Guidance on the Investigation of Bioavailability and Bioequivalence" (16) with the objective to define when data of BA and BE studies are necessary for approval of dosage forms of systemically acting drugs. With a view to biowaiver, this guidance also refers to the possibility of using in vitro as a substitute for in vivo BE studies with pharmacokinetic assessment.

It should be noted that in both guidances BCS-based biowaivers do not apply to food effect BA studies or pharmacokinetic studies other than those designed to test for BE.

The basic approach in both guidances is the classification of drug substance according to the Biopharmaceutics Classification System (BCS), together with the assessment of in vitro drug product dissolution (1,2,14). The underlying justification for BCS-based biowaivers is the assumption that for highly soluble, highly permeable drugs formulated as *rapidly* dissolving IR-dosage forms, no BA problems are expected. Hence, in vivo BE studies can be waived if the dissolution profiles of test and reference product are similar when the dissolution testing is performed according to the guidance (at three pH values within the physiologically relevant range).

The initial step in the evaluation of possible BCS-based biowaivers is the classification of the drug intended for orally administration as follows (17):

Class 1: High solubility–high permeability
Class 2: Low solubility–high permeability
Class 3: High solubility–low permeability
Class 4: Low solubility–low permeability

In order to assure a consistent classification of drug substances according to the classes mentioned above, both guidances provide detailed definitions of the terms solubility and permeability. According to both guidances, a drug is regarded as highly soluble when the highest dose strength is soluble at $37 \pm 1°C$ in 250 mL or less of aqueous media in the physiological pH range 1–7.5 (14). Therefore, solubility profiles should be established by usage of the dose-solubility ratio (ratio of highest dose strength in milligrams and measured solubility in milligram per milliliter). The European CPMP note for guidance (16) recommends the use of buffer solutions at pH 1, 4.6, 6.8. FDA's CDER requires a profiling with higher resolution centered around the pK_a of the drug substance. The use of USP buffer solutions at pH 1, $pK_a - 1$, pK_a, $pK_a + 1$, and 7.5 is recommended. The CDER advises a minimum of three replicate experiments under each pH condition. In order to assure the solubility results at a given pH, the pH should be verified after addition of the drug substance and throughout the entire solubility experiment. Whenever necessary, the pH must be adjusted to the prescribed pH. The concentration of the saturated solutions should be determined using a validated and stability-indicating assay. To establish high solubility, the determined dose-solubility ratio may not be greater than 250 mL at any pH value investigated. In order to avoid influences by counterions or osmotic pressure, mixtures of hydrochloric acid and sodium hydroxide solutions may be used to adjust the pH value. In these cases, it is particularly important to repeatedly check the pH value of the medium during the course of the solubility determination (see 10.2.). Solubility experiments at early phases, mainly with new chemical entities may be performed using different amounts of drug substance and equal volumes of media. This procedure may be needed to level out the influence of impurity on the solubility, especially if a stability-indicating assay has not yet been established (10)

The permeability of the drug substance can be determined by different approaches such as pharmacokinetic studies in humans (fraction absorbed or mass balance studies) or intestinal permeability studies (in vivo intestinal perfusion studies in humans or suitable animal models or in vitro permeation studies using excised intestinal tissue or epithelial cell culture monolayers like CaCo-2 cell line). In order to avoid misclassification of a drug subject to efflux transporters such as P-glycoprotein, functional expression of such proteins should be investigated. Low- and high-permeability model drugs (e.g., antipyrine or metoprolol with designated high permeability and e.g., hydrochlorothiazide with low permeability) should be used as internal standard additionally to zero permeability markers such as PEG 4000 to assure system suitability at each set of experiments. The interlab variability of CaCO-2 results is remarkably high (18), so absolute values of permeability cannot be compared across labs. Alternative cell cultures such as jejunum cell lines may be advantageous (19). The stability of the drug substance in intestinal fluids should be demonstrated for those techniques, which measure the clearance of a drug from the perfusion fluids in the small intestine, since it is necessary to clearly demonstrate that the loss of drug from the perfusion solution arises from drug permeation rather than degradation.

In addition to drug substance properties, which will normally be investigated during the R&D period of pharmaceutical development, the dissolution characteristics of the oral dosage form under consideration also have to be investigated. In general, the guidances will allow biowaivers for pharmaceutical test forms such as tablets, capsules, and oral suspensions, except those that are intended to result in drug absorption from the oral cavity, e.g., sublingual or buccal tablets. It should be noted that waivers of BE studies will only apply to essentially similar products (16). Under certain very restricted circumstances, e.g., tablets vs. capsules, the concept of essential similarity may also be applied to different IR-formulations containing the same active ingredient(s).

Further, both guidances state that BCS-based biowaivers only apply to *rapidly* dissolving IR forms. Unfortunately, a precise definition on what authorities may define as rapidly dissolving IR form is only given in the CDER guidance. The criterion stated here is drug release of not less than 85% within 30 min using either the basket or paddle apparatus and 900 mL dissolution media with the following pH conditions: (i) 0.1 N hydrochloric acid solution (HCl) or simulated gastric fluid (SGF) according to USP, (ii) buffer solution pH 4.5, and (iii) buffer solution pH 6.8 or simulated intestinal fluid (SIF) according to USP. The guidance also specifies the rotational speed, which should be 100 rpm for basket and

50 rpm for paddle. The use of proteolytic enzymes in SGF or SIF needs to be justified. A potential example would be to avoid artifacts when aging results in some cross-linking of gelatin in capsule shells, which in turn hinders in vitro dissolution in the absence of enzymes. In contrast, the CPMP guidance asks for rapid dissolution within the range of pH 1–8 with recommended media at pH 1.0, 4.6, and 6.8.

After classification of the three main pre-requisites—solubility, permeability (BCS-class 1 drug substance), and evaluation of the required dissolution characteristics (rapidly dissolving IR drug product)—the next crucial step is the comparison of the in vitro dissolution performance between the reference and test drug product. This could be the innovator and a generic version in the case of a biowaiver for an ANDA application, or might be the approved product vs. a version that has undergone a scale-up or post-approval change (SUPAC). The recommended dissolution media and procedure are identical to those prescribed for the classification of the dissolution characteristic of the reference drug product (see above). In general, a minimum of 12 dosage units should be evaluated to support a biowaiver request. Samples should be collected at a sufficient number of intervals to obtain dissolution profiles that can be compared using the f_2 similarity factor (see also Chapters 8 and 9). The CDER guidance recommends sampling intervals of 10, 15, 20, and 30 min (see Chapter 11.3.4. for exceptions to the need for profiling).

The pre-requisites for BCS-based biowaivers are summarized in Figure 1.

Biowaiver for Compositionally Proportional Drugs

In addition to BCS-based biowaivers, comparative dissolution testing has also been used to waive in vivo BE requirements for different strengths of a dosage form. Waiver of in vivo studies for different strengths of a drug product can be granted according to 21 CFR Part 320.22(d) (2) when the following pre-requisites are fulfilled:

i. the drug product is in the same dosage form but in a different strength; and

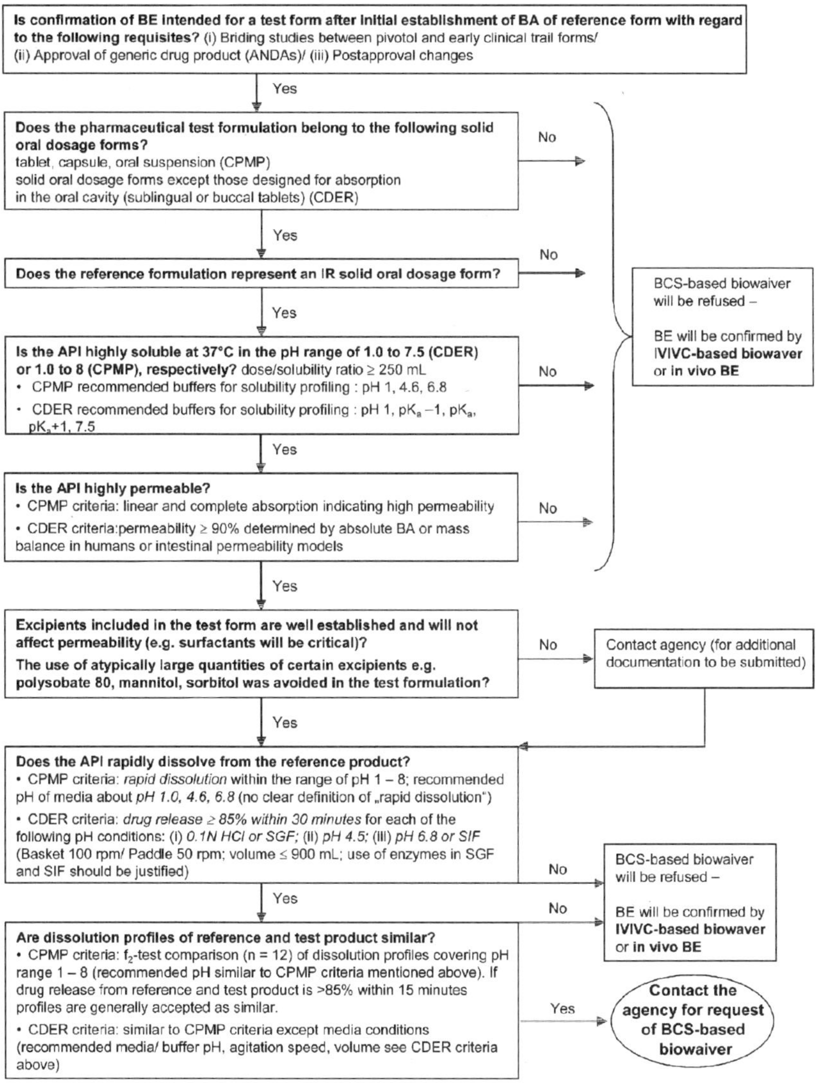

Figure 1 Prerequisites for BCS-based biowaivers according to CDER and CPMP guidelines.

ii. the different strength is proportionally similar in its active and inactive ingredients to the product strength, for which the same manufacturer has conducted an appropriate in vivo study.

The term proportional similarity is implied in the FDA Guidance for Industry—Bioavailability and Bioequivalence Studies for Orally Administered Drug Products (20). Characteristics of proportional similarity are given in Figure 2. If the requirements for proportional similarity are met, dissolution profile comparison of either IR forms or MR forms is then performed according to the scheme shown in Figure 3. For MR products, it is mandatory that the in vivo BA study has been carried out with the highest strength of the current form, whereas for IR dosage forms, data on clinical safety and/or efficacy and linear elimination kinetics may be sufficient to permit application of the biowaiver even for a new product, which has a higher dose strength. For the dissolution profile comparison of IR dosage forms, dissolution profiling using the established dissolution method may be sufficient if it can be shown that the dissolution is not dependent on the pH of the medium. Otherwise, dissolution profiling should be performed for each product in USP buffer solutions at pH 1.2, 4.5, and 6.8.

For MR forms representing MR-beaded capsules, in which the dosage strength is only determined by the number of API-containing beads, dissolution profiling using the established method is sufficient for each product strength. For MR tablets dissolution profiling in USP buffers pH 1.2, 4.5, and 6.8 is required.

Waivers Based on IVIVC in General or When Compositional Changes Are Minor

Additional criteria for waiver of evidence of in vivo BA/BE are given in 21 CFR 320.22 (d)(3). For certain solid oral dosage forms (other than a delayed or extended-release dosage forms), a waiver for the submission of in vivo evidence of BA/BE is possible if the drug product has been shown to meet the requirements of an in vitro dissolution test, which in turn has been shown to correlate with in vivo data. A biowaiver may also be addressed to a reformulated solid oral dosage form identical to another drug product except for color, flavor, or preservatives for which the same manufacturer has obtained approval, if BA data are available for the approved

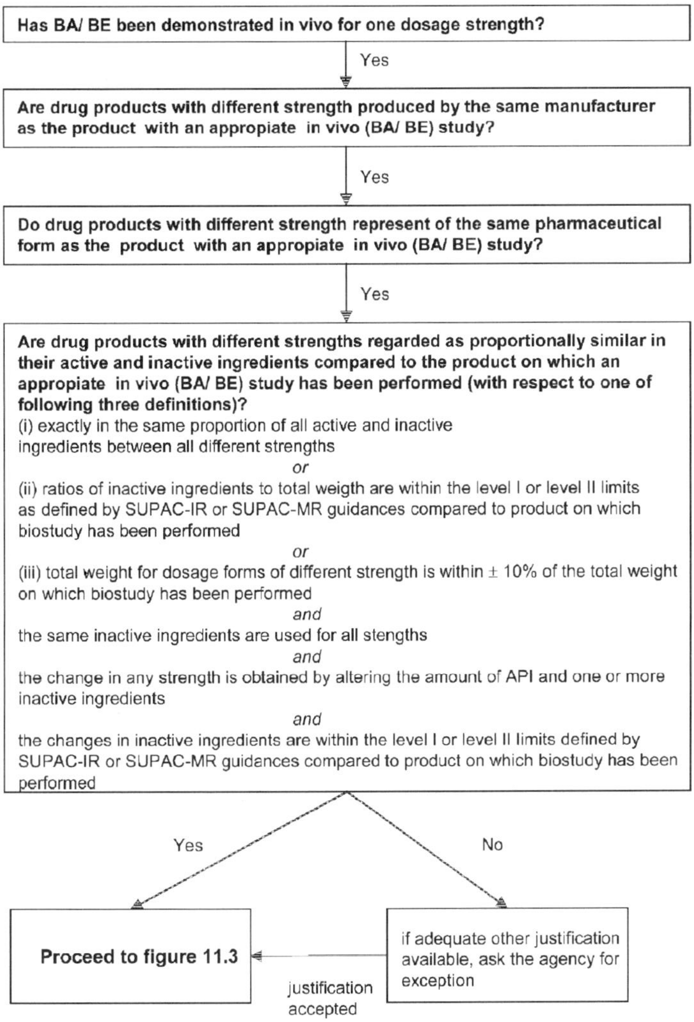

Figure 2 Prerequisites of preapproval waivers of in vivo studies for solid oral dosage forms with different strength supported by in vitro dissolution data.

drug product and both drug products meet an appropriate in vitro test approved by FDA (21 CFR Part 320.22 (d) (4)).

In both cases, dissolution profiling should be performed according to the established method and the similarity of dissolution profiles should be evaluated.

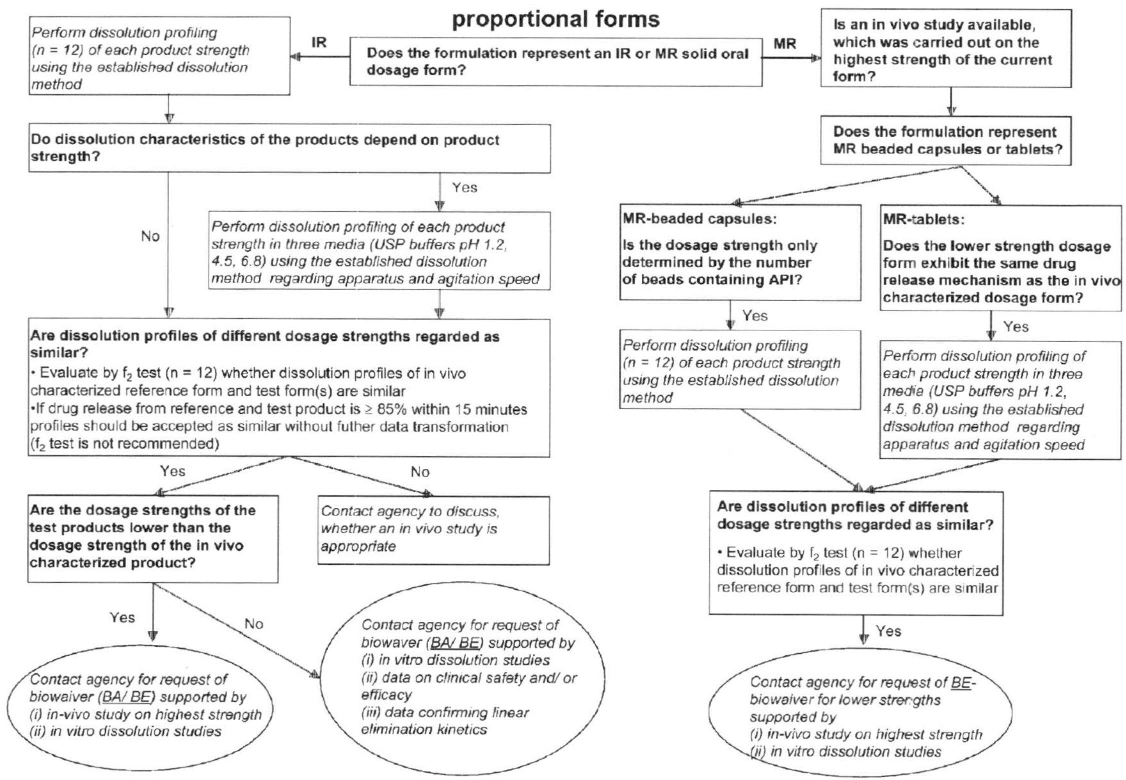

Figure 3 Prerequisites of preapproval waivers of in vivo studies for solid oral dosage forms with different strengths supported by in vitro dissolution data.

How Should Dissolution Profile Similarity Be Assessed?

The method for the evaluation of similarity of dissolution profiles depends on dissolution characteristic of the reference and test drug product. If both formulations (average value of $n = 12$ each) dissolve at least 85% of label claim within 15 min, dissolution profiles are generally assumed as similar and no further testing or data analysis is required.

For formulations not meeting the criterion for very fast release of drug substance, similarity of profiles may be evaluated by model-independent or model-dependent methods as stated in the Guidance for Industry—Dissolution Testing of IR Solid Oral Dosage Forms (1,2).

The most common approach for the comparison of dissolution profiles is model-independent approach using the similarity factor f_2. The pre-requisites for using the f_2-test are the following:

i. dissolution profiles of the two products with $n = 12$ units per product have to be compared;
ii. the mean dissolution rates at each time interval are to be used for the calculation of similarity factor;
iii. dissolution testing of reference and test forms should be conducted under exactly same conditions with the same sampling time intervals;
iv. for SUPAC changes, the reference batch should be the most recently manufactured (pre-change) batch. Alternatively, reference data may derive from the last two or more consecutively manufactured pre-change batches;
v. a minimum of three time intervals should be included in the analysis;
vi. only one time interval with more than 85% dissolved API for test and reference may be included in the analysis;
vii. the coefficient of variation should be not more than 20% for earlier time intervals (e.g., 15 min). Other time points should have a coefficient of variation of not more than 10% (if the intra-batch variation at later

time intervals is more than 15% (CV), a multi-variate model independent approach is more suitable).

The similarity factor is calculated according to the following algorithm:

$$f_2 = 50 \log\left\{\left[1 + \left(\frac{1}{n}\right)\sum_{t=1}^{n}(R_t - T_t)^2\right]^{-0.5} \times 100\right\}$$

where f_2 is the similarity factor, n the number of considered time intervals, R_t the arithmetic mean of dissolved API (% of label claim) from reference product at time interval t, and T_t arithmetic mean of dissolved API (% of label claim) from test product at time interval t. f_2 values of not less than 50 indicate the equivalence of the two dissolution profiles.

Alternative methods and algorithms may be used, such as the model-independent approach to compare similarity limits derived from multi-variate statistical differences (MSD) combined with a 90% confidence interval approach for test and reference batches (21). Model-dependent approaches such as the Weibull function use the comparison of parameters obtained after curve fitting of dissolution profiles. See Chapters 8 and 9 for further discussion of these methods.

SUPAC: Dissolution Profile Comparison Supporting Post-approval Changes

Using the BCS as the basis, the SUPAC guidelines provide a tool-set for proving product sameness after certain changes in the composition, the manufacturing process, or of the manufacturing site without requiring in vivo BE testing.

For IR forms, the SUPAC-IR guidance (15) distinguishes between the following classes of change:

i. changed components or composition of ingredients (levels 1–3);
ii. site changes (levels 1–3);
iii. changes in batch size (levels 1–2);
iv. changes in manufacturing equipment (levels 1–2);
v. changes in manufacturing process (levels 1–3).

Beside additional chemistry documentation, dissolution data are required to support continued approval of the drug product after the intended changes are introduced. Detailed definitions, according to which changes may be assigned to a specific "level," are given in Ref. 15. Depending on the level, different requirements are set for the data that need to be submitted to the agency (in this case, FDA). For all level 1 changes, dissolution data according to the application requirement are sufficient. For higher levels of change, more comprehensive investigations are required. In this context, the guidance distinguishes three cases (Cases A–C), which define in detail how comprehensive the required dissolution testing must be, as well as the acceptance criteria. Details for conducting dissolution testing are given in Figure 4.

Analogously, the SUPAC-MR guidance (1,2) defines level of changes for:

i. change in components and composition of excipients, which do not control the drug release (levels 1–3);
ii. change in components and composition of release-controlling excipients (levels 1–3 with separate requirements for narrow and non-narrow therapeutic drugs);
iii. site changes (levels 1–3);
iv. changes in batch size (levels 1 and 2);
v. changes in manufacturing equipment (levels 1 and 2);
vi. changes in manufacturing process (levels 1–3).

Figure 5 depicts the dissolution test requirements according to case and level. Again, in addition to chemistry documentation, dissolution data are required to support approval of intended changes. For all level 1 changes, dissolution data for the changed drug product (test) and the biobatch (for which BA has been established) or a marketed batch according to the application requirement are requested. For level 2 changes, multi-point dissolution testing of pre- and post-change drug product under varied test conditions (media for controlled release and agitation for delayed release) is required.

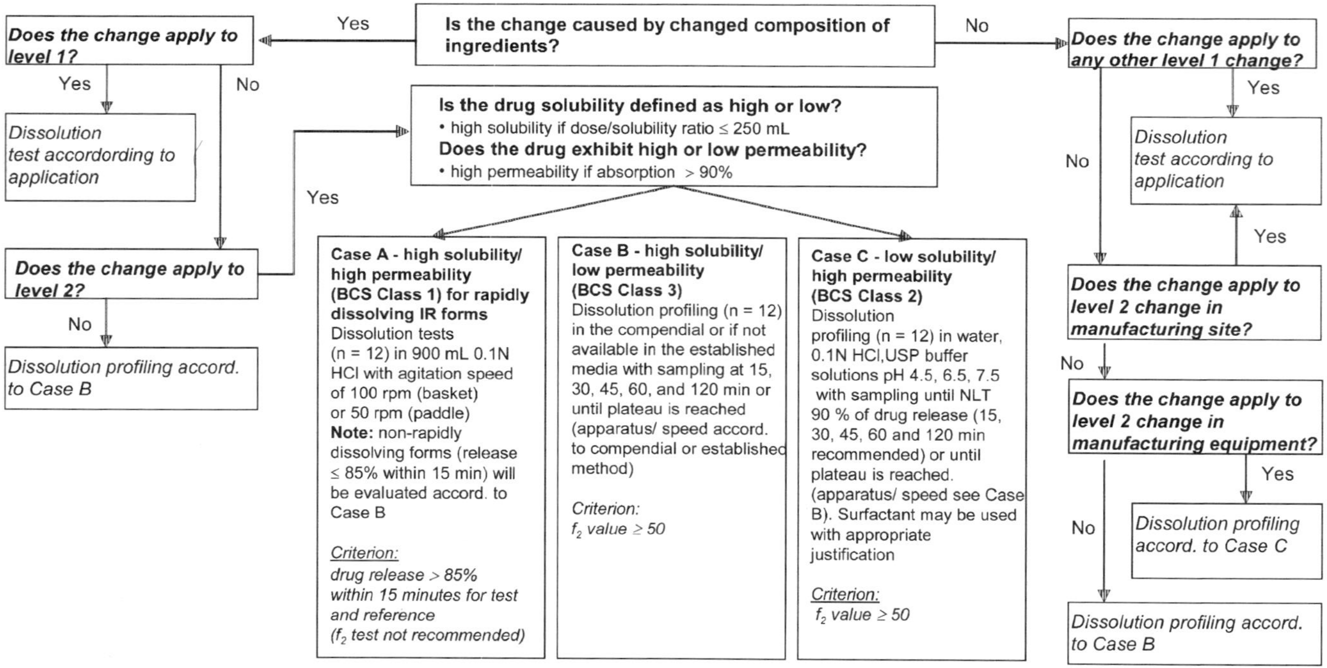

Figure 4 Postapproval changes of IR forms supported by in vitro dissolution data according to SUPAC-IR guidance.

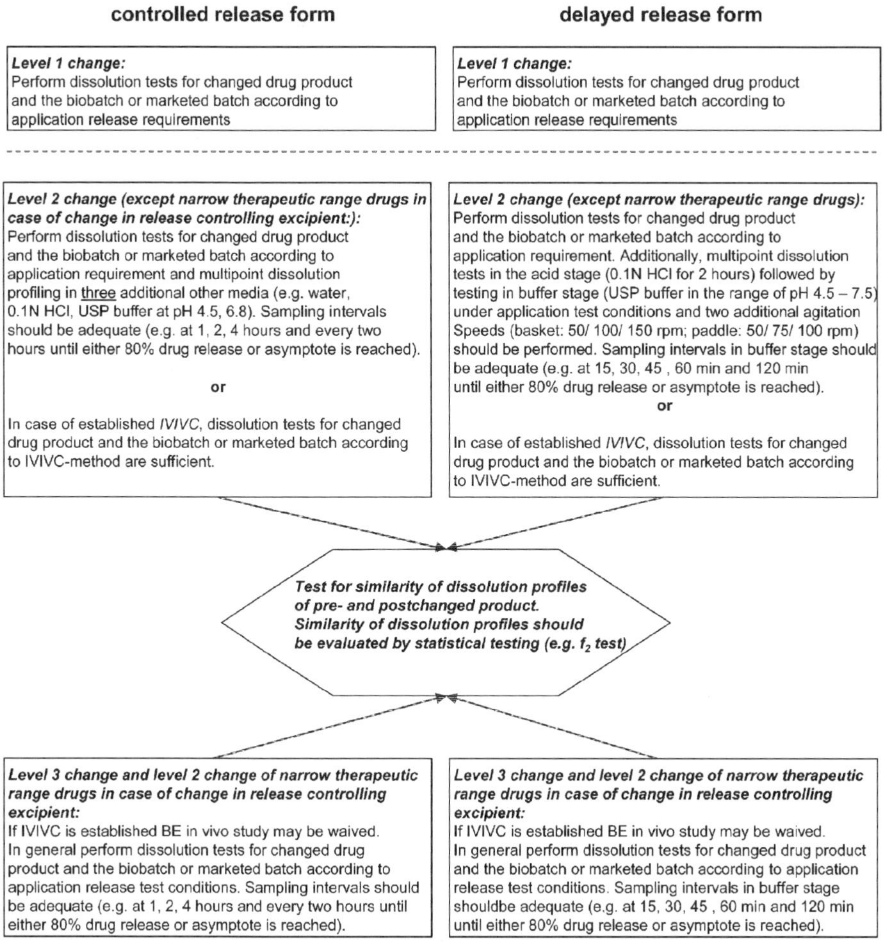

Figure 5 Postapproval changes of MR forms supported by in vitro dissolution data according to SUPAC-MR guidance.

In general, if an IVIVC method has been established, the requirement for additional dissolution test conditions is waived in favor of multi-point dissolution testing according to the in vitro method with which the IVIVC has been established. For level 3 changes, multi-point dissolution testing according to application-release test conditions is required in addition to in vivo BE. If IVIVC is available, this requirement is reduced to comparison of dissolution profiles of test and reference drug product. Methods for establishing IVIVC is described in detail in Chapter (1088) "In vitro and In vivo evaluation of dosage forms" of the USP (see also Chapter 10).

IVIVC: IN VIVO VERIFICATION OF IN VITRO METHODOLOGY—AN INTEGRAL PART OF DISSOLUTION METHOD DEVELOPMENT

As it is often very difficult to quantify therapeutic performance with pharmacodynamic and clinical studies, pharmacokinetic studies are usually the most suitable tool to describe the performance of the drug product in vivo. Once a relationship between the plasma concentration of the drug or active moiety and the therapeutic effect has been established, BA may be considered to be the perfect surrogate parameter for efficacy and/or safety of a drug product.

However, the number of studies that can be performed in humans is limited by both ethical (unnecessary exposure of human volunteers to risks) and economical factors. Therefore, in vitro testing may be invoked as a "surrogate of the surrogate" provided that a linear relationship between relevant in vivo and in vitro exists, i.e., an IVIVC.

The design of pharmacokinetic studies that need to be conducted for product approval is a function of how much is known about the active drug moiety, its clinical pharmacokinetics, and the biopharmaceutical properties of the dosage form, and regulatory requirements. As a minimum,

1. a single-dose crossover study, and/or
2. a multiple-dose, steady-state study using the highest strength are required to characterize the product (USP (1088), (1090) FDA ABBE-Guidance).

According to USP Chapter (1088) the term IVIVC refers to the establishment of a rational stochastical relationship between a biological property, or a parameter derived from a biological property produced by a dosage form, and a physicochemical property or characteristic of the same dosage form. The biological properties most commonly used are one or more pharmacokinetic parameters, such as $c_{\max}$, $t_{\max}$, or AUC, obtained following the administration of the dosage form. The in vitro dissolution behavior of an active pharmaceutical ingredient from a dosage form under a given set of conditions expressed as percent of drug released is the most commonly used physicochemical property. The relationship between the two properties, biological and physicochemical, is to be expressed quantitatively.

An FDA interpretation of IVIVC has been cited as: "To show a relationship between two parameters. Typically relationship is sought between in vitro dissolution rate and in vivo input rate. This initial relationship may be expanded to critical formulation parameters and in vivo input rate" (22).

The both interpretations, the ultimate goal of an IVIVC is clearly to establish a meaningful, ideally linear, relationship between the in vivo behavior of a dosage form and its in vitro performance, according to which the subsequent in vivo behavior can be adequately predicted by in vitro testing. Although the evolution of the IVIVC may be based in conventional IR dosage forms, the concepts are most applicable toward the development and support of MR dosage forms. It must be emphasized that IVIVC for either IR or MR dosage forms are only feasible when the release-controlling mechanism of the dosage form is the principal determining factor for the rate and extent of the drug absorption.

In order to obtain an in vitro–in vivo relationship two sets of data are needed. The first set is the in vivo data, usually entire blood/plasma concentration profiles or a pharmacokinetic metric derived from plasma concentration profile (e.g., $c_{\max}$, $t_{\max}$, AUC, % absorbed). The second data set is the in vitro data (e.g., drug release using an appropriate dissolution test). A mathematical model describing the relationship between these data sets is then developed. Fairly obvious, the in vivo data are fixed. However, the in vitro drug-release profile is often adjusted by changing the dissolution testing conditions to determine which match the computed in vivo-release profiles "the best," i.e., results in the highest correlation coefficient.

Unlike most IR dosage forms, MR products cannot be characterized using a single-time point dissolution test in routine QC. For IVIVC purposes, dissolution profiles must be generated in any case, irrespective of whether the release is IR or MR. The most information-rich IVIVCs are generated when both the in vitro and in vivo data are expressed as profiles (Level A correlation, with correlation between the in vitro dissolution profile and deconvoluted in vivo release on a point-to-point basis). In this case, the IVIVC relationship may be regarded as a calibration function allowing interpolation and being reversible. Typically, not only the batch of interest is studied, but also two "side-batches," i.e., those which are prepared similarly to the batch of interest but which have enough differences to generate in vivo and in vitro results that are clearly distinguishable form those of the product (batch of interest). One of these side-batches should release faster than the batch of interest, the other slower i.e. their behavior should bracket the behavior of the product itself.

Some considerations should be taken into account before attempting IVIVC for solid oral dosage forms:

- the permeability through the gut wall and hence verification that the uptake process is not the rate-limiting step to absorption;
- the release of the active pharmaceutical ingredient from the dosage form (for IR products often limited by drug solubility) is the rate-limiting step for the invasion kinetics;
- the elimination rate of the active pharmaceutical ingredient is independent of dosage form in the therapeutically relevant range.

A higher degree of correlation may be expected with MR formulations, since release from the dosage form is purposely intended to be the rate-limiting step to absorption in these formulations.

The techniques available for evaluating in vivo dissolution rate can be divided in two categories: indirect and direct.

The indirect techniques involve a mathematical treatment of observed conventional plasma, blood, and urine drug concentrations with time. The conclusions drawn depend on the assumption made for the mathematical model. Typical indirect techniques include numerical deconvolution, compartmental modeling (Wagner–Nelson, Loo–Riegelmann), and statistical moments. There are marked differences in the quality of the correlation obtained with each procedure. Thus, these methods are discussed in terms of the advantages of each along with the resulting potential utility as a predictive tool for the pharmaceutical scientist. The recognition and utilization of deconvolution techniques as well as statistical moment calculations represented a major advance over the single-point approach (c_{max}, t_{max}, AUC) in that these two methodologies utilize all of the dissolution and plasma level data available to develop the correlations.

Intubation techniques have been used extensively to appraise the absorption rate in the stomach, duodenum, jejunum, ileum, and colon (23). These methods can be adapted to provide direct evaluation of the dissolution rate in different segments of the GI tract.

Correlation Levels

Three correlation levels have been defined and categorized in descending order of the ability of the correlation to reflect the entire plasma drug concentration–time curve that will result from administration of a dosage form. The relationship of the entire in vitro dissolution curve to the entire plasma level curve defines the correlation.

Level A Correlations

This level provides the most information-rich correlation. It represents a point-to-point relationship between in vitro dissolution and the in vivo input rate of the drug from the dosage form. A linear regression of dissolution and absorption at common time point is established. In such a correlation, the linear relationship of absorption vs. dissolution with a slope of one, an intercept of zero, and a coefficient of determination of one demonstrates superimposable data. The mathematical description for both curves is the same. A y-intercept of the linear correlation plot below zero often reflects a lag-time in the absorption, whereas a positive y-intercept may require additional evaluation.

In the case of a successful Level A correlation, an in vitro dissolution curve can serve as a surrogate for in vivo performance. Therefore, a change in manufacturing site, method of manufacture, raw material supplies, minor formulation modification, and even product strength using the same formulation can be justified without the need for additional human studies.

When linear regression does not yield a good correlation, application of a non-linear function may be feasible (see Chapter 10). The parameter estimates for higher-order or polynomial equations may prove to be more difficult to interpret than for a linear relationship. Nevertheless, this approach may be preferable to using lower-order levels of correlation (B or C) for evaluating the relationship between dissolution and absorption data.

Level B Correlations

Level B utilizes the principles of statistical moment analysis. The mean in vitro dissolution time is compared to either the mean residence time or the mean in vivo dissolution time. Like correlation Level A, Level B utilizes all of the in vitro and in vivo data, but unlike Level A it is not a point-to-point correlation because it does not reflect the actual in vivo plasma level curve. It should also be kept in mind that there are a number of different in vivo curves that will produce similar mean residence time values, so a unique correlation is not guaranteed.

Level C Correlation

This category relates one dissolution time point ($t50\%$, $t90\%$, etc.) to one pharmacokinetic parameter such as $c_{\max}$, $t_{\max}$, or AUC. It represents a single point correlation and does not characterize the shape of the plasma level, which is critical to defining in vivo performance, especially for MR products. Since this type of correlation is not predictive of in vivo product performance, it is generally only useful as a guide in formulation development or as a production QC procedure, unless a multiple Level C correlation can be established.

For MR formulations the in vitro dissolution conditions, which achieve an optimal IVIVC, will be those which possess the discriminatory power to detect the effect of critical manufacturing variables on drug release. An investigation of the dependence of the formulation on pH and surfactants is recommended in media of various compositions. A dependence on dissolution equipment, and range of equipment settings should also be considered in the investigations.

Setting Specifications According to USP Level A IVIVC

Dissolution specifications are limits for the percent of drug released at specific times during the release process. All formulations that meet these limits can be assumed to perform similarly. The specification limits for dissolution testing can be established in case of a Level A correlation by preparing at least of two formulations having significantly different in vitro behavior. One of the batches should show a more rapid release and the other a slower release behavior than the biobatch. The upper and lower-dissolution limits are then selected for each time point established from the BA/BE study of the biobatch. The dissolution curves defined by the upper and lower limits are convoluted to the plasma level curves that result from administration of these formulations. In case that the resulting plasma level data fall within the 95% confidence intervals obtained in the definitive BA/BE study, these ranges can be considered to be acceptable.

Deconvolution

An acceptable set of plasma level data is established both for a batch of material demonstrating a more rapid release and for one demonstrating a slower release than that of the biobatch. These may be selected by using the extremes of the 95%

confidence intervals or ±1 standard deviation of the mean plasma level. In the case of a Level A correlation, these curves are then deconvoluted, and the resulting input rate curve is used to establish the upper and lower-dissolution specifications at each time point. Batches of product must be made at the proposed upper and lower limits of the dissolution range, and it must be demonstrated that these batches are still acceptable by performing a BA/BE study.

Setting of specifications for IVIVC on Level B is more of a challenge. A procedure has been described requiring homomorphic dissolution profiles on the in vitro side and BA data for at least three formulation variables on the in vivo side using interpolation (24). Extrapolation of Level B IVIVC is considered to be very questionable, so one is limited to interpolation within the established limits of the IVIVC. For Level B or C correlations, additional BA/BE will be needed if the IVIVC is to be extended to different types of formulations and/or different brands.

Unfortunately, most of the correlation efforts to date with IR dosage forms have been based on the correlation Level C approach, although there also have been some efforts employing statistical moment theory (Level B). Level A correlation approach is often difficult with IR dosage forms because of the need to sample intensively in the absorptive region of the in vivo study. Thus, Levels B and C are the most practical approaches for IR dosage forms, even though they are not as information-rich and therefore more limited in their application.

Establishing IVIVC for a certain drug product may be of advantage in one or more of the following ways:

i. as a surrogate to bioequivalency studies by SUPAC;
ii. to support and/or validate the use of dissolution testing and specifications as a QC tool;
iii. to predict the in vivo performance of a formulation based on in vitro dissolution data.

In summary, the role of dissolution testing as a surrogate for BE studies in humans has assumed increasing importance in the regulation of drug products. It is more than likely that in the coming years, the application of biowaivers based either on BCS-type principles and/or on IVICS will become even more important.

REFERENCES

1. FDA. Guidance for Industry—SUPAC-MR: Modified Release Solid Oral Dosage Forms—Scale-up and Postapproval Changes: Chemistry, Manufacturing, and Controls; In Vitro Dissolution Testing and In Vivo Bioequivalence Documentation. Food and Drug Administration, Center for Drug Evaluation and Research, 1997.
2. FDA. Guidance for Industry. Dissolution Testing of Immediate Release Solid Oral Dosage Forms. Rockville: Center for Drug Evaluation and Research, 1997.
3. Stippler E. Biorelevant Dissolution Test Methods to Asses Bioequivalence of Drug Products. Frankfurt: Institute for Pharmaceutical Technology, J. W. Goethe University, 2004:414.
4. WHO. WHO Expert Committee on Specifications for Pharmaceutical Preparations. Geneva: World Health Organization, 2004.
5. USP. US Pharmacopeia & National Formulary. Rockville: United States Pharmacopeial Convention Inc., 2004.
6. EDQM. Dissolution Test Stage 4. Strasbourg Cedex: Concil of Europe, 2001.
7. Hoffer JD, Gray V. Examination of selection of immediate release dissolution acceptance criteria. Dissolution Technol 2003; 2:16–20.
8. Layloff T, Nasr M, Baldwin R, Caphart M, Drew H, Hanig J, Hoiberg C, Koepke S, MacGregor JT, Mille Y, Murphy E, Ng L, Rajagopalan R, Sheinin E, Smela M, Welschenbach M, Winkle H, Williams R. The U.S. FDA regulatory methods validation program for new and abbreviated new drug applications. Pharm Technol 2000.
9. Glomme A, März J, Dressman JB. Comparison of a new, miniaturized shake-flask solubility method with automated potentiometric acid/base titrations and calculated solubilities. J Pharm Sci 2004. In press.

10. James KC. Solubility and Related Properties. New York, Basel: Marcel Dekker Inc., 1986.

11. Avdeef A. pH-metric solubility. 1. Solubility-pH profiles and Bjerrum plots, Gibbs buffer and pK_a in the solid state. Pharm Pharmacol Commun 1998; 4:165–178.

12. Polli JE, Lawrence XY, Cook JA, Amidon GL, Borchardt RT, Burnside BA, Burton PS, Chen ML, Conner DP, Faustino PJ, Hawi AA, Hussain AS, Joshi HN, Kwei G, Lee HL, Lesko LJ, Lipper RA, Loper AE, Nerurkar SG, Polli JW, Sanvordeker DR, Taneja R, Uppoor RS, Vattikonda CS, Wilding I, Zhang G. Summary workshop report: biopharamceutics classification system—implementation challanges and extention opportunities. J Pharm Sci 2004; 93(6):1375–1381.

13. Krämer J. The biopharmaceutics classification system—an overview of the current status in relation to IR and MR dosage forms. 1st International Conference on Bioavailability, Bioequivalence and Dissolution Testing, London, 2002.

14. FDA. Guidance for Industry. Waiver of In Vivo Bioavailability and Bioequivalence Studies for Immediate-Release Solid Oral Dosage Forms Based on a Biopharmaceutics Classification System. Rockville: Center for Drug Evaluation and Research, 2000.

15. FDA. Guidance for Industry—Immediate Release Solid Oral Dosage Forms—Scale-up and Postapproval Changes: Chemistry, Manufacturing, and Controls, In Vitro Dissolution Testing, and In Vivo Bioequivalence Documentation. Food and Drug Administration, Center for Drug Evaluation and Research, 1995.

16. EMEA. Note for Guidance on the Investigation of Bioavailability and Bioequivalence. CPMP/EWP/QWP/1401/98. CPMP, The European Agency for the Evaluation of Medicinal Products; 2001.

17. Amidon GL, Lennernas H, Shah VP, Crison JR. A theoretical basis for a biopharmaceutic drug classification: the correlation of in vitro drug product dissolution and in vivo bioavailability. Pharm Res 1995; 12(3):413–420.

18. Möller H. Developing a standardized protocol and data base for in vitro permeability measures and its results. International

Workshop on the Biopharmaceutics Classification System: Scientific and Regulatory Aspects in Practice, London, 2001.

19. Tam KY. Potential of Using Cell Based Technology for Predicting Bioavailability. Amsterdam: Dissolution, Bioavailability & Bioequivalence, 2003.

20. FDA. Guidance for Industry. Bioavailability and Bioequivalence Studies for Orally Administered Drug Products—General Considerations, Food and Drug Administration. Center for Drug Evaluation and Research, 2003.

21. Tsong Y, Hammerstrom T, Sathe P, Shah VP. Statistical assessment of mean differences between two dissolution data sets. Drug Inform J 1996; 30:1105–1112.

22. Cardot JM, Beyssac E. In vitro/in vivo correlations: scientific implications a standardisation. Eur J Drug Metab Pharmacokinet 1993; 18(1):113–120.

23. Lennernas H. Human intestinal permeability. J Pharm Sci 1998; 87(4):403–410.

24. Krämer J. In: Role of in Vitro Dissolution Test. Tokyo: Bioavailability, Bioequivalence and Pharmacokinetic Studies, 1996.

12

Dissolution Method Development: An Industry Perspective

CYNTHIA K. BROWN
Eli Lilly and Company, Indianapolis, Indiana, U.S.A.

INTRODUCTION

In today's pharmaceutical industry, dissolution testing is a valuable qualitative tool that provides key information about the biological availability and/or equivalency as well as the batch-to-batch consistency of a drug. Therefore, a properly designed dissolution test is essential for the biopharmaceutical characterization and batch-to-batch control of the drug product. During drug development, dissolution testing is used to select appropriate formulations for in vivo testing, guide formulation development activities, and assess stability of the drug product under various packaging and storage requirements. For the dissolution test to be a useful drug characterization tool, the methodology needs to be able to discriminate between different degrees of product performance and thus, the collection of a multi-time point dissolution profile is useful. At present, almost all solid oral dosage forms require dissolution testing as a quality control check before a product is introduced into the market place. For the dissolution test to be a useful quality control tool, the methodology should be simple, reliable and reproducible, and ideally be able to discriminate between different degrees of product performance (1).

Dissolution testing is also used to identify bioavailability (BA) problems and to assess the need for further bioequivalence (BE) studies relative to scale-up and post-approval changes (SUPAC), where it can function as a signal of bioinequivalence (2,3). The issuance of the Food and Drug Administration (FDA) guidance document, *Waiver of In Vivo Bioavailability and Bioequivalence Studies for Immediate-Release Solid Oral Dosage Forms Based on a Biopharmaceutics Classification System*, allows dissolution testing to be used as a surrogate for in vivo BE testing under certain circumstances (4). The Biopharmaceutics Classification System (BCS) is a scientific framework for classifying drug substances based on their aqueous solubility and intestinal permeability. When combined with the dissolution of the drug product, the BCS takes into account three major factors that influence the rate and extent of drug absorption from immediate-release solid oral dosage forms: dissolution, solubility, and intestinal permeability (5). Based on the BCS framework, drug manufacturers may request waivers from additional in vivo studies (biowaivers) if their drug product meets certain criteria. In addition, the FDA's guidance on BA and BE (6) allows biowaivers for additional strength(s) of immediate-release as well as modified-release drug products based on formulation proportionality and dissolution profile comparison.

These changes in BE requirements that move away from the in vivo study requirement in certain cases and rely more on dissolution test results, emphasize the significance of dissolution test applications. In all cases where the dissolution test is used as a BE test, a link with a bioavailable product is established. With the advances in dissolution testing and the increased understanding of the scientific principles and mechanisms of dissolution testing, a clear trend has appeared where the dissolution test is not solely a traditional quality control test but may also be used as a surrogate to the in vivo BE test (7).

For the dissolution test to be used as an effective drug product characterization and quality control tool, the method must be developed with the various end uses in mind. In some cases, the method used in the early phase of product and formulation development could be different from the final test procedure utilized for control of the product quality. Methods used for formulation screening or BA and/or bioequivalency evaluations may simply be impractical for a quality control environment. It is essential that with the accumulation of experience, the early method be critically re-evaluated and potentially simplified, giving preference to compendial apparatus and media. Hence, the final dissolution method submitted for product registration may not necessarily closely imitate the in vivo environment but should still test the key performance indicators of the formulation.

To facilitate the development of appropriate dissolution tests several regulatory, pharmacopeial, and industrial organizations have issued dissolution-related guidelines that provide information and recommendations on the development and validation of dissolution test methodology, the establishment of dissolution specifications, and the regulatory applications of dissolution testing (8–16). This chapter describes a systematic approach for the development of a dissolution method. The information is organized and presented in sections that follow the chronological sequence of the method development process. These include the assessment of relevant physical and chemical properties of the drug, determination of the appropriate dissolution apparatus, selection of the dissolution medium, determination key operating parameters, method optimization, and validation of the methodology.

PHYSICAL AND CHEMICAL PROPERTIES

The first step in the development of a new dissolution test is to evaluate the relevant physical and chemical data for the drug substance. Knowledge of the drug compound's physical–chemical properties will facilitate the selection of dissolution medium and determination of medium volume.

Some of the physicochemical properties of the active pharmaceutical ingredient (API) that influence the dissolution characteristics are:

Ionization constants (pK_a),
Solubility as a function of pH,
Solution stability as a function of pH,
Particle size,
Crystal form, and
Common ion, ionic strength, and buffer effects.

Two key physicochemical API properties to evaluate are the solubility and solution-state stability of the drug substance as a function of pH. Knowledge of the pK_a (or pK_a's) is useful because it defines the charge of the molecule in solution at any given pH. Ideally, the drug substance's solubility in the dissolution medium should not be the rate-limiting factor for the drug substance's dissolution from the drug product. Hence, the dissolution rate should be characteristic of the release of the active ingredient from the dosage form rather than the drug substance's solubility in the dissolution medium. When adjusting the composition of the medium to insure adequate solubility for the drug substance, the influence of surfactants, pH, and buffers on the solubility and stability of the drug substance need to be evaluated. The solution-state stability of the API must also be considered in the design of a dissolution test because the molecule's stability in various dissolution media may limit the pH range over which the drug product's dissolution can be evaluated. Typically, the drug's solution stability should be determined at 37°C for 2 hr for immediate-release formulations and twice the designated testing time for sustained-release formulations (17).

During the initial stages of a drug product's development, a dissolution test should facilitate the formulation development and selection. During this phase of the drug development process bioavailability data is usually not available. In the absence of BA, the dissolution medium selection should be based on the physicochemical properties, the formulation design, and the intended dose. The BCS provides a good framework for determining if the dissolution of the drug will be the rate-limiting factor in the in vivo absorption process. Hence, the pH solubility of the drug and the intended dose are essential parameters to consider early in the dissolution method development process.

Once you have a good understanding of the physical–chemical properties of the drug substance, the key properties of the dosage form, i.e., type, label claim, and release mechanism, need to be considered. The most appropriate dissolution testing apparatus and dissolution medium can be selected based on the physical–chemical properties of the drug substance and the key properties of the dosage form. Dosage forms can be designed to provide immediate release, delayed release, or extended (controlled) release. Determining the type of release and anticipated site of in vivo absorption will facilitate the selection of dissolution media, testing apparatus, and test duration.

DISSOLUTION APPARATUS SELECTION

The choice of apparatus is based on knowledge of the formulation design and practical aspects of dosage form performance in the in vitro test system. Dissolution testing is conducted on equipment that has demonstrated suitability, such as described in the 2003 United States Pharmacopeia (USP) under the general chapters of *Dissolution* and *Drug Release* (10,11). The basket method (USP Apparatus 1) is routinely used for solid oral dosage forms such as capsule or tablet formulations at an agitation speed of 50–100 rpm, although speeds of up to 150 rpm have been used. The paddle method (USP Apparatus 2) is frequently used for solid oral dosage forms such as tablet and capsule formulations at 50 or 75 rpm. The paddle method is also useful for the testing of oral suspensions at the recommended paddle speed of 25–50 rpm. The reciprocating cylinder (USP Apparatus 3) has been found to be especially useful for bead-type modified-release dosage forms. The flow-through cell (USP Apparatus 4) may offer advantages for some modified-release dosage forms, especially those that contain active ingredients with limited solubility. Additionally, the reciprocating cylinder or the flow-through cell may be useful for soft gelatin capsules, bead products, suppositories, or poorly soluble drugs. By design, both the reciprocating cylinder and the flow-through cell allow for a controlled pH change of the dissolution medium throughout the test, which allows the apparatus to be easily utilized for physiological evaluations of the dosage form during development. The paddle over disk (USP Apparatus 5) and the cylinder (USP Apparatus 6) have been shown to be useful for evaluating and testing transdermal dosage forms. The reciprocating holder (USP Apparatus 7) has been shown to have application to non-disintegrating oral modified-release dosage forms, as well as to transdermal dosage forms.

In general, compendial apparatus and methods should be used as a first approach in drug development. To avoid unnecessary proliferation of equipment and method design, modifications of compendial equipment or development and use of alternative equipment should be considered only when it has been proven that compendial set up does not provide meaningful data for a given dosage form. In these instances, superiority of the new or modified design has to be proven in comparison to the compendial design.

Table 1 outlines the current status of scientific development for the dissolution or release testing from various dosage forms and recommends, where possible, the dissolution apparatus of "first choice" (13). Refer also to Chapter 2 for further description of the USP apparatus.

DISSOLUTION MEDIUM SELECTION

For batch-to-batch quality testing, selection of the dissolution medium is based, in part, on the solubility data and the dose

Table 1 Apparatus Recommended Based on Dosage Form Type

Type of dosage form	Release method
Solid oral dosage forms *(conventional)*	Basket, paddle, reciprocating cylinder, or flow-through cell
Oral suspensions	Paddle
Oral disintegrating tablets	Paddle
Chewable tablets	Basket, paddle, or reciprocating cylinder with glass beads
Transdermals—patches	Paddle over disk
Topicals—semisolids	Franz cell diffusion system
Suppositories	Paddle, modified basket, or dual chamber flow-through cell
Chewing gum	Special apparatus [European Pharmacopoeia (PhEur)]
Powders and granules	Flow-through cell (powder/granule sample cell)
Microparticulate formulations	Modified flow-through cell
Implants	Modified flow-through cell

range of the drug product in order to ensure that sink conditions are met. The term sink conditions is defined as the volume of medium at least greater than three times that required to form a saturated solution of a drug substance. A medium that fails to provide sink conditions may be justifiable if it is shown to be more discriminating or if it provides reliable data which otherwise can only be obtained with the addition of surfactants. When the dissolution test is to indicate the biopharmaceutical properties of the dosage form, it is more important that the test closely simulate the environment in the GI tract than necessarily produce sink conditions for release. Therefore, it is not always possible to develop one dissolution test or select one dissolution medium that ensures batch-to-batch control as well as monitoring the biopharmaceutical aspects of the drug product.

The dissolution characteristics of oral formulations should be evaluated over the physiologic pH range of 1.2–6.8 [1.2–7.5 for modified release (MR) formulations]. During method development, it may be useful to measure the pH before and after a run to see if the pH changes during the test, especially if the buffer capacity of the chosen medium is low. Selection of the most appropriate medium for routine testing is then based on discriminatory capability, ruggedness, stability of the analyte in the test medium, and relevance to in vivo performance where possible.

For very poorly soluble compounds, aqueous solutions may contain a percentage of a surfactant (e.g., sodium lauryl sulfate, Tween 80 or CTAB) that is used to enhance drug solubility. The need for surfactants and the concentrations used should be justified. Surfactants can be used as either a wetting agent or, when the critical micelle concentration (CMC) is reached, to solubilize the drug substance. The surfactant's CMC depends upon the surfactant itself and the ionic strength of the base medium. The amount of surfactant needed for adequate drug solubility depends on the surfactant CMC and the degree to which the compound partitions into the surfactant micelles. Because of the nature of the compound and micelle interaction, there is typically a linear dependence between solubility and surfactant concentration above the CMC. If a compound is ionizable, surfactant concentration and pH may be varied simultaneously, and the combined effect can substantially change the solubility characteristics of the dissolution medium. Table 2 lists dissolution medium selection criteria as defined in regulatory, industry, and compendial guidances.

The BCS describes the classification of compounds according to solubility and permeability (6). Biorelevant medium is a term used to describe a medium that has some relevance to the in vivo dissolution conditions for the compound. Choice of a biorelevant medium is based on a mechanistic approach that considers the absorption site, if known, and whether the rate-limiting step to absorption is the dissolution or permeability of the compound. In some cases, the biorelevant medium will be different from the test conditions chosen for the regulatory test and the time points are also likely to be different. If the compound dissolves quickly in the stomach and is highly permeable, gastric emptying time may be the rate-limiting step to absorption. In this case, the dissolution test is to demonstrate that the drug is released quickly under

Table 2 Recommended Dissolution Medium Composition and Volume for Rotating Basket or Rotating Paddle Apparatus

Guidance or compendial reference	Volume	pH	Additives
Federation International Pharmaceutique (FIP) (23)	500–1,000 mL; 900 mL historical; 1,000 mL recommended for future development	pH 1–6.8; above pH 6.8 with justification—not to exceed pH 8	Enzymes, salts, surfactants with justification
United States Pharmacopeia (USP) (10–12)	500–1,000 mL; up to 2,000 mL for drug with limited solubility	Buffered aqueous solution pH 4–8 or dilute acid solutions (0.001 N HCl to 0.1 N HCl)	Enzymes, salts, surfactants balanced against loss of discriminatory power; enzymes can be used for cross-linking of gelatin capsules or gelatin-coated tablets
World Health Organization (WHO) (16), European Pharmacopoeia (PhEur) (14), Japanese Pharmacopoeia (JP) (15)	Determined per product	Adjust pH to within ±0.05 units of the prescribed valued	Determined per product
FDA (8,9)	500, 900, or 1,000 mL	pH 1.2–6.8; higher pH justified case-by-case—in general not to exceed pH 8	Surfactants recommended for water poorly soluble drug products—need and amount should be justified; enzymes use need case-by-case justification; utilized for the cross-linking of gelatin capsules or gelatin-coated tablets

typical gastric (acidic) conditions. On the other hand, if dissolution occurs primarily in the intestinal tract (e.g., a poorly soluble, weak acid), a higher pH range (e.g., simulated intestinal fluid with a pH of 6.8) will be more appropriate (18).

The fed and fasted state may also have significant effects on the absorption or solubility of a compound. Compositions of media that simulate the fed and fasted states can be found in the literature (19) (see also Chapter 5). These media reflect changes in the pH, bile concentrations, and osmolarity after meal intake and therefore have a different composition than that of typical compendial media. They are primarily used to establish in vitro–in vivo correlations during formulation development and to assess potential food effects and are not intended for quality control purposes. For quality control purposes, the substitution of natural surfactants (bile components) with appropriate synthetic surfactants is permitted and encouraged because of the expense of the natural substances and the labor-intensive preparation of the biorelevant media.

KEY OPERATING PARAMETERS

Media: Volume, Temperature, Deaeration

As shown in Table 2, the recommended volume of dissolution medium is 500–1000 mL, with 900 mL as the most common volume when using the basket or paddle apparatus. The volume can be raised to between 2 and 4 L, depending on the concentration and sink conditions of the drug, but proper justification is expected.

The standard temperature for the dissolution medium is $37 \pm 0.5°C$ for oral dosage forms. Slightly increased temperatures such as $38 \pm 0.5°C$ have been recommended for dosages forms such as suppositories. Lower temperatures such as $32 \pm 0.5°C$ are utilized for topical dosage forms such as transdermal patches and topical ointments.

The significance of deaeration of the medium should be determined on a case-by-case basis, as air bubbles can interfere with the test results and act as a barrier to dissolution if present on the dosage unit or basket mesh. Additionally, air bubbles can cause particles to cling to the apparatus and vessel walls. On the other hand, bubbles on the dosage unit may increase the buoyancy and lead to an increase in the dissolution rate, or decrease the dissolution rate by decreasing the available surface area. Consequently, the impact of medium deaeration may be formulation dependent, such that some formulations will be sensitive to the presence of dissolved air in the dissolution while other formulations will be robust. To determine if deaeration of the medium is necessary, a comparison between dissolution data generated with non-deaerated medium vs. dissolution data generated with deaerated medium should be performed.

The following deaeration method is described as a footnote in the 2003 United States Pharmacopeia (USP) under the general chapter *Dissolution* (10). The USP deaeration method requires heating of the medium, followed by filtration, and drawing of a vacuum for a short period of time. Other deaeration methods such as room temperature filtration, sonication, and helium sparging are described in literature (20,21) and are routinely used throughout the industry. The deaeration method needs to be clearly characterized, since the method chosen might impact the dissolution release rate (13). It should be noted that dissolution tests using the flow-through cell method could be particularly sensitive to the deaeration of the medium. Media containing surfactants are not usually deaerated after the surfactant has been added to the medium because of excessive foaming. In some laboratories, the base medium is deaerated prior to the addition of the surfactant.

Sinker Evaluation

Currently, the Japanese Pharmacopoeia (JP) is the only pharmacopeia that requires a specific sinker device for all capsule formulations. The USP recommends a few turns of a nonreactive material wire when the dosage form tends to float (12) (see Chapter 2 for illustrations of the Japanese and USP sinkers). Because sinkers can significantly influence the dissolution profile of a drug product, detailed sinker descriptions and the rationale for why a sinker is used should be stated in the written procedure. When comparing different sinkers (or sinkers versus no sinkers), a test should be run concurrently with each sinker. Each sinker type should be evaluated based on its ability to maintain the dosage at the bottom of the vessel without inhibiting drug release.

Sinkers can significantly influence the dissolution profile of a drug. Therefore, the use of sinkers should be part of the dissolution method validation. If equivalent sinkers are identified during the sinker evaluation and validation, the equivalent sinkers should be listed in the written dissolution test procedure. When a dissolution method utilizes a dissolution sinker and is transferred to another laboratory, the receiving laboratory should duplicate the validated sinker design(s) as closely as possible.

Analytical Detection

For determination of the quantitative step in the dissolution method, information regarding the spectral, chromatographic, electrochemical, and/or chemical characteristics of the drug substance should be considered. The quantitative method needs to provide adequate sensitivity for the accurate determination of the analyte in the dissolution medium. Since formulations are likely to change during product development, it is usually advantageous to use high-performance liquid chromatography (HPLC) detection procedures. However, because of the ease of automation and faster analysis time, UV detection methods are more desirable for the routine quality control testing of products.

Filtration of the dissolution sample aliquot is usually needed prior to quantitation. Filtration of the dissolution samples is usually necessary to prevent undissolved drug particles from entering the analytical sample and dissolving further. Also, filtration removes insoluble excipients that may otherwise cause a high background or turbidity. Prewetting of the filter with the medium is usually necessary. Filters can be in-line, at the end of the sampling probe, or both. The pore size can range from 0.45 to 70 μm. The usual types are depth, disk, or flow-through filters. However, if the excipient interference is high, or the filtrate has a cloudy appearance, or the filter becomes clogged, an alternative type of filter or pore size may need to be evaluated.

Adsorption of the drug(s) to the filter needs to be evaluated. If drug adsorption occurs, the amount of initial filtrate discarded may need to be increased. If results are still unsuitable, an alternative filter material should be sought. Centrifugation of samples is generally not recommended, as dissolution can continue to occur during centrifugation and there may be a concentration gradient in the supernatant. A possible exception might be compounds that adsorb to all common filters.

Sampling Time Points and Specifications

Key operating parameters that may change (or be optimized) throughout a product's development and approval cycle are dissolution sampling time points and dissolution limits or specifications by which the dissolution results should be evaluated. The results generated from the dissolution test need to be evaluated and interpreted based on the intended purpose of the test. If the test is used for batch-to-batch control, the results should be evaluated in regard to the established limits or specification value. If the test is being utilized as a characterization test (i.e., biopharmaceutical evaluations, formulation development studies, etc.) the results are usually evaluated by profile comparisons.

For immediate-release dosage forms, the dissolution test duration is typically 30–60 min, with a single time point specification being adequate in most cases for routine batch-to-batch quality control for approved products. Typical specifications for the amount of active ingredient dissolved, expressed as a percentage of the labeled content (Q), are in the range of 75–80% dissolved. A Q value in excess of 80% is not generally used, as allowances need to be made for assay and content uniformity ranges. Since the purpose of specifying dissolution limits is to ensure batch-to-batch consistency within a range that guarantees comparable biopharmaceutical performance in vivo, specifications including test times are usually established based on an evaluation of dissolution profile data from pivotal clinical batches and confirmatory BA batches (8).

When the test is utilized as a characterization tool (i.e., biopharmaceutical evaluations, formulation development studies, etc.) the results are usually evaluated by profile comparisons. In this case, the product's comparability and performance are evaluated by collecting additional sampling time points. For registration purposes, a plot of the percentage of the drug dissolved vs. time should be determined. Enough time points are to be selected to adequately characterize the ascending and plateau phases of the dissolution curve. According to the BCS referred to in several FDA guidance documents, highly soluble and highly permeable drugs formulated with rapidly dissolving products need not be subjected to a profile comparison if they can be shown to release 85% or more of the active ingredient within 15 min. For these types of products, a one-point test will suffice. When an immediate-release drug product does not meet the rapidly dissolving criteria, dissolution data from multiple sampling time points ranging from 10 to 60 min or longer are usually collected.

So-called infinity points can be useful during development studies. To obtain an infinity point, the paddle or basket speed is increased significantly (e.g., 150 rpm) at the end of the run and the test is allowed to run for an extended period of time (e.g., 60 min), and then an additional sample is taken. Although there is no requirement for 100% dissolution in the profile, the infinity point can provide data that may provide useful information about the formulation characteristics during the initial development.

For an extended-release dosage form, at least three test time points are chosen to characterize the in vitro drug-release profile for the routine batch-to-batch quality control for approved products. Additional sampling times may be required for formulation development studies, biopharmaceutical evaluations, and drug approval purposes. An early time point, usually 1–2 hr, is chosen to show that there is little probability of dose dumping. Release at this time-point should not exceed values expected according to the mechanism of release and the intended overall-release profile. An intermediate time point is chosen to define the in vitro-release profile of the dosage form, and a final time point is chosen to show essentially complete release of the drug. Test times and specifications are usually established on the basis of an evaluation of drug-release profile data. For products containing more than a single active ingredient, drug release is to be determined for each active ingredient. Extended-release specifications are addressed in the USP under the general chapter *In Vitro and In Vivo Evaluation of Dosage Forms* (12) and the FDA's guidance document *Extended Release Oral Dosage Forms: Development, Evaluation, and Application of In Vitro/In Vivo Correlations* (9).

METHOD OPTIMIZATION

When human BA data are available from several formulations, the dissolution test should be re-evaluated and optimized (if needed). The goal of dissolution method optimization is to identify in vitro test conditions that adequately discriminate critical formulation differences or critical manufacturing variables. During the method optimization process, the biostudy formulations are tested using various medium compositions (e.g., pH, ionic strength, surfactant composition). The effect of hydrodynamics on the formulations should also be evaluated by varying the apparatus agitation speed. If a non-bioequivalent batch is discovered during a bioequivalency study and the in vivo absorption is dissolution rate limited (BCS Class 2), the dissolution methodology should be optimized to differentiate the non-bioequivalent batches from the bioequivalent batches by dissolution specification limits. This would ensure batch-to-batch consistency within a range that guarantees comparable biopharmaceutical performance in vivo. Once a discriminating method is developed, the same method should be used to release product batches for future clinical studies.

VALIDATION

Once the appropriate dissolution conditions have been established, the method should be validated for linearity, accuracy, precision, specificity, and robustness/ruggedness. This section will discuss these parameters only in relation to issues unique to dissolution testing. All dissolution testing must be performed on a calibrated dissolution apparatus meeting the mechanical and system suitability standards specified in the appropriate compendia.

Linearity

Detector linearity should be checked over the entire range of concentrations expected during the procedure. The ICH recommendation for range of dissolution methods is ±20% of the specification limits (22). For example, if the specification for an immediate-release tablet is "no tablet less than 80% in 45 min," then the range to be checked would be from 60% to 100% of the tablet's label claim. For controlled or extended-release product, the range should be extended to include values 20% less than the lowest specification limit to values 20% higher than the upper specification limit. Typically, the concentration range is divided into five evenly spaced concentrations. Linearity testing of the dosage form should cover the entire range of the product.

Linearity is evaluated by appropriate statistical methods such as the calculation of a regression line by the method of least squares. The linearity results should include the correlation coefficient, *y*-intercept, slope of the regression line, and residual sum of squares as well as a plot of the data. Also, it is helpful to include an analysis of the deviation of the actual data points for the regression line to evaluate the degree of linearity.

Accuracy

Accuracy samples are prepared by spiking bulk drug and excipients in the specified volume of dissolution fluid. The concentration ranges of the bulk drug spikes are the same as those specified for linearity testing. If the dosage form is a capsule, the same size and color of capsule shell should be added to the mixture. The solutions should be tested according to the parameters specified in the method, i.e., temperature, rotation speed, filters, sampling mode, and detection mode. If accuracy solutions are prepared at five concentrations levels across the range, aliquots can be collected at the sampling interval(s) specified in the method and analyzed according to the quantitative method procedure. An alternative approach is to collect at least three sampling aliquots from the low-, middle-, and high-accuracy solutions.

Precision

According to the dissolution method, precision is determined by testing at least six aliquots of a homogenous sample for each dosage strength. The precision should be assessed at each specification interval for the dosage form. The precision can be determined by calculating the relative standard deviation (RSD) of the multiple aliquots from each solution.

Two unique sample tests (e.g., different analysts, instruments, reagents, and standard preparations) performed within the same laboratory would establish the method's intermediate precision. If the dosage form requires the use of a sinker, the sinker specified in the method should be used in precision testing.

Specificity

The dissolution analysis method must be specific for the bulk drug substance in the presence of a placebo. A mixture of dissolution fluid and the excipients (including the capsule shell if applicable) should be tested to specificity. Stability of the drug in the dissolution medium should be considered since the dissolution test exposes the drug to hydrolytic media at 37°C for specified time spans. Simply monitoring the UV spectra of the solutions is not sufficient in determining degradation since many degradation products will have the same UV spectrum as the parent compound. Therefore, specificity testing should be confirmed by analyzing accuracy samples with a selective analysis mode such HPLC. If the capsule shell interferes with the bulk drug detection, the USP allows for a correction for the capsule shell interference. Corrections $>25\%$ of labeled content are unacceptable (10).

Robustness/Ruggedness

Robustness testing should determine the critical parameters for a particular dissolution method. By subjecting each dissolution parameter to slight variations, the critical dissolution parameters for the dosage form will be determined. This will facilitate method transfer and troubleshooting. Robustness testing should evaluate the effect of varying media pH, media volume or flow rate, rotation speed, apparatus sample position, sinkers (if applicable), media deaeration, temperature, and filters. Ruggedness of the methods should be evaluated by running the method with multiple analysts on multiple systems. If the analysis is performed by HPLC, the effect of columns and mobile conditions should also be addressed.

AUTOMATED SYSTEMS

Validation of automated systems must demonstrate a lack of contamination or interference that might result from automated transfer, cleaning, or solution preparations procedures. Equivalency between the results generated from the manual method and the data generated from the automated system should be demonstrated. Since sensitivity to automated dissolution testing may be formulation related, qualification and validation of automated dissolution equipment needs to be established on a product-by-product basis (8,13) (see also Chapter 12 for a more detailed description of automation issues).

CONCLUSIONS

Regulatory changes in BE requirements (that move away from the in vivo study requirements in certain cases and rely more on dissolution testing) emphasize the significance of dissolution test applications. A clear trend has appeared with the advances in and increased understanding of the scientific principles and mechanisms of dissolution testing. The dissolution test is not solely a traditional quality control test but may also be used as a product characterization test that can serve as a surrogate to the in vivo BE test. For the dissolution test to be used as an effective drug product characterization and quality control tool, the method must be developed with the final application for the test in mind. A properly designed dissolution test can be used to characterize the drug product and assure batch-to-batch reproducibility for consistent pharmacological and biological activity.

Therefore, the development and validation of a scientifically sound dissolution method requires the selection of key method parameters that provide accurate, reproducible data that are appropriate for the intended application of the methodology. It is important to note that while more extensive dissolution methodologies may be required for bioequivalency evaluations or biowaivers (i.e., multiple media, more complex dissolution media additives, and multiple sampling time points), it is also essential for the simplified, routine quality control dissolution method to discriminate batch-to-batch differences that might affect the product's in vivo performance.

REFERENCES

1. Dressman JB, Amidon GL, Reppas C, Shah VP. Dissolution testing as a prognostic tool for oral dug absorption: immediate release dosage forms. Pharm Res 1998; 15(1):11–22.
2. U.S. Department of Health and Human Services, Food and Drug Administration, Center for Drug Evaluation and Research. Guidance for Industry: SUPAC IR: Immediate-Release Solid Oral Dosage Forms: Scale-Up and Post-Approval Changes: Chemistry, Manufacturing, and Controls, In Vitro Dissolution Testing, and In Vivo Bioequivalence Documentation, November 1995.
3. U.S. Department of Health and Human Services, Food and Drug Administration, Center for Drug Evaluation and

Research. Guidance for Industry: SUPAC MR: Modified-Release Solid Oral Dosage Forms: Scale-Up and Post-Approval Changes: Chemistry, Manufacturing, and Controls, In Vitro Dissolution Testing, and In Vivo Bioequivalence Documentation, October 1997.

4. U.S. Department of Health and Human Services, Food and Drug Administration, Center for Drug Evaluation and Research. Guidance for Industry: Waiver of In Vivo Bioavailability and Bioequivalence Studies for Immediate-Release Solid Oral Dosage Forms Based on a Biopharmaceutics Classification System, August 2000.
5. U.S. Department of Health and Human Services, Food and Drug Administration, Center for Drug Evaluation and Research. Guidance for Industry: Bioavailability and Bioequivalence Studies for Orally Administered Drug Products—General Considerations, March 2003.
6. Amidon GL, Lennernäs H, Shah VP, Crison JR. A theoretical basis for a biopharmaceutic drug classification: the correlation of in vitro drug product dissolution and in vivo bioavailability. Pharm Res 1995; 12(3):413–420.
7. Shah VP. Dissolution: a quality control test vs. a bioequivalence test. Dissolution Technol 2001; 8(4):6–7.
8. U.S. Department of Health and Human Services, Food and Drug Administration, Center for Drug Evaluation and Research. Guidance for Industry: Dissolution Testing of Immediate Release Solid Oral Dosage Forms, August 1997.
9. U.S. Department of Health and Human Services, Food and Drug Administration, Center for Drug Evaluation and Research. Guidance for Industry: Extended Release Oral Dosage Forms: Development, Evaluation, and Application of In Vitro/In Vivo Correlations, September 1997.
10. 2003 United States Pharmacopoeia, USP 26, National Formulary 21. General Chapter 711 Dissolution; United States Pharmacopeial Convention: Rockville, MD, 2002:2155–2156.
11. 2003 United States Pharmacopoeia, USP 26, National Formulary 21. General Chapter 724 Drug Release; United States Pharmacopeial Convention: Rockville, MD, 2002:2157–2164.

12. 2003 United States Pharmacopoeia, USP 26, National Formulary 21. General Chapter 1088 In Vitro and In vivo Evaluation of Dosage Forms; United States Pharmacopeial Convention: Rockville, MD, 2002:2334–2339.

13. Siewert M, Dressman JB, Brown CK, Shah VP. FIP/AAPS guidelines for dissolution/in vitro release testing of novel/special dosage forms. Dissolution Technologies 2003; 10(1):6–15.

14. European Pharmacopoeia 4th Edition 2002. General Chapter 2.9.3. Dissolution Test for Solid Dosage Forms; Directorate for the Quality of Medicines of the Council of Europe: Germany 2001:194–197.

15. The Japanese Pharmacopoeia 14th Edition 2001. General Test 15. Dissolution Test; Society of Japanese Pharmacopoeia: Japan, 2001:33–36.

16. The International Pharmacopoeia 3rd Edition, Vol. 5, 2003. Tests for Dosage Forms: Dissolution Test for Solid Oral Dosage Forms; World Health Organization Geneva: Spain 2003:18–27.

17. Skoug JW, Halstead GW, Theis DI, Freeman JE, Fagan DT, Rohrs BR. Strategy for the development and validation of dissolution tests for solid oral dosage forms. Pharm Tech 1996; 20(5):58–72.

18. Galia E, Nicolaides E, Hörter D, Löbenberg R, Reppas C, Dressman JB. Evaluation of various dissolution media for predictin in vivo performance of class I and II drugs. Pharm Res 1998; 15(5):698–705.

19. Dressman JB. Dissolution testing of immediate-release products and its application to forecasting in vivo performance. In: Dressman JB, Lennernäs H, eds. Oral Drug Absorption Prediction and Assessment. New York: Marcel Dekker, 2000:155–181.

20. Diebold SM, Dressman JB. Dissolved oxygen as a measure for de- and reaeration of aqueous media for dissolution testing. Dissolution Technol 1998; 5(3):13–16.

21. Rohrs BR, Stelzer DJ. Deaeration techniques for dissolution media. Dissolution Technol 1995; 2(2):1, 7–8.

22. International Conference on Harmonisation, ICH Harmonised Tripartite Guideline, Validation of Analytical Procedures: Methodology, November 1996.

23. Aiache JM, Aoyagi N, Blume H, Dressman JB, Friedel HD, Grady LT, Gray VA, Helboe P, Hubert B, Kopp-Kubel S, Kramer J, Kristensen H, Langenbucher E, Leeson L, Lesko L, Limberg J, McGilveray I, Muller H, Quershi S, Shah VP, Siewart M, Suverkrup R, Waltersson JO, Whiteman D, Wirbitzki E. FIP guidelines for dissolution testing of solid oral products. Dissolution Technol 1997; 4(4):5–14.

13

Design and Qualification of Automated Dissolution Systems

DALE VONBEHREN

Pharmaceutical Development and Quality Products, Zymark Corporation, Hopkinton, Massachusetts, U.S.A.

STEPHEN DOBRO

Product Testing and Validation, Zymark Corporation, Hopkinton, Massachusetts, U.S.A.

FUNCTIONAL DESIGN OF AN AUTOMATED DISSOLUTION APPARATUS

Introduction to Automated Dissolution

Dissolution is becoming one of the most commonly automated functions in the modern pharmaceutical development and quality assurance (QA) laboratory. To the experienced dissolution analyst the reasons seem obvious. Dissolution methods are time-consuming and require a significant amount of labor. Beyond the cost of labor, the true cost of increased regulatory requirements and documentation can be better managed through automation. Additionally, the increased pressure to deliver improved return to shareholders is driving various efficiency improvements relating to various aspects of pharmaceutical development and manufacturing, including dissolution analysis.

Speed to market with the best formulation is critical to the long-term profitability of a new chemical entity (NCE). Intracompany facilities are competing as sites of excellence for finished dosage form manufacturing. Skilled labor is expected to do more, faster, by way of improving overall efficiency, scale up new products faster and assimilating ever-increasing regulatory requirements. Companies are competing for skilled labor as well as retail sales. To meet these demands, world-class efficiency and technology is required. Improved precision and lower per test cost can allow more samples to be tested with an improved resolution to detect smaller changes over shorter periods of time. Automated dissolution can help enable these goals. This chapter is intended to assist the reader to introduce automated dissolution systems tailored to the specific needs of a given company and product profile.

Automating the Manual Method

Before describing the various considerations that go into designing a fully automated dissolution apparatus, it may be worthwhile to discuss automation in general for pharmaceutical applications.

Automation at its basic level can be expressed simply with the statement that "analyses that were traditionally manually performed are now performed mechanically through computer-controlled robotics or workstations." Designers typically have a strong desire to exactly reproduce the manual process. In reality, minor changes to the manual approach must be made in order to make the automated process reliable and efficient.

A simple example relating to dissolution is sampling. In the manual world, samples would be taken with a syringe with a long tube or cannula at the end. The cannula may then be replaced with a filter and the medium expressed through the filter and collected into a test tube. Trying to reproduce the exact manual movements of the analyst exactly with an automated process would be very difficult. For example, matching the exact timings and the pressure applied to the syringe can be more difficult than might at first meet the eye. Furthermore, such a system would be extremely expensive: throughput would be slow and lead to a high cost per sample.

To make automation more practical we take shortcuts which approximate the manual approach. In the above example, the cannula might be located on a drive mechanism that lowers to a programmed location. A pump of some sort (possibly a syringe) could aspirate the sample through longer tubing and convey it directly to a filter-dispensing apparatus. The sample would be conveyed through long tubing to a sample collection device where it would pass through a needle to finally fill the tube (Fig. 1).

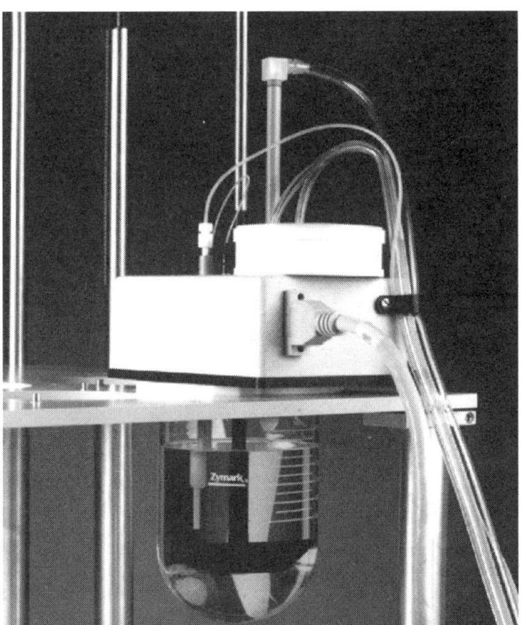

Figure 1 Vessel head for sampling.

The function of the two approaches is identical but the way the task is performed is different. While the reproduced manual method may be expensive it does bring a major benefit. Since it exactly reproduces the manual method, the perceptual barrier to implementation should be relatively low. Implementation may be limited to verifying that the manual steps are accurately reproduced. Additionally, there is no need to formally validate the original chemistry since the procedure reproduces what is already performed manually.

Making the method more automation-friendly requires verifying the suitability of certain steps. As an example, the filtering step is different in that the sample pulled though with a peristaltic pump vs. pushing with a syringe. Equivalence of the two approaches needs to be demonstrated if results of both are to be used interchangeably (Fig. 2).

Demonstrating equivalence of the two approaches does not infer that one is right and the other wrong. One of the unique attributes of dissolution analysis is that there is no right or wrong approach as long as tests can be validated. It is a relative method that is a function of the apparatus and

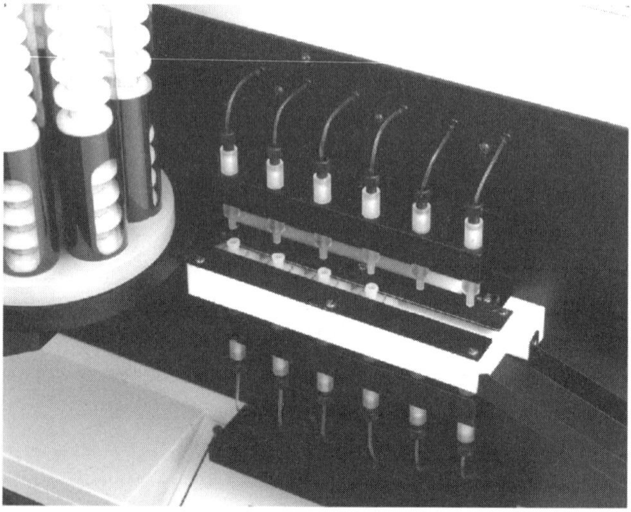

Figure 2 Automated filter assembly.

where everything about it can effect the outcome of the test. Methods are validated to correlate with bioavailability or to discriminate differences between samples for QA purposes. Whether the method is automated or manually performed is inconsequential from a technical perspective. Typically, however, methods are first developed manually so that the suitability of the automated method must be proven to claim equivalence.

The challenge of designing an automated system is to provide an automation-friendly approach that can improve on the efficiency of the manual process (automated or otherwise) while not diverging too far from the manual basics. Each aspect of the analysis that diverges from the traditional approach increases the risk that the system will not be compatible with industry standard hardware and the analogous approach it uses. Compatibility is a critical requirement considering the trend toward global manufacturing. Intercompany facilities, contract laboratories, and governmental agencies need to be as standardized as possible. This is especially important with dissolution analysis since the subtleties of the agitation characteristics have not yet been quantitatively defined.

Regulatory Considerations

In addition to the seemingly obvious concerns of method equivalencies, there is the need to meet local regulatory requirements. In the United States and countries that export to the United States, compliance to Food and Drug Administration (FDA) requirements is mandatory. Other countries' regulations may require a different level of compliance. Fortunately there are forces at work in the industry to harmonize these requirements as much as possible. While this is a slow process, regulatory agencies, the International Conference on Harmonization (ICH) and the Compendia [represented by the European, Japanese, and the United States Pharmacopoeia (USP)] have been making progress.

Prior to designing an automated system, it may be worthwhile to understand the regulatory climate and the official acceptability of automated dissolution analysis. A study of 18 of the most important regulations in the pharmaceutical industry (excluding 21CFR11) was conducted in an attempt to assess the overall acceptability of automation. Of the 18 reviewed documents only three contained a direct reference to automation. The *USP* prominently mentions automated dissolution and at times makes contradictory statements. Interestingly, similarity to the "official" method (re. manual) is mentioned.

One of the most important references to guide us in designing automated apparatus can be found in *USP* (1).

"Automated procedures employing the same basic chemistry as those assay and test procedures given in the monograph are recognized as being equivalent in their suitability for determining compliance. Conversely, where an automated procedure is given in the monograph, manual procedures employing the same basic chemistry are recognized as being equivalent in their suitability for determining compliance."

Here the *USP* makes a very bold statement that if the same basic chemistry is used the method should be considered equivalent in suitability. The authors' interpretation is that an automated method can remain compliant. This is a somewhat drastic statement when thinking about how much a method's physical characteristics can be modified from the original method for the convenience of automation while maintaining the same physical chemistry. *USP* (2) goes on to state:

"...Also, according to these regulations [21 CFR 211.194(a)(2)], users of analytical methods described in the *USP* and the *NF* are not required to validate accuracy and reliability of these methods, but merely verify their suitability under actual conditions of use..."

In other words, if an automated method can be considered equivalent in suitability in determining compliance, and if a compendial method does not require validation, then does it follow that an automated method using the same basic chemistry does not require validation of the original chemistry? This puts automation closer to the same category of change as training new analysts or moving to a new laboratory. At first glance this seems to be a very sweeping proclamation.

Within *USP* (2) the concern over physical differences in apparatus are addressed. Especially with dissolution, it is clear that the physical apparatus is critical to obtaining results and that some sort of test is necessary to verify suitability.

"If automated equipment is used for sampling and the apparatus is modified, validation of the modified apparatus is needed to show that there is no change in the agitation characteristics of the test."

In the practical world, results not only need to be reproducible but also transferable. This requirement helps assure that differences in apparatus for the purpose of automation do not interfere with the method and demands a validation to demonstrate equivalency. Designs which diverge from the strict *USP* and industry convention run the risk of developing a system that cannot be validated at the specific method level. The authors have personally observed cases where extremely subtle changes in apparatus resulted in a failure to demonstrate suitability.

The FDA has also focused specifically on automated dissolution. FDA (3) has stated its acceptance of automated dissolution, however, it specifically refers to *USP* described devices. Presumably this guidance excludes non-USP-compliant apparatus.

"Dissolution methodologies and apparatus described in the *USP* can generally be used either with the manual sampling or with automated procedures."

The FDA (4) casts doubt on the wisdom of straying too far from the established analytical method.

"Use of unusual automated methods of analysis, although desirable for control testing, may lead to delay in regulatory methods validation because the FDA laboratories must assemble and validate the system before running samples. To avoid this delay, applicants may demonstrate the equivalency of the automated procedure to that of a manual method based on the same chemistry."

From these selected references and others, we have confidence that the FDA and *USP* accept automation in general, and automated dissolution in particular. The references confirm the importance of maintaining the same basic chemistry and adhering to compendium design as closely as possible. This is not only a regulatory consideration but also one of practicality. It is extremely important that methods can be successfully transferred to other sites and apparatus (automated or otherwise). With this information we may proceed with our functional design of an automated dissolution system.

Preliminary Requirements

Intended Use

What work will be performed on the system? What are the needs of the analyst serve? What function is being performed? All these and other questions need to be considered.

So far the discussion has revolved around completely automated dissolution. Meaning that media is prepared, dispensed into the vessels, tablets dropped, sampled, filtered, collected or read, and lines and vessels washed. This series of events must be reproduced multiple times without human intervention after it has been initiated (Fig. 3).

This seemingly simple series of events does not address all the requirements. If the device is preparing media does that mean it prepares a buffer to be diluted or only degasses the premixed media? When media is dispensed, is there a need to perform a preliminary dispense to assure removal of the previous media? If samples are to be read on-line is dilution required prior to reading? Systems intended for method development (MD) will have many different requirements than one intended for QA. The value of the automation to the user may be very different for each of these two areas. In fact the MD user may not appreciate the need to automate more than one run at a time and will prefer a semiautomated system, since the MD user may have many different experiments to perform that may be labor intensive. Just a few

Figure 3 Fully automated custom system under construction.

formulations may need to be tested by many methods and conditions. No less of a challenge, QA department requirements may need to run many different samples efficiently with a single method (Fig. 4).

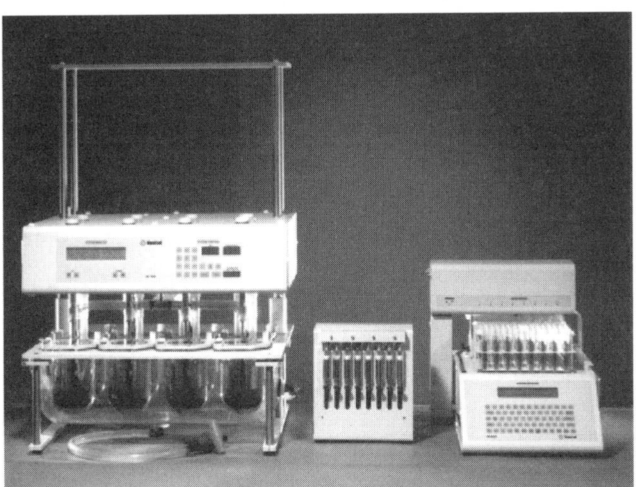

Figure 4 Semiautomated system.

MD requires that the analyst be able to develop methods that can discriminate atypical samples or can be used for correlation to bioavailabilty or serum drug levels. In vivo–in vitro correlation should be established if possible. These objectives require a high degree of flexibility and may become very involved. Taking readings quickly to understand the initial release characteristics or release throughout a range of media pH may be important for the developer. The developer may only want to work with one vessel with a lead candidate or an early prototype that is in short supply or run "quick-and-dirty" tests for preliminary approximation. The effect of various other media components may be evaluated as well. The addition of various other components addressing the physiology at the site of application (e.g., enzymes, bile salts) at key intervals may also be of benefit in MD.

QA requires the efficient analysis of many samples to support routine production release and stability programs. Methods are typically established in the analytical development group. Efficiency and convenience issues, including the speed of media preparation and the relative convenience of data handling and documentation, are important here. While compliance is important in all aspects of the pharmaceutical industry, QA functions must approach compliance perfection. Depending upon the facility, the automated apparatus may be tailored to specific methods with fixed configurations. Dissolution methods may be routine enough that a custom system, optimized for productivity, may be justified. Compliance of USP and use of industry standard apparatus is important to maintain compatibility with other company laboratories or in the case contract laboratory services are required.

The following Table lists features which may be more appealing to QA or development functions, some being obviously of interest to both groups (Table 1).

21 CFR 11 Compliance

21CFR11 is a U.S. regulation requiring security of electronic records and electronic signature requirements. It applies to

Table 1 Feature Comparison Method Development and Quality Assurance

Features of interest to method development
Media modification during run
Short reading intervals
Independent control of vessels
Different drugs/strengths in different vessels
Adjustable sampling height
Change paddle/basket speed during run
pH measurement/adjustment
Alternative vessel sizes
Fiber optic UV measurement
Other continuous measurement
Advanced chemometric capabilities
Data export for nonroutine calculations
Directly compare runs
Long duration runs
Sample dilution or reagent addition
User defined report format
Features of interest to quality assurance
Full compendium compliance
Convenient media preparation and handling
Flexible bracketing of standards
Automatically prepare and run calibration curves
System suitability
Flexible use of blanks with sampling
Multiple component analysis
Comprehensive cleaning
Compatible with industry standard accessories
Centralized networked database
Data output for LIMS
On-line LC capability
Run different methods within a batch of multiple samples
Last minute change to the batch order

any electronic data required by the FDA that is stored to a durable media. Primary attributes include password and log on/off requirements, audit trail, access rights, data security, and integrity. Compliance to this regulation is required for doing business in the United States. Similar regulations are being harmonized by the ICH. To ensure that the product design complies with the regulation, we recommend interpreting each of the individual requirements, and then listing necessary product attributes.

Compendial Requirements

The authors have divided compendial requirements into three different types. For convenience, the *Eur. Ph.* or *USP* may be referenced since they have been harmonized and are identical or nearly identical in all requirements.

- Specifications are the requirements that include a quantifiable tolerance (e.g., *Distance from inside bottom of the vessel and basket is* $25\,mm \pm 2\,mm$ or *rotational speed* $\pm 4\%$). Since these specifications are absolute it is fairly easy to assure compliance.
- Descriptive requirements do not provide quantifiable tolerance and can be somewhat subjective in interpretation. (e.g., Basket free of significant wobbles or sample from a zone midway from the paddle and top of the media.)
- Method requirements play a significant role in the design of the automated system. In the *USP* method specific requirements are included in the individual monographs. Nonmonographed drug products may have also specific requirements described at the general method level. (e.g., media exchange for an enteric method or drug sequestering.)

User Requirements

The individual user is one of the major considerations. The input of those who will use the system day-in and day-out is critical to the design. The role of the user in the design will vary based on the specifics of the automation project. In the case where the system will be customized, the user must have input on almost every aspect. This will allow the resulting method to approximate the manual or current approach as closely as possible. Off the shelf systems (semi and fully automated) likewise are very dependent upon user input; however, a careful balancing act has to be performed. Our experience is that there are about as many ways to run and calculate standards, controls, blanks, and samples as there are users. The challenge of the developer is to either try to build in as many features as considered reasonable, or to standardize on a specific architecture that will appeal to the most users.

We also must recognize that users are the true experts in performing the analysis manually. Their input is very valuable in capturing the function needs to accomplished. It is the developer's task to turn that valuable information into the nuts and bolts of how the task will be accomplished on an automated basis. On-going contact with users (or in the authors' case, customers) is important to determine which features are appreciated and which features are not.

Extent of Automation

When starting to develop functional requirements we often observe the tendency to want to automate everything. In fact, there must be trade-offs between cost and benefits. Yes, it is possible to automate just about everything; however, the increased time and expense may not be worthwhile. Previous experience on the part of the developer can be very helpful. This is also an area where standardized workstations or modular approaches to automation are useful.

Semiautomated—Generally systems that perform sampling, filtration, and UV reading or collection are termed semiautomated systems. They are generally simple to set up and operate with a much lower overall cost and can provide short walk away periods during which samples are taken. Generally, procedures such as media preparation, dispensing, and clean out are not performed by semiautomated systems. Most of the dissolution tester manufacturers as well as other automation technology companies offer semiautomated systems. Purchasing a system from a dissolution tester company can assure compatibility of a discrete system designed to work together. Automation companies can provide custom integration of the apparatus (tester and UV) that you already own and are using, to help lower cost and provide better assurance of equivalent results with your manual approach.

Fully Automated Systems

Fully automated systems typically automate the entire process including some aspects of media preparation, media dispensing, tablet drop, sample removal, filtration, and analysis. More or fewer functions can be added to the design, based on the benefit to the user. Media may be fully prepared by mixing a concentrate, heating, and degassing. Media can be dispensed initially and additional (different) media can be added within a method. Some methods require media removal, which can also be automated. Analysis can be performed in a straightforward manner, for example using flow-through cells with UV detection, or, with simple sample collection. The automated analysis can, however, be more complex when dilution of samples is required, reagents have to be added or samples sequenced for subsequent HPLC analysis.

Fully automated systems can be purchased off the shelf or fully customized. Customized systems offer exactly what the customer wants and needs, for example a system might be optimized for one high volume product. Off the shelf systems are available that are fully integrated systems with components designed by the provider. As with the semiautomated systems modular approaches are also available primarily through automation companies. Modular approaches allow the use of standard industry apparatus that the user already owns and uses (Fig. 5).

Finalizing Requirements

The preliminary requirements discussed above are very broad in nature. In order to realize a specific product, we must be very detailed with our specifications, so that critical features function correctly. The debate regarding the appropriate level of detail will never end. One rule to go by is, if an attribute matters, then specify it. If the attribute does not matter, then allow the engineer flexibility in the design. A common failing at this step of the process is that specifications tend to tell the engineer how the function is to be accomplished rather than what function is required.

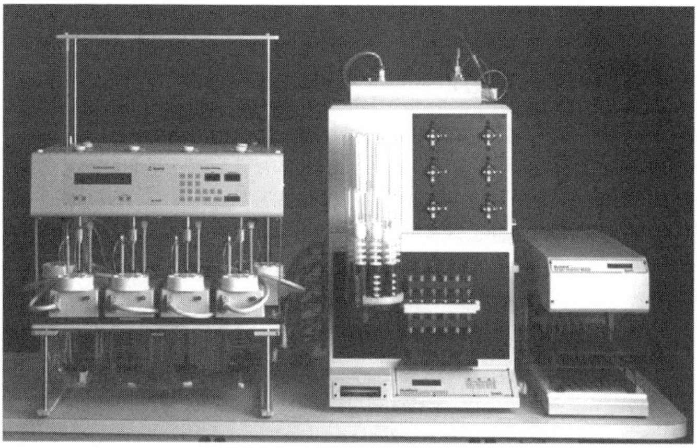

Figure 5 Fully automated dissolution system.

Functional Requirement Specification

The functional requirement specification (FRS) and its nearly identical twin, the user requirement specification (URS), is a list of functions and features the device should process. If there are specific needs the customer (user) has then this is the place to include it. The level of specificity may be dependent on the experience the end-user has with dissolution. An experienced dissolution scientist will be sensitive to issues such as cross-contamination or the importance of timing etc. Critical specifications need to be clearly stated since the FRS serves as the starting point of the test plan (discussed in the next section).

When considering any particular function, it is important to break it down to the smallest components possible, and determine which are important to specify. Let us look at media dispensing as an example of the required level of detail:

A. Prior to dispensing media, the containers, lines, pump, and vessels will be rinsed to effectively remove media from the prior dissolution run.

   1. The volume of rinse shall be user-selectable from 0 to 500 mL.

2. The rinse medium will be de-ionized water at room temperature.
3. The rate shall be fixed at 50 mL/min.
4. The rinse volume shall be recorded in the sample data base.
5. The rinse will leave the media fill lines full of media.

B. Media will be preheated

1. User-selected temperature.
2. User-selected tolerances.
3. How much media will be preheated.
4. Preheat setting will be selectable to 0.1°C.
5. Media must heat from 20 to 37°C in less than 5 min.

C. Media will be degassed

1. De-gas after media heating.
2. De-gas using vacuum approach.
3. De-gasing should result in less than 1 mg/L dissolved O_2 in the vessel after dispensing.

D. Initial dispensing of media

1. The user may select one of five media to dispense.
2. Media will be dispensed to the vessels when the specified conditioning temperature is achieved.
3. Media will be dispensed with media contact surfaces composed of a material compatible with 1.0 N HCl and buffers with pH of < 11.
4. Volume is user selectable from 20 to 1000 mL.
5. Six vessels must be filled to 1000 mL within 3 min from the time at which the media achieve the preselected temperature.
6. Volume must be used to calculate results and included in the sample data.
7. Media volume required for tubing dead volume and flush volume must be accommodated in the total overall volume.

8. Volume of dispensed medium must be within $\pm 1\%$ of the selected amount.

E. Supplemental dispensing of media

1. After a user-specified period of time, additional media may be added to the vessel.
2. This may be the same or one of the other four media available for selection.
3. Volume is user-selectable from 20 to 1000 mL with selections that cause overflow (>1025 mL) not allowed.
4. Supplemental media dispensed must be within $\pm 1\%$ of the selected amount.
5. Total updated media volume must be included in the sample data, and the calculation of results.
6. Supplemental media must be able to be added repeatedly at least eight times during a run.

In the example above, it appears that all the bases are covered and they may well be, depending upon the analyst's needs. It is easy to overlook valuable functions that we may expect without further thought. In this case, we have not specified that media should be preheated while a previous method is running. The sequence of events has not been well characterized. Here, it can cause a delay in run time for the batch is media is not heated prior to completion of the prior batch.

The following considerations have been assembled to help assure that meaningful FRS is constructed that might best fit the users needs. This list is intended to help provide areas of consideration and should not be considered all-encompassing.

A. Custom system or a generic workstation?
B. Quality assurance or development or both?
C. Level of flexibility required

1. USP type I, II, III, and IV?
2. Single method or several similar methods.
3. Diverse methods within a USP type.

4. Diverse methods within multiple USP types.

D. Desired degree of automation

1. Semiautomated sampling to sample collector.
2. Fully automated sampling with media dispenses.
3. Media preparation (mix, heat, and de-gas?).
4. On-line sample collection or on-line analysis?
   a. LC, UV, fluorescence.
   b. Collection after UV analysis.
   c. Dilution or further sample preparation.
5. Continuous loop analysis.

E. Run options

1. Enter values for calculations at run time (e.g., standards).
2. Baseline measurement.
3. Timer start delay.
4. Tablet drop stagger.
5. Staggered reading time.
6. Reading at time zero.

F. How many different media are to be used?

1. Dispensing specifications.
2. Volumes.
3. Supplemental media addition.
4. Sample loss replacement.

G. Sampling

1. Minimum sample frequency.
2. Sample volumes.
3. Dead volume.
4. Flush volume.
5. Precision.
6. Sample height.
7. Sample filtration
   a. Choice of filtration frequency.
   b. Type of filter (fixed cannula, interchangeable).

H. Data processing
   1. Single or multiple components.
   2. Advanced mathematical functions.
   3. Multiple standards.
   4. Bracketing of standards (multiple modes).
   5. System suitability and other controls.
   6. Mode of calculation.
   7. Standard curve fitting.
   8. Comparative features.
   9. Data reporting.
      a. Types.
      b. Graphic display.
      c. Comparisons.

I. Networking
   1. Shared client server or workstation.
   2. Multiple workstation database support.
   3. Data export to LIMS.
   4. Export spread sheets.

J. Compliance
   1. 21 CFR11 compliant.
   2. Data fulfill GMP compliance.
   3. USP compliant.
   4. EU safety compliance.

K. Utilities
   1. Calibration.
   2. Validation.

L. Device compatibility
   1. Bath, UV, LC, diluter injector.
      a. Develop custom devices.
      b. Use off-the-shelf devices.

The FRS or URS should be agreed to before the design requirements are started. Whereas the FRS/URS describes what functions are to be included with the product, the design requirements describe how the functions will be provided.

Until now we have not discussed hardware vs. software. We have only discussed functional requirements without differentiation. In reality, most any functional requirements will be comprised of both hardware and software design requirements. Differentiation of hardware and software attributes becomes more important in developing the design requirements as a means to meeting the functional requirements. There are specific product functional requirements that are largely software focused (e.g., 21CFR11 compliance) however, a distinction should not be made in terms of functional requirements. The software could be designed many ways and yet remain compliant to 21CFR11.

Developing design requirements is the role of the project manager, mechanical, and software engineers. It is important, however, that design reviews with the entire project team be conducted to assure that the functional requirements will be met. Because of the detail and level of expertise typically required, separate software and hardware design specifications are developed. Eventually a prototype is constructed by the engineers. This would start the testing process to be discussed in the next section.

Hopefully this discussion has provided food for thought in developing your own automated dissolution capabilities. The following section relating to testing and qualifications will help the user assure that the intended functionality is indeed delivered.

SYSTEM QUALIFICATION

Introduction

System qualifications are quality checks. They are a part of the validation of a product. Validation is defined as, "Establishing documented evidence which provides a high degree of assurance that a specific process will consistently produce a product meeting its predetermined specifications and quality attributes (5). A product that is validated is considered to be of much higher quality than one that is not validated. Automated dissolution systems need to be validated as a requirement of their use in regulated laboratories.

Automated dissolution testing for pharmaceutical dosage forms involves processing samples that are related to the manufacture and control of a product destined for human or veterinarian use. As such, the system must comply with the current good manufacturing practice (cGMP) regulations (6). 21 CFR 211.68 states that when "certain data, such as calculations performed in connection with laboratory analysis, are eliminated by computerization or other automated processes" validation data shall be maintained. Thus the requirement of validation is established. Systems that are designed to store data electronically or allow for electronic signatures must also adhere to "21 CFR Part 11: Electronic Data and Electronic Signatures."

Types of Qualifications

There are several types of system qualifications. The quality of a system is dependent not only on the qualifications that are done following the system's development, but also on the qualifications that are done as part of the system's development.

System qualifications include development reviews, development testing, and instrument qualifications. Development reviews occur as part of the design process and include such things as functional specification reviews, design document reviews, and code reviews. Development testing is the work that is performed to demonstrate that the product meets its specifications prior to the equipment being available for delivery to customers. Development testing includes unit testing, integration testing, system testing, and regulatory compliance testing. Instrument qualifications are the tests that are performed after the equipment is installed in a laboratory for use. Instrument qualifications include installation qualification, operational qualification, and performance qualification.

Development Reviews

Specification, design, and code reviews are the earliest form of system qualifications. Quality cannot be tested into the product; quality must be built into the product. Reviewing specifications is the most efficient and least expensive way to eliminate defects. As the product development cycle progresses, it becomes more and more expensive to find and correct defects. During each phase of the product development cycle, there are important quality checks that can be performed. Design reviews and code reviews are important quality checks that are performed during product development.

Development Testing

Development testing encompasses a wide range of testing to verify and validate the product. There are several major types of testing that can occur, which include unit testing, integration testing, system testing, and regulatory compliance testing. The terminology used to categorize these types of testing can vary. The major types of testing can then further be broken down into many subcategories of types of testing.

Unit testing is the testing of the individual "units" of software. Unit testing verifies the functionality of algorithms and code modules. This type of testing is generally performed using software-debugging tools within the environment on the developer's computer. Each path of the code can then be tested, including error paths that are impossible to intentionally produce, during integration and system testing. The developer of the code or another developer on the project team often performs this type of testing. More often than not, minimal documentation is created for this type of testing.

Integration testing is the next level of testing after unit testing and involves testing the combined functionality of different code modules and pieces of the system. Typically, both developers and QA personnel perform this type of testing. The developers will test first to make sure that the combined modules perform correctly according to their design specifications. The quality personnel will follow with testing that verifies that the integrated modules perform the functions as specified in the requirements documentation.

System testing includes beta testing and applications testing. Beta testing is testing that is performed by actual customers. Customers are given a product to try out, often with a well-defined plan of testing based on how they plan to use the system. Applications testing is testing that is performed by the manufacturer, which simulates how customers will use the product. For automated dissolution, applications testing involve running actual chemistry on the equipment to evaluate proper performance. More information is given on applications testing in a later section.

Regulatory testing is testing the product for compliance to regulations. Often these regulations are governmental, including CE for Europe, and CSA for Canada. These regulations are imposed upon the manufacturer that wants to claim compliance, which can be a condition in order to sell into certain countries. Sometimes these regulations are from an independent quality organization such as underwriters laboratory (UL) in the United States. Manufacturers will work to comply with these regulations in order to compete in a specific marketplace. For manufacturers of automated dissolution equipment, regulations that are imposed on their customers by agencies such as the FDA in the United States are also an important consideration. These pharmaceutical manufacturers must comply with good manufacturing practices (GMP, 21 CFR Parts 210 and 211) and the electronic records and electronic signatures regulation (21 CFR Part 11). The supplier of automated dissolution equipment must supply compliant-ready devices in order to be competitive. More specifics of part 11 testing are provided in a later section.

Application Testing

A key aspect of producing a good product is making sure not only that the design meets specification, but also that the design meets the needs of the customer. Checking the design against the specification is often referred to as verification. Checking that the design meets the customer needs is often referred to as validation. Application testing involves testing the equipment by using it exactly as the customer will use it. Although the best way to determine if the product will meet the needs of the customer is to allow the customer to test the product, this type of testing, which is also known as beta testing has its limitations. The product manufacturer performs the most prevalent form of application testing.

Beta testing can provide important feedback to the manufacturer, but it is limited in a few key ways. Beta testing often occurs late in the development of the product because the product must be in good working shape before exposing it to customers. At this point in the development cycle it is often difficult to make any major changes to the product. Another limitation of beta testing is that the customer often has a very limited amount of time to test the product. The customer is often left on their own to complete the beta tests. This not only often leads to delays in completing the tests, but also allows the customer to stray from the desired tests of the manufacturer. A third limitation of beta testing is the difficulty in communicating the results of the testing. This difficulty can arise from the fact that the testing is performed in a different location, the information gets passed through many people, and many times the information is interpreted only from written messages.

The application testing that is performed by the manufacturer is key to the characterization of the product's capabilities. When a manufacturer of automated dissolution testing equipment designs their product, the process to be automated is broken down into the individual functions that are performed. These functions include dispensing, dropping tablets, aspirating samples, and calculating results. Each of these functions could then be verified as operational according to the specification of that function. These functions could even be integrated and tested as a system. The system at this point could pass all of its specifications, but will it satisfy the needs of the customer? This question cannot be answered without application testing. Application testing involves running real chemistry on the system to validate that it will perform with chemistry similar to what the customers will be using.

21 CFR Part 11 Testing

Automated dissolution equipment in most cases must be compliant with the FDA electronic records and electronic signatures regulation (21 CFR Part 11). The requirements of the regulation include use of validated systems, secure storage of records, computer generated audit trails, system and data security via limited access privileges, and the use of electronic signatures.

As with any set of requirements, the product must be tested to verify that the system can meet the requirements. Compliance to the regulation is achieved not only through features in the product, but also through practices and procedures that are instituted by the users of the equipment. The manufacturer of the equipment can thus only provide a compliant-ready product. The users of the equipment can then achieve compliance by configuring and operating the equipment in a manner that meets all the requirements of the regulation.

In order to provide a compliant-ready product, the manufacturer must make sure that the features required for compliance are built into the product. For verification purposes, a requirements traceability matrix should be created to match the appropriate tests for each of these requirements. An excerpt of an actual matrix is show in the following table (Table 2).

Instrument Qualifications

Instrument qualifications are the tests that are performed after the equipment is installed for use in a laboratory. Instrument qualifications include installation qualification, operational qualification, and performance qualification. These tests verify that the equipment is installed, operates, and performs according to the manufacturer's specifications. Each of these types of qualifications is defined in more detail in the following sections.

Installation Qualification (IQ)

IQ is defined as documented verification that all key aspects of the hardware and software installation adhere to appropriate

Table 2 Requirements Traceability Matrix

11.50 Signature manifestations	Test case description	Test no.
(a) Signed electronic records shall contain information associated with the signing that clearly indicates all of the following: (1) the printed name of the signer; (2) The date and time when the signature was executed and (3) The meaning (such as review, approval, responsibility, or authorship) associated with the signature.	Signed electronic records include printed name, date and time of signing, and the meaning.	3.14
(b) The items identified in paragraphs (a)(1), (a)(2), and (a)(3) of this section shall be subject	Electronic signatures are secure	3.15
	All signature information is included on reports that are displayed as well as printed.	3.14

codes and approved design intentions and that the recommendations of the manufacturer have been suitably considered. The IQ consists of checks to verify that the hardware and software have been installed properly. Component version numbers, electrical connections, and fluid path connections are checked during IQ.

The following activities may be performed to qualify the installation of an automated dissolution system:

- System qualifiers' identification.
- Verification of site preparation procedures.
- Environmental condition verification as recommended by manufacturer (space, electricity, water, gases, temperature, humidity, etc.). System location documentation.
- Complete listing and identification of components to be installed on the system, to include system components and peripheral device identification (HPLC, UV, etc.).

- Sales order identification and compliance.
- Reference document identification (operating manuals, maintenance manuals, validation certificate).
- Manufacture data review.
- Verification of installation procedures, including plumbing and electrical connections.
- Verification of correct software installation (proper software versions loaded).
- Application of power to the instrument to ensure that all modules power up and system initializes properly.

Operational Qualification (OQ)

OQ is defined as documented verification that the system or subsystem performs as intended throughout representative or anticipated operating ranges.

For automated dissolution systems, OQ testing can include testing balance functionality, testing the functionality of individual components including bath communication, sample cannulae, waste cannulae, thermistor communication, tablet dispensers, sensors, valves, pumps, filter dispenser and holder, and testing fluid pathways.

Performance Qualification (PQ)

PQ is defined as documented verification that the system performs its intended function in accordance with the system specification while operating in its normal environment.

For the purposes of instrument qualification, the PQ involves testing the equipment for overall system functionality. For dissolution equipment, these tests verify that the equipment can perform the entire dissolution process. A sample method should be observed to run properly. This can include running actual chemistry and analyzing the data results.

Instrument Qualification Design Considerations

When designing instrument qualifications for automated dissolution systems, some key considerations are determining the functions to validate, cost, testing using equipment diagnostics, integration of different manufacturer's equipment, protocol format, and scope.

Functions to validate

The cornerstone of validation and qualification is testing to a set of specifications. Without specifications, proper qualifications cannot be performed. For an automated dissolution system, the specifications originate from a few sources, which include the USP, the manufacturer's FRS, and the manufacturer's detailed design specifications, which may include HDS and SDS.

Functions to validate on automated dissolution systems may include bath operation, balance operation, media dispensing operations, media removal, sampling operations, media replacement, thermistor operation, robot operation, sample timing, sequence, and dilution.

Cost

Equipment manufacturers are faced with the challenge of qualifying all the functionality of complex equipment at the customer's lab while keeping the costs at a reasonable level. There is an expectation that the cost to qualify laboratory instrumentation be only a small fraction of the cost of the equipment itself. However, there are costs associated with both developing the qualification protocols and executing the qualification protocols.

Testing using equipment diagnostics

Equipment manufacturers design diagnostic routines into the equipment to make troubleshooting hardware and chemistry issues as simple as possible. A question arises as to how much of the qualification testing can be performed using equipment diagnostics. Using diagnostics to qualify the instrument can make the testing quicker and therefore less expensive, but it must also accurately represent the functions as they would be used when the system is operating.

Integration of different manufacturer's equipment

It is often the case that laboratories combine the use of equipment from more than one manufacturer into systems that need to be qualified. Each individual device must be qualified for the functionality of that device. Sometimes one manufacturer will sell and qualify other manufacturers' devices that connect to their equipment. In this case, one company is responsible for the instrument qualifications of the entire integrated system. It is usually required that each manufacturer qualify its own device, and that following the qualification of the individual devices, the manufacturer that supplies the interface must then qualify the interfaces between the devices.

The order of instrument qualification can be important, as checking the specification of one device may rely on an attached device being calibrated and functioning properly. In the case of automated dissolution testing, the bath should be calibrated and qualified prior to the qualification of the device that pulls samples from the bath. By performing the qualification in this order, it is not possible to fail the qualification for pulling samples due to a problem with the bath. The bath should be calibrated and qualified first to make sure that it is functioning properly, and then the device that pulls samples can be qualified. Additionally, it can be very difficult to diagnose a qualification failure of one piece of equipment that is caused by a specification failure of another piece of equipment.

Protocol format

While there are many different formats that can be used for instrument qualifications, there is a minimum amount of information that needs to be provided as part of the testing.

The level of detail put into the protocol depends on many factors including the level of expertise of the operator who will execute the testing, how often the testing will be performed, and the complexity of the product. Cost is always a driving factor, so time should be reduced wherever it can without sacrificing quality. A greater amount of detail should be put into the protocol if the level of expertise of the operator who will be executing the testing is low rather than high. If the testing will be performed on a very regular basis, the level of detail within the protocols could be streamlined. In this scenario, it would make sense to make the test protocols concise and reference separate documents for the methods and procedures. This would allow for unneeded duplication of the method and procedure sections in each testing documentation package. If a product is very complex and many settings must be configured for operation, it is required that the detail in the protocols not only have instructions for all of the settings that must be made, but that the protocols include checks throughout the protocols to make sure that proper configuration is made for the testing that is performed. The checks throughout would help to avoid getting to the end of a lengthy test only to find that one of the settings was configured improperly.

A typical protocol may include the following sections:

- The *objective* will state the purpose for the test and the specific module(s) to be tested. Prerequisite protocols will be listed.
- The *scope* will state the specific operations and/or functions to be examined by the procedure.
- The *overview* will provide general information describing interpretation of results.
- The *required materials* will include any operational prerequisites required to perform the test such as reagents and disposables.
- The *acceptance criteria and data evaluation* will describe the acceptance criteria or expected results for the tests. This may include a comparison of the observed response with an expected response or statistical analysis.
- The *test procedure* will include a detailed description of each of the test steps. This will include manual setup steps, system operations, and human operations. It will include tables as necessary. It will also detail each of the procedural steps, the acceptance

criteria or expected results, and give a space to enter the raw test data. As a test script is utilized it will become part of the qualification observation log. The test script and the error log can be referred to as the observation log. The error log will contain all information about unexpected responses or unacceptable results.

- The *results* will be summarized.

Scope

How much testing should be performed during instrument qualification? This is not always an easy question to answer. There are usually innumerable configurations and settings that can be made to the instrumentation. The manufacturer must do his/her best to determine the best way to test the major functions of the system while operating the equipment over the range of settings that the customer will most likely use. The number of tests that will be executed over a range of setting types must also be determined, as well as how many replications will be performed at each of the determined settings. Also to be determined are which systems options will be enabled for testing and how many permutations of the system options will be tested. Use of different equipment peripherals leads to many different system configurations that can occur. The question is how to qualify a specific customer configuration while at the same time keeping a reasonable cost on the creation and delivery of instrument qualification. More and more manufacturers of dissolution equipment face this dilemma as the development of open systems proliferates.

Instrument Qualification Execution

Prior to execution, the site preparation and document approval must take place. The equipment manufacturer will provide detailed site preparation requirements for the system. It is the responsibility of the customer to prepare the site as per the documented requirements. The operator will verify the site preparation during the testing of the installation qualification protocols. The customer, with the appropriate signatures, must approve the protocol documentation package for use prior to execution. Sometimes it is planned that the protocols will not be followed exactly. In these cases, deviation reports, which are planned changes to a protocol or test plan prior to the start of testing, must be written and approved as well. Deviation reports are used primarily due to observed failures (such as known protocol errors) or due to customer specific situations (improper hot water temperature may necessitate not using that option).

A trained operator then executes the protocols. If events or data that do not match the expected results are observed, then an error log must be written. The error log details the issue and its resolution. Proper retesting of a failed protocol can then occur. Following completion of the execution of the protocols, customer signoff is again required.

Instrument qualifications should be executed on a scheduled basis that can be determined with the help of manufacturer's recommendations. Automated dissolution systems that are used regularly are typically re-qualified every six months to one year. Re-qualification is also recommended for other reasons including moving equipment or replacing parts. Below is a typical system re-qualification policy.

RE-QUALIFICATION POLICY

Installation Qualification Execution Frequency

- Upon initial system installation.
- When equipment is physically moved to another location. Definition of another location must be made within the company's SOP. Manufacturer recommends a re-qualification if equipment is lifted during the move or if the environmental conditions are different in the new location.

Operational Qualification Execution Frequency

- Upon initial system installation following IQ.

- When system components are upgraded or serviced depending on the extent of the work performed.
- On a regular time interval as determined by use. Manufacturer recommends at least once per year.

PQ Execution Frequency

- Upon initial system installation following IQ and OQ.
- When system components are upgraded or serviced depending on the extent of the work performed.
- On a regular time interval as determined by use. Manufacturer recommends at least once per year.

The appropriate amount of testing that needs to be performed in the laboratory is open to interpretation. It is the responsibility of the company using the equipment to determine if the equipment is suitable for its own use. Government regulations and guidelines do not dictate to the company the appropriate amount of testing that must be performed on the equipment. The manufacturer can share documentation created during the development of the product. During product development, much testing is performed. Ideally, a comprehensive set of documented test results that match up to all of the product requirements and specifications is available for review. With reference to the manufacturer's testing documentation, the company that uses the equipment can justify not repeating the same tests.

The manufacturer often creates an instrument qualification plan and provides installation, operational, and performance qualifications to be executed in the customer's laboratory. The company using the equipment must determine if the manufacturer supplied instrument qualifications is comprehensive enough to be sure that the equipment is installed, operating, and performing correctly. If they feel it is not, they may choose to perform more tests themselves.

SUMMARY

System qualifications are important for producing a high quality product. These tests occur throughout the entire life cycle of the product, including development. Qualifying the specifications is the least expensive way to remove product defects, as they are discovered very early in the process and can be corrected with a few pen strokes. Qualifications are important quality checks during the development of a product to help find defects before the product reaches the market place. After the product has shipped, instrument qualifications are used to validate that it is installed, operating, and performing correctly. Routine qualifications are performed to regularly check the equipment.

REFERENCES

1. USP28. General Notices.
2. USP28. Validation of Compendial Methods.
3. FDA Guidance. Dissolution Testing of IR Solid Oral Dosage Forms (Appendix A), Apparatus August 1997.
4. FDA Guidance. Submitting Samples and Analytical Data for Methods Validation Appendix C, B. Automated Methods. February 1987.
5. FDA Guidelines on General Principles of Process Validation. May 1987.
6. cGMP are defined in Title 21 of the Code of Federal Register, Parts 210 and 211.

14

Bioavailability of Ingredients in Dietary Supplements: A Practical Approach to the In Vitro Demonstration of the Availability of Ingredients in Dietary Supplements*

V. SRINI SRINIVASAN

Dietary Supplements Verification Program (DVSP), United States Pharmacopeia, Rockville, Maryland, U.S.A.

* The approach outlined in this chapter reflects the collective thinking of the USP Council of Experts (formerly known as USP Committee of Revision) with whom the author has had the privilege of working closely over the past 16 years.

Since the U.S. Congress passed Dietary Supplement Health and Education Act in October 1994, the landscape of the dietary supplement industry has changed in the United States dramatically. In fact, as early as the late 1980s, the U.S. Pharmacopeia's elected Council of Experts (then known as the USP Committee of Revision) was evoking great interest in the development and establishment of public standards for the multitude of multivitamin and multivitamin–mineral combination products as well other nutritional supplement products marketed in the United States.

The U.S. Pharmacopeia's interest in dietary supplements was triggered by Prof. Ralph Shangraw who conducted studies (1) on the use of calcium salts as fillers for tablets and capsules and noted that, in addition to not dissolving, in many cases the calcium salt tablets took as long as 4–6 hr even to disintegrate. Shangraw made the same observations when testing multivitamin–mineral combination and single vitamin preparations. In recognition of the impact of these findings on consumer confidence in the dietary supplements, the U.S. Pharmacopeia initiated work to establish public standards for multivitamin–mineral combinations as well as single vitamin and mineral and other dietary supplement preparations. These standards address performance i.e., disintegration/dissolution as well as content uniformity requirements for oral solid dosage forms of these preparations.

The commonly accepted definition of bioavailability is the proportion of the nutrient that is digested, absorbed, and available for metabolism via the normal pathways (2). Bender (3) refines the definition further by stating that the bioavailability should be defined as "the proportion of a nutrient capable of being absorbed and becoming available for use or storage; more briefly, the proportion of a nutrient that can be utilized." Thus, it is not enough to know how much of a nutrient is present in a dietary supplement; the more important issue is how much of the amount present is bioavailable. It is important that the nutrient or dietary ingredient of concern contained in a dietary supplement is present in an absorbable form. A common tenet regarding bioavailability of dietary supplements is that the dietary ingredient or nutrient must be in solution in order to be absorbed into the body. In order to assure that this condition is achievable, it is essential that all oral solid dosage forms of dietary supplements must meet in vitro test requirements for both disintegration and dissolution.

In developing appropriate performance standards for a given solid oral dosage form, the intended use of the product must be taken into consideration. Drug products are taken for the treatment, cure, and alleviation of disease states, while dietary supplements, as the name implies, are intended to supplement a diet that may be deficient in certain nutrients, thereby preventing certain disease states and/or maintaining health status. However, formulation development and manufacturing technology involved in the preparation of dietary supplements are essentially the same as those in the manufacture of drug products. Nevertheless, there are certain fundamental differences, which distinguish dietary supplements from drugs, which must be considered in the context of development of standards for dietary supplements:

1. Nutritional supplements are consumed for prevention of diseases and maintenance of a state of well-being.
2. Nutrients enter into biological processes that are not characterized by a well-defined dose–response relationship. Therefore, in many cases, the dietary supplement itself is not expected to exhibit a characteristic dose–response curve.
3. Another difference from drug therapy is that the dosing interval of a nutritional supplement is often not a critical parameter for a positive outcome. This lack of a strong dose–response relationship is an important consideration in setting of standards for dietary supplements and is in stark contrast to the situation for drug products.
4. Further, nutritional supplements provide benefits that are not expressed well by scalar measurements distributed over periods of a few hours, such as pharmacokinetic profiles after single administration.

Much longer periods are involved (typically weeks to months) and benefits may be qualitative and variable rather than expressable as a quantifiable outcome.

5. Interactions between foods and dietary supplements are complex and measurement of nutrient absorption presently lacks the precision of characterization generally achieved with drug bioavailability.

Thus, while the *content uniformity* requirement for drug products is an acknowledgment of the existence of a well-defined dose–response curve and thus the need to establish a suitable dosing interval, such a requirement was at first not considered appropriate for dietary supplements based on the lack of dose–response curves for these products. As an alternative, it was suggested that a *weight variation* requirement could be used to provide an assurance that the article was indeed manufactured under good manufacturing practices and this requirement was adopted by the U.S. Pharmacopeia early in 1991 for judging the quality of nutritional supplements. However, the current thinking of U.S. Pharmacopeia's Expert Committees on Dietary Supplements is that content uniformity is indeed a very important attribute for dietary supplement products from both consumer and good manufacturing practices point of view. This change in appraisal of the situation for dietary supplements has resulted in major revisions to the requirements for dosage uniformity of dietary supplements (4). The proposal, which requires *content uniformity* as a measure of performance characteristics, takes into consideration the analytical burden this would bring to bear on multivitamin–mineral combination products. Thus, the proposal calls for a hierarchy of index vitamins and index minerals to determine content uniformity in multi-ingredient dietary supplements. This approach simplifies the content uniformity determination to a practical level but makes the assumption that if the content uniformity of ingredients present in lesser amounts can be demonstrated, the rest of the components will also be evenly distributed.

Compliance with the content uniformity requirements for vitamins and minerals in multivitamin–mineral combination products may be determined by measuring the distribution of a single index vitamin or a single index mineral present in the product. Folic acid is the index vitamin when present in a multivitamin formulation. For formulations that do not contain folic acid, cyanocobalamin is the index vitamin. If neither folic acid nor cyanocobalamin is present in the formulation, the index vitamin is vitamin D and in the absence of vitamin D, the index vitamin is vitamin A. If none of the above four vitamins is present in the formulation, the vitamin labeled in the lowest amount is used as the index for content uniformity.

With regard to minerals, copper is the index mineral when present in the formulation and in its absence zinc becomes the index mineral. If neither copper nor zinc is present, the index mineral is iron and in the absence of all these minerals, the element labeled as present in the lowest amount is the index mineral. While this approach may not be ideal, it does represent a significant improvement over the weight variation requirement that guided the industry through the 1990s.

In spite of the lack of clearly defined dose–response curve, a dietary supplement formulated into tablet or capsule is expected to disintegrate in the stomach within a reasonable time to release the active ingredient or nutrient. This disintegration will then facilitate further dissolution in the biological fluids prior to gastrointestinal absorption. Because nutritional supplements are formulated and manufactured using essentially the same technology as drugs, in vitro dissolution is considered appropriate as a surrogate for in vivo absorption for oral solid dosage forms of multivitamin–mineral products.

An in vitro dissolution procedure is very useful:

- To assist in formulation development.
- In predicting the in vivo performance of the product.
- In assuring equivalence between the pilot batch and scale-up batch.
- In assuring performance characteristics when formulation change occurs.

- To help differentiate between commercially available preparations.
- To serve as a quality control tool to assure consistency in batches produced.

APPROACH TO IN VITRO DISSOLUTION IN DIFFERENT CATEGORIES OF DIETARY SUPPLEMENTS

Multivitamin–Mineral Combination Dietary Supplements—Indexing of Vitamin and Minerals

In a typical multivitamin–mineral combination product consisting of 10–15 ingredients, it is neither practical nor necessary to require in vitro demonstration of each and every vitamin and mineral. Consequently, in a unique approach to establishing in vitro dissolution for multivitamin–mineral combination products, an index vitamin and an index mineral are identified as markers for dissolution. In an attempt to account for the many different permutations of vitamins and mineral combinations, a hierarchy of index vitamins and index minerals was arrived at and specified (5). Table 1 shows the hierarchy of index vitamins and minerals specified for demonstration of dissolution requirement in the nutritional supplements monographs in USP 25-NF20.

Riboflavin (vitamin B2) was chosen as the number one index vitamin because among the so-called "water-soluble vitamins," it is the least soluble in water. If riboflavin is demonstrated to dissolve within the specified time, it is assumed that all other water-soluble vitamins will have also

Table 1 Hierarchy of Index Vitamins and Minerals

Index vitamin	Index mineral
Riboflavin (B2)	Iron
Pyridoxine (B6)	Calcium
Niacin or niacinamide	Zinc
Thiamine (B1)	Magnesium
Ascorbic acid (C)	

dissolved. In the absence of riboflavin, pyridoxine (vitamin B6) becomes the index vitamin if present. Where a formulation contains neither riboflavin nor pyridoxine, then niacin or niacinamide, if present, becomes the index vitamin.

In view of the reported growing importance ascribed to folic acid deficiency in the prevention of various disease conditions, such as neural tube defects, megaloblastic anemia, colon cancer, and colorectal cancer, a dissolution requirement is specified for folic acid when it is present in multivitamin–mineral combination products. Currently, the dissolution standard required in the official articles of dietary supplements (including vitamin–mineral combination products) places folic acid outside the index vitamin hierarchy. Therefore, a mandatory dissolution test for folic acid is required that is independent of and in addition to the mandatory index vitamin test for multivitamin preparations containing folic acid.

Table 2 contains the currently official (USP24-NF19) issolution conditions and requirements for multivitamin–mineral combination products labeled as USP, while Table 3 illustrates the USP dissolution requirements, according to the combination of vitamins or minerals present.

In contrast to the dissolution criteria used for water-soluble vitamins, the hierarchy for index minerals is based on their importance in public health. For example, iron was chosen as the number one index mineral because iron deficiency is the most prevalent condition in the United States and because iron is present in almost all the multivitamin–mineral combination products currently available on the

Table 2 Recommended Dissolution Test Conditions for Multivitamin–Mineral Combination Products Labeled as USP

Medium	0.1 N Hydrochloric Acid, 900 mL[a]
Apparatus 1	100 rpm (for capules)
Apparatus 2	75 rpm (for tablets)
Duration	1 hr

[a]For formulations containing 25 mg or more of the index vitamin, riboflavin, the same conditions are recommended, expect for the volume, which is increased to 1800 mL.

Table 3 USP Dissolution Requirements According to the Combination of Vitamins or Minerals Present

USP Class	Combination of vitamins or minerals present	Dissolution requirement
I	Oil-soluble vitamins	Not applicable
II	Water-soluble vitamins	One index vitamin; folic acid (if present)
III	Water-soluble vitamins with minerals	One index vitamin and one index mineral; folic acid (if present)
IV	Oil- and water-soluble vitamins	One index water-soluble vitamin and one
V	Oil- and water-soluble vitamin with minerals	One index water-soluble vitamin and one index element; folic acid (if present)
VI	Minerals	One index element

market. Similarly, calcium was chosen as the next index mineral in view of its importance in the prevention of osteoporosis. As with the vitamins, a similar hierarchical approach based on presence in a given preparation is used to determine the index mineral in a given supplement, i.e., iron, then calcium, then zinc, then magnesium.

Botanical Preparations

In accordance with the provisions of the Dietary supplement Health and Education Act 1994, in the United States botanical dosage forms can be marketed as dietary supplements provided the label makes no medical claim; however, structure–function claim is allowed. In most countries other than the United States, botanical preparations are regulated as drugs thus posing a different set of challenges. This fact must be taken into consideration in standard setting.

In contrast to vitamin and mineral products, which are chemically well-defined, the biopharmaceutical quality and behavior of botanical dosage forms marketed as dietary supplements are often not well documented. In most cases, in vitro/in vivo biopharmaceutical characterization is complicated by the complex composition of botanical dosage forms, extensive metabolism of constituents, and the resulting analytical challenges.

Though predictable biological effects and dosing intervals have yet to be determined through systematic and acceptable clinical trials for the majority of marketed botanical dosage forms, in view of the above-mentioned similarities to drug manufacturing technology, one can argue that content uniformity and dissolution testing requirements should be an integral part of the public standards for these preparations as well.

Such requirements are expected to assure that the dosage form is formulated and manufactured appropriately to ensure that the index or marker ingredients are uniformly distributed and will dissolve in the gastrointestinal tract and be available for absorption. No assumption is made that the marker or index compound selected for demonstration of dissolution is responsible for the purported effect. The test is valuable in that it assures that the formulation technology used is reflective of the state-of-the-art technology, provides a means to evaluate lot-to-lot performance over a product's shelf-life and that excipients used to facilitate transfer of the index or marker ingredients of the botanical to the human system are appropriate.

Botanical preparations differ from vitamin–mineral preparations in the following respects:

1. Since botanicals are natural products (usually extracts), variations in the composition of the chemical constituents due to seasonal variations, crop location, time of harvest, etc. are commonly encountered.
2. Botanical preparations may contain either the powdered part of the plant or an extract derived from the part of the plant, or a mixture of both.
3. Depending on the nature of solvent and manufacturing procedure employed for extraction of the plant material, the quality of extract varies considerably both in composition and the nature of constituents

present. Instability of some constituents may in addition influence the composition of the extract.

4. The different constituents present in the plant may belong to different chemical groups. For example, chamomile contains pharmacologically active essential oils, polyacetylenes, terpenoids, flavonoids, coumarins, and polysaccharides.

Botanical raw materials and their extracts therefore usually contain complex mixtures of several chemical constituents. For a large majority of botanical plant material and extracts of these used as dietary supplements, it is not known with certainty which of the various components is responsible for the purported pharmacological effect. It is generally believed that several constituents act synergistically to provide the purported effect. In actual practice, two or more of the chemical constituents present in the plant material are identified as marker compounds that are characteristic of the plant material to be tested, for identification and monitoring of the stability of the extracts.

Marker constituents of botanical products can be different types.

Active Principles

In some cases, constituents with known clinical activity and these may be called by the name *active principle(s)*. (e.g., Sennosides in Senna Extract).

Active Marker(s)

Constituents that have some known pharmacological activity that contributes to some extent to the efficacy of the product have been identified. These are known as *active markers*. An example of this category is alliin, which is converted to allicin in presence of allinase enzyme, and is present in garlic. These active markers may or may not have clinically proven efficacy in their own right. A minimum content or range for active markers is usually specified in pharmacopeial articles. A quantitative determination of *active marker(s)* during stability studies of botanical dosage forms provides necessary information in arriving at suitable expiration dates.

Analytical Markers

Where neither defined active principles nor active markers are known, certain constituents of the botanical raw material and their extracts are chosen as candidates for quantitative determination. These markers aid in the positive identification of the article to be tested. In addition, maintaining a minimum content or a specified range of the analytical markers helps achieve standardization of the plant extract and arrive at suitable expiration date during stability studies.

Negative Markers

Some constituents may have allergenic or toxic properties that render their presence in the botanical extract undesirable. A stringent tolerance limit for these negative markers may be specified in compendium articles. These markers are considered noxious contaminants and thus outside the scope of discussion in this chapter.

To meet the challenges in the biopharmaceutical characterization of botanical preparations, a Special Interest Group (SIG, of which the author of this chapter is a member), which is a working group of the International Pharmaceutical Federation (FIP) and was established in 1999, is currently working on arriving at suitable recommendations. The FIP group is of the opinion that the Biopharmaceutical Classification System (BCS), which was originally developed for chemically well-defined synthetic organic drug substances, could possibly be extended to cover botanical dosage forms, which contained well-defined and characterized botanical extracts. An initial draft report (6,7) published simultaneously in both *Pharmazeutische Industrie* and *Pharmacopeial Forum* contains the working group's initial recommendations with regard to the biopharmaceutical characterization of herbal medicinal products. For herbal preparations, the entire extract is regarded as the active pharmaceutical ingredient. The working group recognizes the differences in the types of marker compounds present in botanical extracts as outlined above.

One can argue that botanical preparations can be considered pharmaceutically equivalent if they contain an extract (taken here as the active ingredient) prepared by the same solvent extraction procedures, having same specifications, in the same quantity and in the same dosage form. This means that extracts from the same plant material manufactured with different solvents and/or manufacturing procedures are not pharmaceutically equivalent. Further, different dosage forms such as plain-coated tablets, hard gelatin capsules, or soft gelatin capsules containing the same extract are not pharmaceutically equivalent. Even when products are deemed to be pharmaceutically equivalent, this does not mean that they are bioequivalent, since differences in excipients and/or manufacturing process may lead differences in their in vitro dissolution and in vivo absorption characteristics.

Is the BCS that was developed with reference to chemically characterized and well-defined synthetic drug substances relevant for application and or adoption to botanical preparations? (8) If one assumes, as is reasonable, that bioavailability of the "active" component(s) in a botanical dosage form depends on both solubility and permeability, the solubility of the botanical extract could be controlled through appropriate formulation technology and dissolution testing. The applicability of the BCS to botanical preparations will certainly be increasingly researched, debated, and discussed in the coming years.

REFERENCES

1. Shangraw RF. Standards for vitamins and nutritional supplements: who and when. Pharmacopeial Forum 1990; 16:751–758.
2. Faithweather-Tait SJ. Nutr Res 1987; 7:319–325.
3. Bender AE. Nutritonal significance of bioavailability. In: Nutrient Availability: Chemical and Biological Aspects. Dorchester, Dorset: The Royal Society of Chemistry, 1989:3–9.

4. United States Pharmacopeial Convention Inc., <2091> Weight variation of nutritional supplements-proposed revisions to. Pharmacopeial Forum 2002; 28(5):1548–1554.
5. The USP 25-NF 20, General Chapter <2040>. Disintegration and Dissolution of Nutritional Supplements 2002; 2484.
6. Lang F, Keller K, Ihrig M, Oudtshoorn-Eckard J, Moller H, Srinivasan VS, Yu He-ci. Pharmazeutische Industrie 2001; 63(10).
7. Lang F, Keller K, Ihrig M, Oudtshoorn-Eckard J, Moller H, Srinivasan VS. Yu He-ci in FIP recommendations for biopharamceutical characterization of herbal medicinal products. Pharmacopeial Forum 2002; 28(1):173–181.
8. Blume HH, Schug BS. Biopharmaceutical characterization of herbal medicinal products: are in vivo studies necessary? Eur J Drug Metabol Pharmacokinet 2000; 25:41–48.

Index